DIESEL PLANT
OPERATIONS
HANDBOOK

Other Books of Interest

Baumeister & Marks • MARKS' STANDARD HANDBOOK FOR MECHANICAL ENGINEERS
Bralla • HANDBOOK OF PRODUCT DESIGN FOR MANUFACTURING
Brunner • HANDBOOK OF INCINERATION SYSTEMS
Corbitt • STANDARD HANDBOOK OF ENVIRONMENTAL ENGINEERING
Elliott • STANDARD HANDBOOK OF POWER PLANT ENGINEERING
Freeman • STANDARD HANDBOOK OF HAZARDOUS WASTE TREATMENT AND DISPOSAL
Ganic & Hicks • THE MCGRAW-HILL HANDBOOK OF ESSENTIAL ENGINEERING INFORMATION AND DATA
Gieck • ENGINEERING FORMULAS
Grimm & Rosaler • HANDBOOK OF HVAC DESIGN
Harris • HANDBOOK OF NOISE CONTROL
Harris & Crede • SHOCK AND VIBRATION HANDBOOK
Hicks • STANDARD HANDBOOK OF ENGINEERING CALCULATIONS
Juran & Gryna • JURAN'S QUALITY CONTROL HANDBOOK
Karassik et al. • PUMP HANDBOOK
Maynard • INDUSTRIAL ENGINEERING HANDBOOK
Parmley • STANDARD HANDBOOK OF FASTENING AND JOINING
Rohsenow, Hartnett, & Ganic • HANDBOOK OF HEAT TRANSFER APPLICATIONS
Rohsenow, Hartnett, & Ganic • HANDBOOK OF HEAT TRANSFER FUNDAMENTALS
Rosaler & Rice • STANDARD HANDBOOK OF PLANT ENGINEERING
Rothbart • MECHANICAL DESIGN AND SYSTEMS HANDBOOK
Shigley & Mischke • STANDARD HANDBOOK OF MACHINE DESIGN
Tuma • ENGINEERING MATHEMATICS HANDBOOK
Tuma • HANDBOOK OF NUMERICAL CALCULATIONS IN ENGINEERING
Wadsworth • HANDBOOK OF STATISTICAL METHODS FOR ENGINEERS AND SCIENTISTS
Young • ROARK'S FORMULAS FOR STRESS AND STRAIN

DIESEL PLANT OPERATIONS HANDBOOK

Clive T. Jones

McGraw-Hill, Inc.
New York St. Louis San Francisco Auckland Bogotá
Caracas Hamburg Lisbon London Madrid
Mexico Milan Montreal New Delhi Paris
San Juan São Paulo Singapore
Sydney Tokyo Toronto

Library of Congress Cataloging-in-Publication Data

Jones, Clive T.
    Diesel plant operations handbook / Clive T. Jones.
      p.   cm.
    Includes index.
    1. Diesel electric power-plants—Handbooks, manuals, etc.
I. Title.
TK1075.J66   1991   621.31'2133—dc20   90-26979
ISBN 0-07-032814-5

Copyright © 1991 by McGraw-Hill, Inc. All rights reserved. Printed in the United States of America. Except as permitted under the United States Copyright Act of 1976, no part of this publication may be reproduced or distributed in any form or by any means, or stored in a data base or retrieval system, without the prior written permission of the publisher.

1 2 3 4 5 6 7 8 9 0   DOC/DOC   9 6 5 4 3 2 1

ISBN 0-07-032814-5

The sponsoring editor for this book was Robert W. Hauserman, the editing supervisor was Caroline Levine, and the production supervisor was Pamela A. Pelton. It was set in Times Roman by McGraw-Hill's Professional Publishing composition unit.

Printed and bound by R. R. Donnelley & Sons Company.

While the text of this book takes on masculine gender in many cases for convenience, there is no intent on the part of the author to slight or ignore women, whose role in engineering and plant management has seen tremendous growth in recent years.

Information contained in this work has been obtained by McGraw-Hill, Inc. from sources believed to be reliable. However, neither McGraw-Hill nor its authors guarantees the accuracy or completeness of any information published herein and neither McGraw-Hill nor its authors shall be responsible for any errors, omissions or damages arising out of use of this information. This work is published with the understanding that McGraw-Hill and its authors are supplying information but are not attempting to render engineering or other professional services. If such services are required, the assistance of an appropriate professional should be sought.

This work largely owes its creation to John P. Brydon, who generated the seed of an idea, and to Laurence F. Jonson, who provided a rain of encouragement.

CONTENTS

Preface xi
Acknowledgments xiii

Chapter 1. Introduction — 1.1

Chapter 2. The Operator — 2.1

Chapter 3. Engine Stopping, Starting, and Protection — 3.1

Chapter 4. Engine Log Sheets — 4.1

Chapter 5. Engine Fault Identity — 5.1

Chapter 6. Housekeeping — 6.1

Chapter 7. Crankcase — 7.1

Chapter 8. Crankshaft and Torsional Vibration Damper — 8.1

Chapter 9. Bearings — 9.1

Chapter 10. Cylinder Liner — 10.1

Chapter 11. Pistons — 11.1

Chapter 12. Cylinder Head — 12.1

Chapter 13. Cylinder Head Valves — 13.1

Chapter 14. Camshafts and Engine Timing — 14.1

Chapter 15. Fuel Injection Equipment — 15.1

Chapter 16. Turbocharger — 16.1

Chapter 17. Aftercooler — 17.1

Chapter 18. Governor — 18.1

Chapter 19. Filters — 19.1

Chapter 20. Aspirating Air — 20.1

Chapter 21. Exhaust System — 21.1

Chapter 22. Fuel Oil — 22.1

Chapter 23. Cooling System — 23.1

Chapter 24. Lubricating Oil and System — 24.1

Chapter 25. Instrumentation — 25.1

Chapter 26. Foundations, Mounting, and Alignment — 26.1

Chapter 27. Vibration — 27.1

Chapter 28. Engine Preservation and Storage — 28.1

Chapter 29. Seals and Gaskets — 29.1

Chapter 30. Pipes and Valves — 30.1

Chapter 31. Manuals, Parts Books, Workbooks, and Service Bulletins — 31.1

Chapter 32. Engine Overhaul and Maintenance Routines — 32.1

Chapter 33. Spare Parts and Materials — 33.1

Chapter 34. Generator Operation — 34.1

Chapter 35. Generator Overhaul and Upkeep — 35.1

CONTENTS

Chapter 36. Switchgear — 36.1

Chapter 37. Batteries — 37.1

Chapter 38. Compressed-Air Systems — 38.1

Chapter 39. Starting Systems — 39.1

Chapter 40. Engine Turning — 40.1

Chapter 41. Elementary Power Plant Safety — 41.1

Chapter 42. Terminology — 42.1

Chapter 43. Journeyman Notes — 43.1

Chapter 44. Power Plant Control — 44.1

Chapter 45. Annual Inspections of the Entire Power Plant Complex — 45.1

Chapter 46. Records — 46.1

Chapter 47. Measures and Conversions — 47.1

Chapter 48. Equipment Inventory — 48.1

Index follows Chapter 48.

PREFACE

The *Diesel Plant Operations Handbook* has been prepared for the day-to-day reference of the personnel employed in the running of diesel generating plants. Used as intended, the Handbook is a valuable practical guide for the proper operation of any diesel-powered generation complex.

The author has been engaged in the operation, maintenance, and overhaul of diesel power plant main and auxiliary machinery for many years and has drawn upon his considerable experience in providing a Handbook that will assist all of those personnel charged with the responsibility of running a safe, efficient, and economical facility.

The Handbook is also particularly suitable for either the familiarization or training of plant staff at any level or vocation. All references to any particular components are generic in nature so that the entire contents of the Handbook have a direct relevance to any plant where the Handbook is consulted.

The 48 chapters of the Handbook cover the functions of the principal engine components and develop in the operator an early awareness of a machine in difficulty. The Handbook discusses in some considerable detail the five systems upon which all internal combustion engines rely, i.e., air, fuel, exhaust, cooling, and lubrication. Maintenance and overhaul affairs are examined, and suggestions are made throughout the Handbook that will improve the efficiency of the plant in many spheres in a cost-effective manner.

The Handbook will serve the attentive reader as a buffer between the all-too-often inadequate operator's manual provided by the manufacturer and the shop repair manual.

Clive T. Jones

ACKNOWLEDGMENTS

In the preparation of this Handbook numerous publications were consulted, which included:

Diesel Engine Manufacturers' Association, *Standard Practices for Stationary Diesel and Gas Engines*, New York, 1977.
Lister, Eugene C., *Electric Circuits and Machines*, McGraw-Hill, Inc., New York, 1960.
Odberg, Eric and F. D. Jones, *Machinery Handbook*, The Industrial Press, New York, 1964.
Sothern, J. W. M., *Marine Diesel Oil Engines*, vols. I and II, James Munro, Glasgow, 1952.
Westinghouse Electric Corp., Apparatus Repair Division, *Westinghouse Maintenance Hints*, Pittsburgh, 1965.
Wilbur, C. T., and D. A. Wright, eds., *Pounders Marine Diesel Engines*, Butterworth & Co. Ltd., Kent, United Kingdom, 1984.
Williams, D. S. D., *Oil Engine Manual*, Temple Press, London, 1950.

Numerous guidebooks, bulletins, manuals, parts books, and instructional material were also utilized, particularly those issued by Caterpillar Inc., Peoria, Illinois, and their agent in Vancouver, British Columbia, Finning Tractor Ltd.; Electromotive Division of General Motors of La Grange, Illinois, and their Canadian agent, Mid-West Power Products of Winnipeg, Manitoba; Cummins Engine Company, Columbus, Ohio, and their agent in Western Canada, Alberta Governor Ltd., Edmonton, Alberta; and Maintenance Lubrication Services Ltd., Oil Analysts, Edmonton, Alberta.

Reference was also made to the publications of the Waukesha Engine Co., Mirrlees-Blackstone Ltd., GEC-Ruston Ltd., Lucas-CAV, and others.

DIESEL PLANT OPERATIONS HANDBOOK

CHAPTER 1
INTRODUCTION

1.1 THE PURPOSES OF THIS HANDBOOK

This handbook is intended to bridge the gap between the information given in the various operator's manuals and the shop manuals supplied by the equipment manufacturers and thereby help the power plant personnel, particularly the operators, to carry out their duties in a more effective manner. The work done by the operators can be divided into three main activities:

1. Operating the generating plant machinery in the most economic manner possible and appropriately to the prevailing load
2. Maintaining a continuous watch over the entire plant and safeguarding the plant against damage or deterioration
3. Performing the day-to-day maintenance duties that are prescribed by the owner and are in keeping with safe working practices

Diesel power plants have tremendous variations in purpose, equipment, capacity, manning, and management. In discussing the various aspects of operation and upkeep, the terms *operator*, *mechanic*, *planner*, *superintendent*, and *owner* are used frequently. In many instances, however, the operator may also perform the functions of mechanic, planner, and superintendent. Conversely, in larger plants several persons may be assigned to only one of the duties and in almost all plants there is always a certain amount of overlap of functions. The handbook should be read with those variations in mind.

If a diesel-powered generating plant is to operate continuously, reliably, and economically, certain standards of performance must be established by the owner and demanded of both the operators and the equipment. The standards of performance implied in this book are not hard to follow; they are merely statements of what must be done in any diesel plant if the consumer is to get value for his money, if the owner is to get a fair return on his investment, and if the employee is to give service worthy of his wage. On the other hand, the performance standards outlined here are by no means complete, and they never can be, even for a particular plant.

It will be necessary for the owner and the plant superintendent to maintain a constant pressure on the plant personnel to work to the best possible standard of performance and maintain it consistently. The diesel generating plant that has no standard of performance by which to abide will not be able to provide constant and predictably good service.

Another intention of this handbook is to provide a yardstick by which an engine operator can judge his abilities and his progress in learning about his occupation. The book can also be a teaching guide for the plant superintendent. The owner will benefit greatly by hiring the best possible grade of operator, because virtually all the unpredictable engine expenses in manned plants are caused by shortcomings in operator performance. In turn, those shortcomings are caused by lack of training and background experience.

A further intention of the handbook is to effect an improvement in the diesel plant operator's professionalism. In far too many instances, stationary diesel plant operating personnel have been regarded as short-term employees who possess only limited skills, are transient, and are unworthy of much attention in the way of training. It is hoped that this handbook will assist the diesel plant operator in expanding his general knowledge of sound power plant operating practice and thereby encourage him, while expanding his experience, to remain in the industry.

No engine manufacturer produces bad engines, but many engines are unsuccessful and are condemned by their owners, users, or operators as bad because, in truth, they were improperly selected, applied, operated, or maintained. It is quite understandable that any person who is associated with diesel engine operating or upkeep work should form his own opinions on the merits of various engines. As time goes by and a person's experience broadens, that opinion becomes very useful. But no matter what attitudes toward a particular make or model of engine they may have, the operator, the mechanic, and the superintendent can only base their engine care activities on the premise that all of the engines and equipment they are required to work on is suitable for its duty. They must never knowingly evade giving the machinery under their control their best possible attention at all times.

1.2   OPERATOR EDUCATION

The operator's ability to perform effectively will be in direct proportion to the training and experience that he brings to the job or is able to acquire. Diesel engine operating is a respectable and responsible occupation, and it is hoped that this handbook will contribute something to the trade.

To operate a diesel generating plant successfully, a knowledge of purpose and expected performance of the plant under various conditions is necesssary, and so is an understanding of the part played by the many components. By its very nature, a plant operator's job is lonely. For that reason, an operator, once his basic and often all too brief training is complete, will learn nothing more about his job unless he takes the trouble to teach himself. The operator who is prepared to learn more about the nature and operation of the equipment that surrounds him will find other people such as station superintendents and journeymen more than willing to supply further instruction.

Operators should be encouraged to read on shift. Anything may be read as long as it is directly related to the job. Operator's manuals, parts books, procedure manuals, safety codes, *original equipment manufacturer's* (OEM) instruction books, training guides, job-oriented textbooks, and similar literature are provided by the owners, suppliers, and technical schools. If little material is available, it is not unreasonable for the operator to request it. By continuous reference to this reading matter, an operator can give himself a large amount of necessary education. He will find that the more he studies, the more interesting and satisfying his job will become.

Unfortunately, OEM-supplied manuals usually contain only the rudimentary elements of running and servicing the engine, whereas the shop manuals deal only with the rebuilding of the engine or its individual components. Generally, none of these manuals provide much in the way of diagnostic assistance, trouble recognition, or fault avoidance.

Modern engines have many exact and carefully calculated adjustments that are necessary for efficient, smooth, and trouble-free operation. It is therefore essential that any operating adjustment or setting work shall be carried out by properly trained personnel who must use the relevant instruction manual.

There is simply no way a person, even with the best memory in the world, can remember the sequences, clearances, torque values, pressures, temperatures, and all the other vital information for handling an engine. Without there being a continuous reference to the OEM manuals during repair, maintenance, or operation, one may expect starting, running, and upkeep problems. There is a proverb in the engine world that "disregarded manuals and broken engines lie together in the same junk heap."

At least four basic books are essential for engine upkeep: operator's manual, the shop manual, the parts book, and the workbook. The first three are supplied by the OEM, and the fourth is supplied by the owner. They must, naturally, be available to both the operator and the mechanic at all times. In the event of loss or disintegration, their replacement is necessary. For engines of 750 kW at 720 rpm and larger, test sheets also are usually available. Occasionally, some or all of the four basic books will be found combined within one cover.

Any newly hired mechanic or operator will check out the presence of the four basic books as an early priority in his new duties. It is his duty to ask for them if they are not available. It must be understood by all personnel that no matter how many copies of these publications appear to be available, they always remain the property of the engine owner—past, present, or future. Although all manuals and other information sources must be properly looked after, nobody on the plant staff is entitled to the exclusive custody of any particular document. All teaching, guiding, and specifying material must be visible to all those who have good reason to use it.

1.3 CLOTHING

Simply as an elementary form of personal protection anyone occupied within the power plant should observe some basic clothing rules.

1. All power plant personnel should wear some kind of headgear. If the plant regulations require their use, hard hats or bump caps, must be worn. In the absence of such regulations, a cap at least should be worn. A person with longer hair must consider using a hair net also.
2. A long-sleeved garment should be worn so that the arms are fully covered.
3. Hard-toed boots or shoes are recommended, but in no case should rubber-soled footwear be permitted. Jogging shoes and Wellington boots are particularly dangerous.
4. Gloves are regarded as dangerous and as inhibiting the proper performance of operator and journeyman duties. They should be used only when handling very hot or very cold materials or tools. Hands inside gloves are very poor sensors, and they adversely affect an operator's judgment of conditions.

5. Hearing protection and safety glasses must be used if plant regulations demand them, and the use of both is highly recommended in all plants on a voluntary basis.
6. Further, and purely in the interest of personal safety, it is prudent that rings (both finger and ear), wristwatches, necklaces, bracelets, and neckties should be left at home.

Any of the owner's rules regarding attire, protection or safety apply equally to any power plant visitor.

1.4 OEM DEFINITION

The term *OEM* will frequently appear in the text of this handbook; it includes the engine builder, the alternator manufacturer, or any of the companies that provide switchgear, governors, fuel injection apparatus, or any other equipment incorporated in the final product as purchased. The OEMs provide their own service bulletins, specification sheets, parts books, instruction manuals, and other guidance information.

CHAPTER 2
THE OPERATOR

2.1 OPERATOR IN CHARGE

At any diesel power plant there can be only one person in charge of generator operations. That person, known as the *operator in charge* may carry out his duties alone or as one of several persons under that authority of a station superintendent. Whatever the command structure, the person acting as the operator in charge is responsible for the safe and proper starting, running, and stopping of the generator sets.

Under ordinary circumstances, the operator goes about his duties in a routine manner; he is the person trained and assigned to handle the generating equipment and its auxiliaries. The occasion may arise, however, when the operator is not sufficiently experienced to handle an emergency. If there is an outage or some other form of breakdown, it is quite common practice for the station superintendent, electricians, mechanics, or other personnel to attend the plant and make their assistance available. In those circumstances, it must be well understood by all concerned that, no matter who helps out, the operator remains in charge of operations unless he is relieved by the superintendent. If that occurs, the superintendent must tell the others who is then the operator-in-charge.

The occasion may also arise when more than one shift operator is at the plant during an emergency situation. Unless the station superintendent has stated otherwise, the operator in charge is the operator assigned to the shift within the current time frame.

The operator in charge is responsible for the proper and safe conduct of plant generating operations. All workmen performing any task within the environs of the plant must defer to the operator. They should not open any cubicle, remove any floor plates, commence welding, or carry out any other work that may infringe on the operator's normal activity or the plant's power-producing capability without clearly advising the operator of their intentions and obtaining formal clearance. Following such advice, it is the operator's duty to see that the necessary and appropriate flags, tags, or other warnings are applied in accordance with standard safety rules and practices.

At the time of shift change, the incoming operator must be fully briefed by the outgoing operator on any ongoing changes or alterations that have occurred during the preceding shift that may impede the normal operation of the next shift. The operator in charge must be aware of all plant visitors, whatever their rank or business, and he should acquaint them with the site safety regulations applying to such visitors.

Although the primary function of the operator in charge is to produce safe, economic, and uninterrupted power for the consumers, he is also obliged to extend his cooperation to all efforts being made to improve, maintain, or repair the generating facility and its surroundings.

2.2 SHIFT CONDUCT

There should be a continuous dialogue between the shift operator and the maintenance personnel who are working around the plant so that the operator is advised in good time of any intended test running, circuit checking, power isolation, or other activities.

No actions that may have a bearing on the operation of the plant should take place at or about the time of shift change. If necessary, the operator may wish to make arrangements for an additional operator to be on hand if there is likely to be much distraction, caused by extraordinary activity of the shift operator from his regular duties.

It is useful to make available a list of all the items that must receive the operator's attention during a shift. The list should be there for the operator as a guide, but the operator's inspection tours should not be limited to the listed items. Some of the points may require more than one visit during the shift, and it should be understood that the list is open-ended and additional checks will be added as required. It may be found convenient for the operator to break the inspection tour into several sections and handle each section during a different period of the shift. Whatever the routine adopted, the required activities should be conscientiously carried out so that the operator who picks up the next shift can do so with confidence.

Here are some examples of what should receive attention:

1. Review the day service tank, the transfer tank, the bulk storage tank levels, and the tanks themselves for leakage, venting, security, and so on.
2. Inspect the fuel-water separator. Drain any water accumulation, and drain the day service tank for water indication.
3. View the exhaust outlets for smoke, color, and signs of slobber, sparks, and so on.
4. Inspect the guards of the air intake hoods and remove any seed, pollen, frost, or other obstructive accumulation.
5. Review the integrity of the transformer yard fence and outdoor area protection.
6. Check the condition of auxiliary machine belt drives.
7. Inspect the fire extinguisher seals.
8. Test the functioning of the emergency lighting system.
9. Ensure that sufficient quantities of lubricating oil are on hand for both the current and the next shift.
10. Look at all gas cylinders that are permitted to be within the plant and check that they are safe. Remove all unpermitted or excess cylinders.
11. Wipe all instrument faces clean.
12. Drain all air-oil separators and dispose of waste oil in an acceptable manner.

13. Inspect all cable troughs for the presence of lubricating oil, water, or fuel.
14. Replace all manhole covers and floor plates in their proper positions.
15. Park the crane, its hook, maneuvering chains, and control box in safe positions.
16. Clean battery terminals and add electrolyte if needed.
17. Clean drip trays as necessary and record leakage in log book.
18. Inspect the washroom for cleanliness and sufficiency of supplies.
19. Check all service motors for signs of overheating, vibration, and so on.
20. Read the preceding day's log sheets and workbook entries.
21. Drain the compressor water traps and also the air receivers.
22. Familiarize yourself with ongoing work routines and locations, alterations and changing conditions.
23. Inspect the raw water pumping plant and raw water discharges.
24. Ensure that all rotating-machinery guards are in place.
25. Refill air-start lubricators as needed.
26. Check chain-type air filter oil levels.

The list could be extended, and it will vary with the equipage peculiarities of the plant. Worldwide, however, the basic inspection routines will be very similar except for the special attention that regional climatic conditions demand. In a well-run plant there are few surprises because the continuous process of inspection and checking will reveal any intentional discrepancies before they occur. It is good practice to have one day per week or per month scheduled for the more infrequent but necessary routines. In that category will be found such activities as greasing motors, oiling hinges and linkages, and inspecting the property fences.

2.3 ODORS IN THE POWER PLANT

There are many "vital signs" about an engine which will warn an operator or mechanic of trouble. Some of those signs are more obvious than others, but all are worthy of notice. For example, the odors of a power plant change from time to time, and one smell will dominate another. Recognition of odors and their changes will assist in troubleshooting or early fault detection. A person arriving at a plant to commence work will notice an unusual odor much more readily than the person who has been there for some time. For that reason, it is recommended that each of the power plant personnel commence his "smell inspection" before he steps inside the power plant door at the commencement of his shift. That may seem a bit ridiculous, but it is a perfectly valid and extremely useful part of a plant inspection. It is stressed that the smell inspection must be made with a nose that has not been deadened by hours of exposure to diesel fumes.

It is rather difficult to describe the odors one might encounter in or around a diesel power plant, but the personnel should take heed of the sweetish smell of unburned diesel fuel, the hot-chocolate character of overheated insulation, the throat-catching aspect of battery acid, the steamy wet smell of escaping hot coolant, the pungent eye watering of scorching cloth or wood, and the easily recognizable tang of overheated metal or burning paint. These and other smells indicate anomalies and nearly all will demand immediate attention.

2.4 HAZY ATMOSPHERE IN THE POWER PLANT

Upon arrival at the power plant to assume his shift, the operator should take note of the atmospheric conditions in the engine hall. The atmosphere may be clear, or it may be murky. Ideally, the air in the plant should always be clear, but some plants always run with a slight haze. The operator must notice any change in the atmosphere, identify the reason for the change, and act in a way that suits the occasion.

A plant that is permitted to operate with a dirty atmosphere will quickly develop a major housekeeping problem. The walls and ceiling will discolor and become black if the haze is allowed to persist; the ventilation fans will become coated with oil, which will then be thrown off in blobs that spatter the machinery, floor, and walls; the air filters will have a short life; and the turbochargers will require dismantling and cleaning much more often. Visibility in the plant will be reduced as an oily film settles on the lighting fixtures. Many man-hours will have to be diverted to cleaning duties, and the morale and safety management of the plant will be impaired.

Haze may be caused by any of the following:

1. Fumes leaking from the engine exhaust system
2. Broken high-pressure fuel lines spraying semiatomized fuel
3. Overheated machinery evaporating the volatile components of previously deposited paints, varnishes, or atmospheric dirt
4. Fuel or lubricating oil dripping onto and vaporizing from hot surfaces
5. Excessive crankcase pressures that cause lubricating oil mists to escape from the crankcase
6. Smoldering rags
7. Leakage in the waste-heat recovery systems

If the plant appears to be unusually vapor-laden at any time, the operator must make it his business to find the cause and then decide what must be done. All power plant personnel should know that many power plant fires are preceded by smoke, mist, or fumes. A plant with a poor atmospheric condition is a plant in trouble.

2.5 POWER PLANT CLIMATE

The climate within the average diesel power plant is generally unchanging throughout the year, but any noticeable change can be taken as a warning of something amiss. Merely opening the entrance door may serve to warn the operator that there is something wrong with the plant's ventilation system.

If the door is very "heavy" to open and slams shut, the probability is that the ventilation system is closed or shut off and the running engines are being starved of air. Any engine consumes large quantities of air, usually from inside the building. So that adequate amounts of replacement air can enter, manually or barometrically operated louvers are set into the power plant walls, and air filter panels screen the outside air before it passes to the engine. The flow of air will be impeded when the filter panels become dirty, and the atmospheric pressure of the plant interior will be reduced below the external pressure because of the engine air consumption. That will account for the heavy door. The maintenance routine of any plant must allow for a regular inspection, cleaning, and replacement of the filter panels.

In winter it is not uncommon to find the thermostatically operated louvers of wall-mounted radiators out of adjustment, which will cause the radiator fans to exhaust excessive quantities of air from inside the plant. Refer to Figs. 23.8 and 23.9 for the correct arrangement of inlet and outlet louvers. A dirty radiator core can often be the cause of an abnormally cold power plant interior. In a mistaken effort to keep the engine and its cooling system running, the ventilation is fully opened up in the hope that surface cooling of the engine will alleviate the difficulties. The only real solution, however, is to clean the radiator core.

In any nonequatorial plant location, there is no reason why a comfortable working temperature of 24 to 32°C (75° to 90°F) cannot be maintained at all times. Whenever the normal operating temperature of the plant interior moves toward the uncomfortable range, the reason must be found and corrective action must be taken without delay.

If in spite of the engines being in operation, the plant seems to be uncomfortably cold, the cause of the discomfort must be found and rectified. No matter what arrangements are provided for heat rejection or waste-heat recovery, sufficient heat must always be available for the needs of the generating plant as a first priority. The operator must resist the pressures of other waste-heat consumers who wish to take more heat to the detriment of the power-generating facility. Obviously, all reject heat must be utilized whenever possible, but the generating plant priorities cannot be downgraded.

It is suggested that all year-round windows, doors, or other openings in the plant be kept closed at all possible times. Open doors will permit a great deal of dust, insects, pollen, and other airborne debris to enter the plant. Not only will the debris cause premature and unnecessary expensive air filter element changeouts but the generators, switchgear, control gear, the engine itself, the structural steel, and many other hard-to-clean areas will become coated with dirt. That will increase costs immensely, with no offsetting benefit to the owner or the operator.

The simple matter of security demands that the generating plant buildings be kept closed whenever feasible, and environmental considerations suggest that the noise of the plant be kept behind closed doors. A grittiness on the power plant floor caused by open doors may be taken as notification that the engine air filters are going to be plugged sooner than they should be. Just sweeping the floor is not good enough; grit entry must be stopped.

A hot steamy atmosphere may develop in an engine hall when an external coolant leak of some proportion develops. It is unlikely that a coolant leak in a radiator would be manifested in this manner, but a ruptured flexible hose, burst frost plugs, or an overflowing header tank would readily provide such humid conditions.

2.6 MACHINE SOUND LEVELS

The characteristics of the sounds produced by any machine are indicative of the machine's condition and the loads being imposed. That statement is generally true, but it is valid only for the person who is familiar with the particular machine and the generic noises to be expected from the various machine classes. An operator's familiarized ear is a good source of information on a machine's condition.

The listeners's ear must be practiced; if an operator is to become acquainted with the sound pecularities of a particular engine, he must listen to it regularly and frequently. He must use the same listening device each time and listen to the same places on the machinery in much the same way that a vibration-monitoring tool is used. Indeed, some vibration-monitoring instruments are capable of reading sound quantities, but not sound qualities.

Some operators may content themselves with using a hammer shaft rather than a stethoscope for listening to the engine's internal noises, but whatever the method, familiarity is all-important. Of course, when a noise inspection is carried out, the prevailing load, the effects of other machines in operation, and other variable conditions must be taken into account. Changes in engine noise levels and rhythm are a very valuable indication of altering engine conditions, and they deserve the operator's continuous attention.

2.7 DATA PLATES

Virtually every engine has a data plate that shows the engine make, model, and serial number as a bare minimum. Other plates that may be on the engine will indicate the capacity, the horsepower, the rotational speed, the rotational direction, the year of manufacture, the manufacturer's address, the original supplier, the critical speed range, the arrangement number, and the weight. The attached generator will have similar pertinent data.

Data plates will also be found on variety of equipment such as safety valves (make, model, set pressure), pumps (make, model, capacity, rotation speed), and waste-heat boilers (make, model, working pressure, evaporation rate or horsepower), compressors (make, model, serial, CFM rating, and maximum pressure). Each of these plates has the purpose of stating what performance can be expected. Each of the power plant personnel should familiarize himself with the potential capabilities of the plated equipment and take appropriate action when the stated performance becomes unattainable. The action taken may be in the form of log entry, written report, or activity directed by the superintendent.

2.8 ENGINE POSITIONAL DEFINITIONS

Fig. 2.1 shows the accepted engine layout and positional definitions. The front of the engine is the free end; the rear is the end from which the load is driven. The left bank of cylinders is the A bank, and the right bank of cylinders is the B bank as viewed when standing at the free end of the engine.

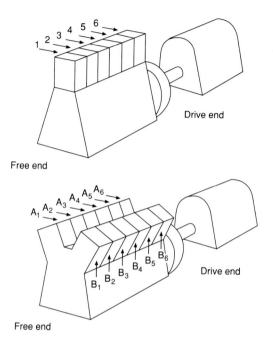

FIGURE 2.1 Standard engine and cylinder identification scheme.

2.9 ROUTINE MAINTENANCE

Small Plants

A typical maintenance routine for a 300-kW generator set in continuous service is outlined below. It is expected that most of the work will be done by the plant operator. The typical overhaul work on the same unit is described under "Scheduled Overhauls." Comparison of the two activities will reveal the great differences between maintenance work and overhaul work.

Daily or Every 8 h (Whichever Occurs First)

- Inspect the lubricating oil dipstick. Take notice of both the level and the appearance of the oil.
- Inspect the air filter element if visible. Note the position of the service indicator.
- Inspect the fuel tank level; check for leaks; drain the water; and read the fuel meter.

- Inspect the governor oil level; add makeup oil as needed; and record the amount.
- Record the kilowatt-hour meter reading and, in conjunction with the fuel meter reading, establish the kilowatts per gallon or per liter performance for the preceding period.
- Wipe down the entire machine with full and due regard for safety. Note any defects.
- Drain the water from the fuel-water separator.
- Clean the drip trays.
- Check the oil level of the pedestal bearing.
- Check the coolant level when possible.
- Do a walk-around inspection of the entire plant including fuel system, cooling system inside and outside, the waste-heat recovery system, ventilation system, exhaust system, switchboards, transformer yard, and outbuildings. Record all noticed defects.
- Perform assigned housekeeping duties.
- Write up the log sheet. Include any unusual events.

Weekly or Every 100 h (Whichever Occurs First)

- Check the electrolyte level in the batteries.
- Check the cleanliness of the radiator.
- Check the tension and condition of the fan belts. (For this operation, the engine must be at rest.)
- Drain the water and sediment from the fuel filter housings.
- Take and record vibration readings from set points on the generating units.
- Exercise all standing units.

Biweekly or Every 250 h (Whichever Occurs First)

- Change the lubricating oil; change the lubricating oil filters; and clean out the filter housing. (The intervals of these activities may be longer or shorter, but the maximum must be in accord with the OEM manual.)
- Examine the oil filter elements for metal presence.
- Examine the air filter element.
- Clean the crankcase breather element.
- Clean the top and terminals of the batteries.
- Check the specific gravity of the battery fluid.

Monthly or Every 500 h (Whichever Occurs First)

- Check the pH of the coolant. Adjust as necessary.
- Grease the fan hub bearings, alternator bearings, and tachometer drive.
- Inspect the generator and exciter brushes; record the sparking and condition.

Bimonthly or Every 1000 h (Whichever Occurs First). It will be seen at this point that the maintenance duties begin to get shared between the operation staff and the journeyman mechanics.

- Take a sample of the lubricating oil for analysis.
- Lubricate the governor speeder motor.
- Test all unit protection devices.[J,O][1]
- Renew worn generator brushes; dress the commutator.[J]
- Clean the slip rings.[J][2]
- Clean and recoat the battery terminals.[J]

Biannually or Every 2500 hours (Whichever Occurs First)

- Adjust valve lash.[J]
- Check valve rotation when applicable.[J]
- Clean out the bottom of the sump.

Annually or Every 5000 h (Whichever Occurs First)

- Carry out scheduled overhaul of the engine.[J]
- Carry out scheduled overhaul of the alternator.[J]
- Carry out scheduled overhaul of auxiliary equipment.(*)
- Check alignment.[J]
- Test all equipment and prove its ability to perform in accordance with specifications.[J,O]

Large Plants

For plants of 1.0 MW or more, the nature of the maintenance work will not be very different from that outlined for a small plant, but the quantity of work will be much greater. It is probable that an operator will be continuously present, and it is expected that he will do the bulk of the work during his shift. In the very large plants, some of the routine maintenance will be assigned wholly to dayworkers of the various trades.

Every Hour. Inspect the lubricating oil dipstick, and take notice of both the level and the appearance of the oil. Check the lubricating oil levels on the governor, the turbocharger(s), the camshaft, or any other component with an independent lubricating oil system. Take at least one full turn on each self-cleaning filter. Write up the log sheet, and note any unusual event.

Every Shift or Every 8 h (Whichever Occurs First)

- Inspect the air filter element if it is visible. Note the position of the indicator.

---

[1]Work requiring the presence of both a journeyman and an operator.
[2]Specialist work handled by a journeyman of the appropriate trade.

- Inspect the daily service fuel tank. Look for leaks, spillage, or water.
- Take and record vibration readings from set points.
- Check the pressure drop across the lubricating oil filters.
- Drain the water from the fuel-water separators.
- Drain the water from the compressor water separators.
- Wipe down the assigned equipment. Working with due regard for safety; log noted defects; and perform designated housekeeping duties.
- Clean all drip trays.
- Check the coolant level when possible.
- Do a walk-around inspection of the entire plant including the bulk fuel system, primary and secondary cooling systems, the waste-heat recovery equipment, the transformer yard, and the storage facilities whether internal or external to the main plant buildings.
- Write up the log sheet and record all unusual events.

Weekly or Every 100 h (Whichever Occurs First)

- Check the electrolyte level in the batteries; clean the battery tops and terminals. Recoat the terminals.
- Check the cleanliness of the radiator and the tension of the fan belts of all standing units.
- Drain water and sediment from the fuel filter housings on standing units.
- Exercise all standing units.
- Lubricate the governor linkage.
- Clean magnetic plugs where fitted and enter comments in logbook.

Monthly or Every 500 hours (Whichever Occurs First)

- Check the pH of the coolant and adjust it as necessary.
- Check all shutdown devices.
- Grease fan hubs, motor bearings, linkages, and soon, as described in standing instructions.
- Change the oil in oil bath air filters. (Frequency will vary with local environment conditions.)
- Change and wash oil-wetted air filter elements.
- Inspect and service brushes, sliprings, and commutators of standing machinery.[J]
- Adjust the tappet clearances of all standing engines.[J]
- Change the lubricating oil in the turbochargers.[J]
- Check the condition of and adjust the compressor drive belts as necessary.
- Record cylinder peak pressures and rack settings.

Bimonthly or Every 1000 h (Whichever Occurs First)

- Take a sample of lubricating oil for analysis.

- Test all engine protection devices.[J,O] Take a set of crankshaft deflections.[J]
- Change the lubricating oil and lubricating oil filter elements. Clean filter housings, including self-cleaning filters. Clean out the sump. (Oil change frequency should be in accordance with OEM recommendation.)
- Examine filter elements.
- Clean engine and turbocharger breather.
- Perform planned segments of overhaul cycle.[J]

Further Planned Work at Set Frequencies

All work scheduled for intervals of more than 1000 h on larger units involves the rework or replacement of components having definite life spans. Although the plant operating staff may assist in it, this further work is normally handled by a specialist journeyman. There are a multitude of engine care activities that must take place frequently but which cannot feasibly be written into a fixed schedule. For example,

   1. *The air filter elements* should be changed when indicated as necessary by the monitoring instrumentation or when it is obvious that the element is loaded. The time required for the filter to plug will vary with the load, the seasonal climatic conditions, and the changes in the local environment (e.g., adjacent dusty roads becoming muddy roads in the wet season).
   2. *The primary fuel filter* will collect more debris immediately after new fuel has been delivered. After the fuel has settled, there will be less sediment movement. The fuel-water separator will also precipitate more water after the bulk supply has been agitated.
   3. *The secondary fuel filter elements* will serve for a longer time if the fuel supply is relatively clean, and the plant operating staff can exercise their judgment accordingly.
   4. *Governor bleeding* will be necessary from time to time to eliminate any hunting or surging. This activity can be carried out only by a journeyman mechanic who is fully conversant with the procedures.
   5. *Battery cleaning* will be necessary at intervals, and the frequency is dependent largely on the care taken to mop up fluid that splashes when the hydrometer is withdrawn from individual cells and when the batteries are topped up. Geographic location also can have an effect on the frequency. In humid areas and coastal locations, the moisture in the atmosphere will foster an increase in the deposits of lead oxide and sulfuric acid on the battery terminals, whereas in very dry atmospheres, water loss will be greater.

2.10 TOOLS FOR SHIFT USE

The operator must have access to various tools if he is to conduct his shift in a satisfactory manner. The tools that he is expected to use must be those best suited to the work to be done. The operator should not be given pipe wrenches, adjustable wrenches (crescent wrenches), and slip-joint pliers of varying designs to use during his shift. None of these "versatile" tools do everything well, and in most cases they ruin the surface of the part or fastener to which they are applied. None of these tools can be recommended, and so the superintendent must make the right tools available to the operator and make the operator responsible for their safekeeping.

2.11 HAND GAUGING OF TEMPERATURE

Temperature gauging by hand is not necessarily conclusive because the temperature prevailing may exceed the hand bearable limit of 50°C (120°F). Even up to 50°C (120°F) it will be sketchy, and considerable practice would be needed to attain a reliable temperature-judging proficiency.

2.12 POWER PLANT DOORS AND OPENINGS

Besides the dust and grit that naturally enters a power plant that is operated with its doors open, there are other considerations. If the power plant doors are kept shut, there is little risk of small animals or birds entering the building. The internal environment is totally inappropriate to these creatures, who tend to take refuge in warm confined places such as generator windings, switchgear centers, cable ducts, and motor control centers (MCCs).

All openings such as pipe entries, cable and pipe trough endings, sprung siding, damaged doors, and distorted door frames are common points of entry, and all of them can be easily and usefully closed for the good of all. Birds will come into the plant through the overhead door or through the vent louvers, and they too will attempt to hide after they are exhausted. Simple wire mesh screens over vent openings easily prevent the entry of birds.

CHAPTER 3
ENGINE STOPPING, STARTING, AND PROTECTION

3.1 STOPPING AN ENGINE

An engine can normally be stopped by returning the fuel pump rack to the zero, or no-fuel, position. The fuel rack return can be accomplished by electrical (solenoid), mechanical, or manual means. Before the final stopping sequence is initiated on any engine, it is essential that it be brought down to idle speed under pressurized lubrication. That allows the turbocharger to run down to its slowest speed with a full flow of oil to the bearings.

If the stopping sequence is initiated at synchronous speed, the engine will come to rest before the turbocharger has stopped rotating. On many engines in which the lubricating oil supply to the turbocharger is pumped from the engine lubricating pump, there is a real danger of turbocharger seizure. Allowing the engine to run at idle speed for about 5 min also permits the internal temperatures to be substantially reduced and the dissipation of residual heat to be facilitated, thereby lessening the risk of coolant afterboil.

Never use emergency shutdown devices for normal shutdown purposes. In many engines, the design of the emergency devices is such that the devices will readily wear out if overused. Shutting the engine down by closing the fuel line valves can be extremely dangerous, and many engines have fuel return lines which cause complications. Before the supply valve has been fully closed, there will be fuel in the engine system, and until all the fuel in the line between the fuel pump and the shutoff valve has been evacuated, the return line valve cannot be closed.

If the fuel return line is closed prematurely, there is a possibility of fuel pump or fuel line rupture. Most of the fuel transfer pumps are of positive-displacement type, and it is not normal to have pressure relief valves fitted in a fuel injection system. Second, shutting the supply fuel off can result in severe damage to the fuel pump plungers. As they become starved of fuel, they lose the lubrication that is provided by the passing fuel and consequently seize. It must also be noted that shutting the fuel off at the supply valve is a very slow way of getting the engine stopped. When an engine is stopped in that manner, it will be necessary to prime the system again before the unit can be restarted.

When an engine fails to start because of a safety shutdown trip or shuts itself down in normal operations, the true cause of the shutdown must be found and rectified before the engine is started up again. If the engine is restarted without the shutdown cause being clearly understood, there is a risk of aggravated damage.

3.2 SHUTDOWN SEQUENCE OF A NORMAL UNIT

1. Complete load transfer to the incoming unit.
2. Trip the breaker of the outgoing unit at no load.
3. Switch out the automatic voltage regulator of the outgoing unit.
4. Lower the engine speed to low idle.
5. Reduce the load governor limit to 50 percent.
6. Allow the turbochargers to decelerate to minimum speed.
7. Place the droop control in the start-up position.
8. Make a walk-around inspection.
9. Stop the engine.

Do not take shortcuts in starting or stopping procedures. Shortcuts become bad habits, and bad habits shorten engine life. Do not make "emergency" starts. A few extra minutes of darkness never harmed a good engine or the community it serves. *Learn to stop an engine first; then learn to start it!*

3.3 STARTING AN ENGINE

In the past, much damage has been done to engines by following improper starting practices and incorrect operating methods. The successful operation of the equipment is greatly dependent on the attitude and knowledge of the plant personnel. Before attempting to start any engine, it is good practice to make a visual inspection of the unit. Check the lubricating oil level; be sure that it has not risen since the engine was last shut down. Check that all of the necessary isolating valves for the fuel and cooling services are open. Check that the governor oil is at the correct level. See that all of the shutdowns have been reset and are in operating positions. Ensure that the prelubrication and preheat facilities are functioning correctly. Be sure that the barring mechanisms are in the OUT position and that all the running gear is clear of obstructions. See that the coolant level is adequate and shows no sign of contamination by either fuel or lubricating oil. Check the shutdown solenoid indicator light. See that the governor controls are in an acceptable position. When the engine is mounted on vibration isolators, be sure that nothing is wedged between the underframe and the floor. Make sure that all persons are standing clear before initiating the start.

Many engines have governors featuring an external load limit setting knob. In every such case the load limit setting must be less than 50 percent before an attempt to start the engine is made. If the load limit setting is too high, the engine can and will immediately go to overspeed after startup. It should be understood by everyone that a governor is there only to control normal operating speed. Overspeed protection is provided separately by either a mechanical or an electrical shutdown device. If the governor is improperly set, overspeed of the engine will result, and only then will the overspeed protection come into play.

If an engine fails to start after 15 s of cranking, allow the starter(s) to cool down for several minutes before repeating the starting procedure. (Use the stand-down time to check the systems and look for a possible problem or the reason for difficult starting.)

ENGINE STOPPING, STARTING, AND PROTECTION

The use of ether as a starting aid is strongly discouraged; it must not be resorted to except under the most extraordinary circumstances. (If an engine has failed to start the first time by using ether, there is danger of uncontrollable runaway if additional ether charges are used.) Ether must never be used by a person who has not been properly trained in its use and in its safe handling. Under no circumstances must ether be used to aid the starting of an engine that is fitted with glow plugs. *If ether is used on a glow plug engine, an explosion will result.*

Jacket water heaters are fitted to diesel engines to maintain the coolant within the jacket of a standing engine at a temperature conducive to easy starting. That temperature should be about 27°C (80°F).

3.4 STARTUP CHECK OF NORMALLY AVAILABLE GENERATOR SETS

1. All covers and guards in place
2. Clear of obstructions
3. Overhead crane clear
4. Battery or air charge at optimum level
5. Annunciator flags in operating position
6. Mechanical overspeed protection device in run position
7. Cooling system controls in operating position
8. Coolant levels correct
9. Lubrication levels correct
10. Fuel level, grade, and temperature correct
11. Fuel lines open and system primed
12. Governor load limit in start position
13. Solenoid indicator light
14. Automatic voltage regulator switch in OFF position

3.5 STARTUP SEQUENCE OF NORMALLY AVAILABLE GENERATOR SETS

1. Perform startup checks.
2. Start the engine.
3. Run at low idle speed.
4. Observe fuel and lubricating pressures.
5. Make a walk-around inspection.
6. Set the droop controls on incoming and running units in the appropriate position.
7. Raise the engine to synchronous speed.
8. Switch on the automatic voltage regulator.
9. Check the voltage.

10. Switch on the synchronizer.
11. Adjust the engine to synchronizing speed.
12. Close the breaker and transfer part of the load.
13. Switch off the synchronizer.
14. Adjust the speed of the synchronized units.
15. Make a walk-around inspection of the oncoming unit.
16. Transfer the remaining load.

3.6 LOAD DIVISION

When generators are operating in parallel, the kilowatt load should be divided between the running units proportionately to their ratings. On ac generators the load can be moved from one unit to another only by speed control, not by manipulation of the voltage regulator rheostat. Alteration of the voltage regulator rheostat will only alter the power factor (PF) and thus the current output of the generators, which will result in undesirable crosscurrents.

When generators are running in parallel, the load carried by a unit running at less than full load can be increased by transfer from another unit merely by operating the governor raise control. The frequency also will rise, and it will be normalized by operating the governor lower control on the second unit which surrenders load.

After the kilowatt load has been allocated in good proportion, the reactive (VAR), or wattless, load also must be proportioned. If the generators are of the same capacity, their amperages should be similar. If there is a difference in the amperages for the same load, crosscurrents are probably present. They should be eliminated by adjusting the voltage regulator rheostats on the respective units. Only minute adjustments of the adjusting knob should be made, and enough time should be allowed between adjustments for the meters to settle. If the generator sets are of different sizes, the load will be divided proportionately and the amperages will differ, but the reactive load (VARS) must be balanced if circulating currents are to be avoided.

When the load distribution between the running units has been established, very little subsequent adjustment will be needed if the load increases or decreases. Thereafter, the load apportionment will be handled by the regulating governor's speed and the reactive load will be handled by the crosscurrent compensation feature of the voltage regulator.

3.7 ENGINE OPERATING TEMPERATURE

The maintenance of the correct operating temperature of the lubricating oil in any engine is of great importance. Under normal circumstances, the lubricating oil temperature is kept approximately the same as the jacket temperatures. That permits efficient lubrication, and any water that may accumulate in the system is able to vaporize and leave the crankcase via the breather. If water remains in the crankcase, it tends to combine with combustion products to evolve sulfur and thus form sulfuric acid. Low oil temperatures will lead to sludge formation, shorten filter element life, decrease pan capacity, and accelerate wear rates.

3.8 IDLE RUNNING

Low idle speed must not be set below 65 percent of the synchronous speed; high idle speed should be about 3 percent above synchronous speed. It is vitally important that the voltage regulator be switched out when the unit is run at anything less than synchronous speed.

3.9 ENGINE PROTECTION

No person should be permitted to start an engine without being fully conversant with all the stopping methods. Each and every person responsible for stopping and starting engines must be fully familiar with all of the shutdown devices attached to the engine in his charge. Each of the shutdown devices has a resetting facility of either the automatic or the manual type.

Alarms are provided to warn the operating personnel that a situation that demands immediate attention has arisen. Shutdowns are provided to protect the machinery from any damage that would occur if the machine continued to operate under unacceptable conditions. The operator must never rely on the alarm system to advise him of a developing situation. Instead, he must exercise the continuous vigilance that will keep the alarms from sounding off.

Before starting, check all shutdown devices for function, reset all those that have tripped, identify why they were tripped, and exercise appropriate caution and take appropriate action before initiating another start. Nearly all engines have protection against:

- Low lubricating oil pressure.
- High cooling water temperatures.
- Overspeed.

The owner should not permit the operation of any engine without these three basic protective devices being in perfect working order. It is imprudent to let an interval of more than 60 days pass without testing the devices, even if the engine has not run all of the 60 days.

Many engines have additional protective devices:

1. Mechanical lubricating oil pressure shutdown
2. Mechanical jacket water temperature shutdown
3. Mechanical overspeed shutdown
4. Excessive vibration shutdown
5. Low coolant level switch
6. High lubricating oil temperature switch
7. Crankcase pressure switch

8. Low lubricating oil level switch

The devices can vary considerably with the individual applications, and some but not necessarily all of them will be found on different generator units. When these devices are fitted, they are subject to the same inspection and proving routines as the three basic devices.

All operators and operating superintendents must understand that under no circumstances are any personnel other than journeyman mechanics to be allowed to change out governors or perform any mechanical work related to the speed-control systems. NEVER permit an engine to be started without the governor linkage being fully and properly connected. This rule applies to any engine or personnel, whether owner-, contractor-, or dealer-supplied.

When test-running an engine after any work on the shutdown devices, the governor, or the fuel system has been done, at least two people must be present. Both of them must understand the stopping methods and be capable of activating the various normal and emergency stopping devices.

Although almost all stationary and marine propulsion engines are fitted with protective shutdown devices, not all are backed with alarm devices. The operator must familiarize himself with the nature and location of all engine, switchgear, fuel, fire, and safety devices. The engine shutdown devices may be of a design that allows them to be manually tripped. In such cases, manual tripping of any unit may take place only under emergency conditions.

When an engine is operating under normal conditions, the wear on the moving parts will be minimal. If the conditions deteriorate, the wear rates will probably accelerate. Ideally, the protective devices will be set to sense the changing condition and shut the machine down before any damage is sustained. If a generating set does incur a protective shutdown, two things must happen: (1) The situation that triggers the shutdown must be normalized and then (2) the equipment must be fully examined for any damage that may have been sustained.

It is foolhardy to restart any unit that has been protectively stopped without making the best possible attempt to see that the engine, its generator, and the auxiliaries are in a safe, operable condition. A few more minutes without power are a lot less costly than a wrecked machine.

3.10 SAFETY SHUTDOWN SETTINGS

Protective devices with factory-sealed settings cannot be adjusted in the field. Without proper calibrating equipment on hand, any adjustment is at best risky and at worst downright dangerous. Uneducated tinkering with protective devices can turn an engine into a fairly lethal machine. No protective device should be manipulated or revised just to allow an engine to run. Any fault in the protection system must be properly rectified before the generator set is put into service.

3.11 LUBRICATING OIL PRESSURE SHUTDOWN

The sole purpose of the low lubricating oil pressure shutdown device is to keep the crankshaft from being damaged if a low lubricating oil condition develops. To operate effectively, the shutdown device must be so set that shutdown takes place while there is still sufficient pressure in the lubricating oil system to serve the crankshaft. It is, therefore, necessary to set the device at a pressure only just below the normal operating pressure. Refer to the OEM manuals for the correct operating oil pressures.

3.12 ENGINE INSPECTION FOLLOWING A LOW LUBRICATING OIL PRESSURE SHUTDOWN

After a shutdown is initiated by the low lubrication oil pressure protection device and before any attempt is made to restart the unit, it is imperative that the engine be thoroughly examined and the cause of the low lubricating oil pressure condition be positively identified and corrected. Experimental starts made before the problem is fully understood will have a ruinous outcome.

The lubricating oil filter elements and their housings must be inspected for metallic debris. If any metal is found, even in minute quantities, an in-depth inspection of the bearings must be undertaken. The sump, or pan, must be dredged, and any debris found must be recognized and its origin established. A visual inspection of the bottom, end, and main bearings must be carried out while particular attention is given to the dryness, discoloration, or other anomalous appearance of individual bearing caps.

No attempt to remove any inspection covers or access plates in order to conduct a crankcase inspection shall be made until at least one hour after the engine has come to rest. There is always a grave risk of a crankcase explosion in a hot engine that has suffered partial seizure.

A set of crankshaft deflection readings will show if there has been any bending or distortion of the crankshaft, and each main and bottom end bearing should be sighted for signs of emerging bearing metal. The engine must be barred over at least two complete turns, preferably with either the indicator cocks open or the injectors removed. The barring should be done by the attending mechanic, who will be familiar with the "feel" of the engine and be aware of any tight or rough spots.

The entire inspection procedure must be carried out and all components must be found to be in good order before the unit can be safely restarted. A hurried check will save neither time nor money, and the consequences may be disastrous if the inspection is skimped.

3.13 JACKET WATER TEMPERATURE SHUTDOWN

The purpose of the high jacket water temperature shutdown device is to save the engine from seizure in the event of overheating of the jacket water. To eliminate that possibility, the high temperature shutdown device must be set to stop the engine before coolant boiling occurs. The temperature at which boiling occurs will depend, in part, on the normal operating pressure of the cooling system and the altitude at which the engine is operated.

The high jacket water temperature shutdown switches should be set to operate at 5°C (10°F) above the normal operating temperature of the coolant leaving the engine. The high jacket water temperature alarm switches should be set to operate at 2.5°C (5°F) above the normal operating temperature of the coolant leaving the engine.

3.14 ENGINE INSPECTION FOLLOWING A JACKET WATER TEMPERATURE SHUTDOWN

The level of the lubricating oil in the sump shall be checked immediately after the engine has stopped and at hourly intervals thereafter. Cracked heads of liners and perhaps heat-damaged liner O-rings may permit coolant leakage into the sump with a consequent rise in level. Do not restart the engine if there is any likelihood of coolant being picked up and circulated by the lubricating oil pump. An attempt must be made to bar the engine around by hand. If the engine is difficult to move, it is likely that some seizure has taken place. The point of seizure must be located, and remedial action must be taken.

Examine the lubricating oil filters for metal splinters, particles, or flakes. At the same time, look for coolant in the filter housings and judge the general condition of the lubricating oil.

3.15 OVERSPEED SHUTDOWN

The purpose of the overspeed shutdown device is to save the engine from damage that would be caused by excessive speed. Some engines have an electrical shutdown switch backed by a mechanical shutdown device; others have only one overspeed protection unit. In all cases, the first-stage or only protection is set at 10 percent above the normal maximum operating speed and the second-stage device, when one is fitted, shuts the engine down at 15 percent above the normal maximum operating speed. When both mechanical and electrical shutdown devices are fitted, it is usual for the electrical device to have the lower and the mechanical setting the higher setting. For example:

Normal RPM	+10%	+15%
900	990	1035
1000	1100	1150
1500	1650	1825

Overspeed protection device shutdown settings must be adjusted only by a journeyman, who must refer to the OEM manual in all instances. Also, he must have the specific permission of the plant superintendent to do such work. When any work has been carried out on either the overspeed shutdown or the low lubricating oil pressure shutdown, at least two people must be present for the startup of the engine.

3.16 OUTAGES AND OVERSPEED TRIPS

There are occasions when outages occur and leave the generator set tripped off the board and stopped with its overspeed protection devices tripped. The cause of the outage is often identified as an overspeed situation, but in virtually every case that diagnosis is erroneous.

If there is a disturbance in the distribution system or a fault in the plant electrical system that can cause the breaker to trip, the instantaneous dumping of the load can cause the generator set to go into overspeed before the governor has time to respond. The overspeed protection acts to cut the fuel off immediately. It is more or less impossible for an engine under load to overspeed, but if it were to, the overvoltage or overfrequency protection devices would initiate a stoppage.

For the inquest that should follow each outage, it is most helpful if correct identification of the fallen flags and their sequence has been made. Together with any other factors that might have influenced the event, that will ensure that lessons will be learned.

3.17 GENERATOR SET INSPECTION FOLLOWING AN OVERSPEED SHUTDOWN

Following an overspeed incident, the field or armature coil windings of the generator and exciter must be inspected for any signs of bare copper, spattered solder, broken ties, or split insulation. The rotor and stator must be inspected to see that there has been no iron-to-iron contact as indicated by bright patches on the lamination packs. A small hand-held magnet passed over the interior of the generator will indicate any iron or steel debris that may have resulted from such contact.

On the engine, the valve clearances must be checked; any increase in clearance will indicate bent valves or pushrods, distorted rocker shafts, or other valve gear damage. The lubricating oil filter elements must be checked for particles of metal, and all of the sump, or pan, should be dredged. The source of debris recovered must be clearly identified. Fragments of piston rings, pieces of or complete snap rings, nuts, bolt heads, broken spring washers, and feathers of bearing metal are to be expected, and each advises of serious damage.

A set of crankshaft deflection readings will show if there has been any bending or distortion of the crankshaft. Each bearing must be inspected for evidence of pounded-out bearing metal.

The engine must be barred over by a skilled person. Either the indicator cocks must be open or the injector must be removed, and any tightness, roughness, or binding felt must be followed up.

3.18 RESTART OF AN ENGINE AFTER A PROTECTIVE SHUTDOWN

Following the shutdown of an engine through the actuation of one of the protective devices, it will be necessary to ascertain and remedy the root cause of the problem. This must not be done in haste, and neglect of any essential checks can nullify the benefit of the protection.

As a general rule, protective devices fail safe (mechanical overspeed shutdowns are an exception), so that when the protection device becomes unable to perform its function, the engine will shut down. Nevertheless, the operator must never assume that an engine has stopped because of a fault in a protection device. Every individual stoppage must be thoroughly investigated, and the engine must remain out of service until the true cause of the stoppage has been identified and rectified. When the rectification work is completed and the operator is satisfied that it is safe to restart the unit, he must reset all of the tripped protection devices.

Although numerous protection devices can be fitted to any engine, the variety of conditions that can develop and damage the machinery are greater in number. It is not feasible to provide an auto-monitoring device for every possible event. Therefore, in any continuously manned plant, the operator must monitor the machinery in every respect, including the areas in which auto protection is provided.

3.19 COMMON ALARM ARRANGEMENTS

1. Basic radiator-cooled engines
    a. Low lubricating oil pressure
    b. High jacket cooling water temperature
    c. Overspeed
2. Heat exchanger or common rail-cooled engines
    a. Low lubricating oil pressure
    b. High jacket cooling water temperature
    c. Overspeed
    d. Low jacket cooling water level
3. Raw-water-cooled engine with aftercooler
    a. Low lubricating oil pressure
    b. High jacket cooling water temperature
    c. Overspeed
    d. Low jacket cooling water pressure
    e. High intake manifold air temperature
4. Raw-water-cooled engine with aftercooler and oil cooler
    a. Low lubricating oil pressure
    b. High jacket cooling water temperature
    c. Overspeed
    d. Low jacket cooling water pressure
    e. High intake manifold air temperature
    f. High lubricating oil temperature

An engine equipped with an electric governor may have a low-voltage alarm on the control battery. A reverse-power alarm or circuit-breaker-tripped alarm may be installed on an ac generating unit. If any other engine function is monitored, the fault indicator will be built into or located next to the control panel. Such indicators are often referred to as *communicators*.

Practically any additional engine function involving speed, temperature, and pressure control can be sensed at extreme limits by a special alarm or shutdown system. The use of such a system will depend entirely on the type and extent of monitoring and automation that the owner required. Nearly every installation will differ in some or all details. No matter what alarms and shutdown devices are provided, none of them relieve the operator in a manned plant of standard watch-keeping responsibilities.

3.20 ALARM AND ADVISORY SIGNALS

In any diesel power plant, numerous audible and visual signals will be provided to advise the operator of events in progress. These signals may warn of an unusual event which must be given priority of attention, or they may merely advise that a routine event of an intermittent nature is in progress. The operator must develop and retain quite different and separate attitudes to the two types of signal, and those attitudes must not be confused.

3.21 NEGATION OF AUDIBLE ALARMS

The habitual removal of one or more fuses to silence protective alarms must not be permitted. If an alarm sounds repetitiously and can be silenced only by fuse removal, the circumstances triggering the alarm must be reviewed. In a well-run diesel power plant, an alarm should be regarded as a recordable event and be logged. If there are so many alarms that they become commonplace, it may be said that the operator does not have proper control of his shift but the lack of control is not necessarily his fault. Alarms are provided only to advise of an extraordinary event in the making. At no time should they be utilized to advise the operator of a routine occurrence.

3.22 EMERGENCY HANDLING

When an emergency situation develops in a power plant, the operator must be able to respond quickly and do the right things immediately to restore a normal condition. He should not run at this or any other time. He should walk swiftly to wherever he has to go, thinking all the time. The flashlight that he uses so often during the routine inspection work should be ready for use. If he has any doubt about the correct course to take, these rules of thumb will help him decide:

1. Never start an engine or any other machine unless it is known to be safe and ready to run.
2. A few extra minutes of preparation in darkness are much more acceptable than a hastily started and subsequently damaged engine or generator.
3. Do not waste time answering the telephone. The people who can or want to help will come to the plant anyway. The customer always gets the power restored as soon as the supplier sees it is safe to do so.
4. Regardless of the emergency, the safety of personnel is paramount. The safety of the plant and equipment comes next in the order of priority.

When an engine cannot be stopped by normal means, emergency methods can be used:

1. Shut the fuel supply line off at the stop valve closest to the engine. *Do not close the fuel return line valve.*
2. Place a piece of heavy polyethylene or closely woven fabric over the air filter(s). Two people will be needed to do this simultaneously on V-form engines. Cloth is not very effective, but a parka, raincoat, or coveralls can be used. Do not attempt to strangle the engine if the air filters are not in place.
3. Inject carbon dioxide (CO^2) into the air intake.

3.23 OVERLOAD

The load applied to an engine is limited in normal operation by:

1. The highest permissible exhaust temperature
2. The highest permissible alternator phase amperage (on any one phase)
3. The highest permissible stator winding temperature

The total load, in kilowatts, may be either greater or less than that stamped on the alternator data plate without exceeding either (1) or (2), but it will certainly not be more than the kVA rating, which also is noted on the alternator data plate.

Alternator overload will occur when the amperage exceeds the OEM specifications. That overload can and may occur even if the kilowatt rating is not exceeded. If there is phase unbalance, the amperage on only one phase may rise above acceptable limits; also, if the voltage falls, the amperage may rise on one or more phases beyond design limits. However, on some larger installations, there will be both undervoltage and overcurrent protection which will protect the alternator from any damage.

For most practical purposes, the operator may take the alternator data plate as the applicable limit. There will be the occasional exception, as when an alternator has been matched to an engine with an output that is insufficient to drive the alternator at rated capacity.

Most engines of greater than 1 MW capacity are fitted with individual cylinder exhaust temperature gauging methods. Midsize units of 500 to 1000 kW often have only exhaust manifold temperature gauges. Units of less than 500 kW generally have no exhaust temperature indication at all.

For reasons of economy, it is essential that any machine selected to run should be kept fully loaded. If the load level falls to the capacity of a smaller unit, that unit must be started and the load then transferred to it.

Note. For practical purposes, kVA × power factor = kW

3.24 SELECTING CAPACITY; ALLOCATING AND DISPATCHING GENERATOR SETS

The selection of generator sets to match the prevailing load is important if reasonable fuel economy is to be achieved. If the most suitable unit is dispatched, there will also be substantial savings in the cost of overhaul parts. Using parts prices prevailing in 1990, the replacement parts costs on an overall basis were about 1.5 cents/kW of capacity operated per hour. For example, if a 500-kW unit were run for 700 h (1 month) when a 300-kW set would have been sufficient, the difference in parts costs would be:

$$(500 \times 700 \times 1.5 \div 100) - (300 \times 700 \times 1.5 \div 100) = \$2,100/month$$

In an efficiently dispatched diesel generating facility the operator will always be ready and prepared to use the most suitable capacity and aim for an almost impossible to achieve 100 percent utilization factor. In a normal diesel generating system, it is not at all unreasonable to aim for an average utilization factor of 85 percent. Any individual plant factor below 85 percent would suggest that there is room for an improved utilization performance. To act as effectively as possible, the operator must be able to exercise initiative by loading the most suitable engine intelligently to its intended maximum without going into an overload situation.

3.25 FUEL ECONOMY

There are several ways in which an operator can obtain the maximum benefit from the fuel consumed in a diesel driven generator set:

1. Ensure that the fuel injector pressure is correct and that the fuel pump rack settings are right.
2. Do not run engines without load unnecessarily.
3. Keep the air filters clean.
4. Be sure that the turbocharger is turning freely and that the correct boost pressure commensurate with the load is being attained.
5. Operate the engine with the correct thermostat all year round, and check the glycol mix ratios regularly.

Most important of all, size the engines to the load, even if it means starting and stopping different engines several times a day (Figs. 3.1 through 3.4). Fuel efficiency equivalents are listed in Table 3.1.

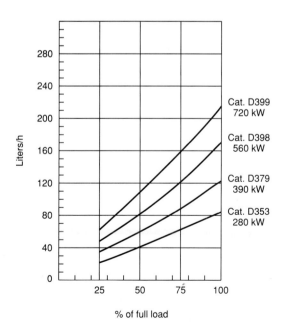

FIGURE 3.1 Typical fuel curve (liters) for a 6.25-in-bore engine. All units are without engine-driven fans. Continuous duty at 0.8 PF.

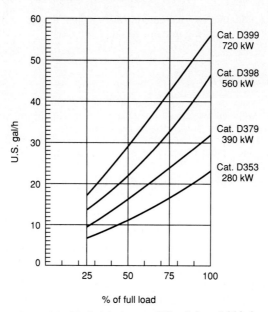

FIGURE 3.2 Typical fuel curve (US gal) for a 6.25-in-bore engine. All units are without engine-driven fans. Continuous duty at 0.8 PF.

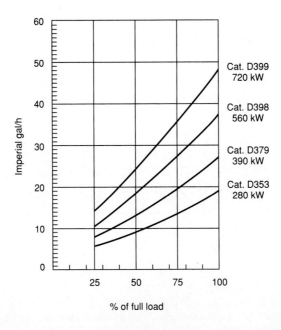

FIGURE 3.3 Typical fuel curve (imp gal) for a 6.25-in-bore engine.

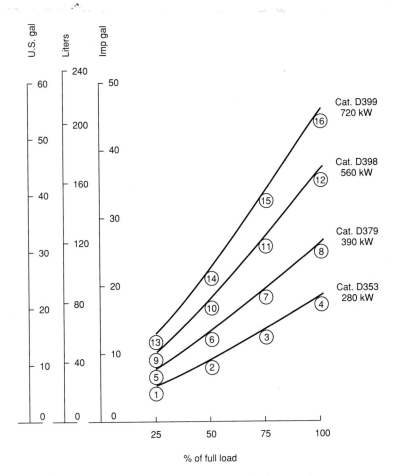

FIGURE 3.4 Typical fuel curve for 6.25-in-bore engine (summary).

3.26 FUEL CONSUMPTION IMPROVEMENT COMPARISON

Consider the difference in fuel consumption when a diesel generating plant formerly operating at 14.75 kW/imp gal (12.65 kW/US gal or 3.25 20kW/L) is improved to 15.25 kW/imp gal (13.10 kW/US gal or 3.35 kW/L).

$$\frac{15.25 - 14.75}{15.25} = 3.27\% \text{ improvement}$$

TABLE 3.1 Engine Fuel Efficiency Equivalents

kW/imp gal	kW/L	kW/US gal	Best possible fuel performance by model, (lb/(BHP-h)
19.0	4.17	15.82	
18.9	4.15	15.73	
18.8	4.13	15.64	
18.7	4.11	15.57	
18.6	4.09	15.47	Mirrlees MB (0.327)
18.5	4.06	15.39	Cat D3603 (0.327)
18.4	4.04	15.32	
18.3	4.02	15.22	
18.2	3.99	15.15	
18.1	3.97	15.07	
18.0	3.95	14.97	
17.9	3.93	14.89	
17.8	3.91	14.81	
17.7	3.89	14.73	
17.6	3.87	14.64	
17.5	3.84	14.57	
17.4	3.82	14.47	
17.3	3.80	14.40	Mirrlees K Major (0.339)
17.2	3.78	14.32	
17.1	3.75	14.23	
17.0	3.73	14.14	
16.9	3.71	14.06	
16.8	3.69	13.98	Ruston RK—Waukesha (0.370)
16.7	3.67	13.90	Cat D3516
16.6	3.65	13.82	
16.5	3.63	13.73	
16.4	3.60	13.64	
16.3	3.58	13.56	Cat D3512
16.2	3.56	13.48	
16.1	3.54	13.39	Cat D3408 (0.370)
16.0	3.51	13.32	
15.9	3.49	13.24	Cat D3406
15.8	3.47	13.15	
15.7	3.45	13.07	
15.6	3.43	12.99	Cat D349 (0.386)
15.5	3.40	12.90	Cat D3306 (0.390)
15.4	3.38	12.81	
15.3	3.36	12.73	
15.2	3.34	12.65	Cat D353
15.1	3.32	12.58	Cat D399
15.0	3.29	12.48	

If the plant has an average output of 1000 kW/h, the savings in fuel oil over a 1-year period are considerable.

$$\frac{1000 \text{ kW} \times 8760 \text{ h}}{15.25 \text{ kW/imp gal}} = 574{,}426 \text{ imp gal}$$

TABLE 3.1 Engine Fuel Efficiency Equivalents)*Continued*)

kW/imp gal	kW/L	kW/US gal	Best possible fuel performance by model, (lb/(BHP-h)
14.9	3.27	12.39	
14.8	3.25	12.32	
14.7	3.23	12.24	Cat D379
14.6	3.21	12.15	
14.5	3.18	12.07	
14.4	3.17	11.98	
14.3	3.16	11.90	
14.2	3.13	11.82	DD71 Series (0.435)
14.1	3.11	11.74	
14.0	3.08	11.65	
13.9	3.05	11.57	
13.8	3.03	11.48	
13.7	3.07	11.41	
13.6	2.99	11.33	
13.5	2.97	11.23	
13.4	2.94	11.14	
13.3	2.92	11.07	
13.2	2.90	10.99	
13.1	2.87	10.90	
13.0	2.85	10.82	
12.9	2.83	10.74	
12.8	2.81	10.64	
12.7	2.79	10.56	
12.6	2.77	10.49	
12.5	2.75	10.40	
12.4	2.73	10.32	
12.3	2.70	10.24	
12.2	2.68	10.15	
12.1	2.66	10.07	
12.0	2.64	10.00	
11.9	2.62	9.90	
11.8	2.59	9.81	
11.7	2.57	9.73	
11.6	2.55	9.66	
11.5	2.53	9.56	
11.4	2.50	9.49	
11.3	2.48	9.39	
11.2	2.46	9.32	
11.1	2.44	9.24	

1.0 imp gal = 4.546 L = 1.2009 US gal
1.0 US gal = 3.79 L = 0.8333 imp gal
1.0 L = 0.219 imp gal = 0.263 US gal
    Source: OEM data sheets.

but

$$\frac{1000 \text{ kW} \times 8760}{14.75 \text{ kW/imp gal}} = 593{,}898 \text{ imp gal}$$

which represents a difference of 19,472 imp gal (23,366 US gal or 88,519 L). The

TABLE 3.2 Typical Fuel and Output Ratios for Caterpillar 6.25-in-Bore Engines

Engine*	Ref. point	kW/US gal	kW/L	kW/imp gal
Cat. D353, 280 KW	1	10.67	2.85	12.91
	2	12.72	3.35	15.26
	3	12.72	3.35	15.26
	4	12.33	3.25	14.79
Cat. D379, 390 KW	5	10.42	2.75	12.50
	6	12.18	3.21	14.61
	7	12.31	3.25	14.77
	8	12.22	3.23	14.66
Cat. D398, 560 KW	9	11.29	2.97	13.54
	10	12.78	3.37	15.33
	11	12.80	3.38	15.56
	12	12.58	3.33	15.09
Cat. D399, 720 KW	13	11.25	2.96	13.50
	14	13.33	3.51	15.99
	15	13.17	3.47	15.80
	16	13.99	3.45	15.70

*Continuous at 0.8 PF without fan.

3.27 percent improvement in fuel efficiency can, in most cases, be initially obtained simply by matching the generating equipment more appropriately to the prevailing load. Equipment operating at an average utilization of better than 75 percent would be expected to show a fuel efficiency of at least 15.000 kW/imp gal if it is in good repair.

CHAPTER 4
ENGINE LOG SHEETS

4.1 ENGINE LOG SHEET PURPOSE

The importance of the engine log sheet cannot be overemphasized. Without such a record, it is almost impossible to plot the history or performance of any given engine. The scope and complexity of the log sheet to be used in conjunction with any particular installation will vary with the size of the unit, the nature of engine duty, and the requirementsa of the owner.

As a general rule, it can be said that the larger the engine the more frequent and extensive the data to be logged. Some log sheets require a report every hour, others every 2-h, and still others need only be completed at the end of a shift or even daily. Whatever the frequency demanded, the operator must make the effort to provide clearly written, accurate figures.

Many figures have to be written down, and to the operator the task may appear to be tedious and without purpose. However, when the entry items are reviewed, the real usefulness of the logged informational bridge between the operator and the owners is evident. A carefully devised log sheet will show trends in engine performance and give some clues to why the performance of the engine deviates. All of the information columns have individual purposes, and it is important that the data inserted be accurate. The operator will in time learn to analyze his own log sheets and act accordingly when there is an indication that something is awry.

4.2 LOG DATA CATEGORIES

It will be noticed that readings required for the log fall into distinct categories. In one category, there are readings that remain more or less constant; in the second category are readings that vary depending upon the load; in a third are readings that are cumulative.

In the constant category are lubricating oil pressures and temperatures, primary fuel pressure, cooling water pressures and temperatures, lubricating oil levels, and voltage and frequency. All of the constant readings are regulated automatically with the exception of the lubricating oil and coolant levels. A short time after startup, say about 30 min, the pressures and temperatures will have reached their normal operating status. Thereafter, the operator must recognize any departure from the norm in any of the constant readings.

In the variable category are kilowatt loading, phase amperage, boost pressure and exhaust temperatures. All these will rise or fall with the load imposed on the generator set. In the cumulative category are the fuel consumption, kilowatt production, engine hours, engine starts, and breaker trips.

4.3 LOG ABUSE

There will be occasions when the shift operator will find it more important to attend to a situation rather than collect log data. If the hourly readings are not taken, the operator should mark the vacant log spaces "Readings not taken because...." and make a concise notation of the interfering event in the remarks column. Under no circumstances should the log be falsified by inserting readings that were not actually observed. The practice of "log flogging" is not uncommon, but it is a grossly misleading and useless activity. The entire purpose of the log is defeated if meaningless data are recorded; similarly, no operator should take it upon himself to write in blended numbers in an effort to camouflage anomalies. It is incumbent upon each operator to preserve the integrity of his trade.

4.4 LOG SCANNING

It will be seen that the daily engine log sheet plays a very useful part in engine health monitoring, but its function is not complete without the participation of the plant superintendent and the journeymen mechanics, and electricians. They must be in the habit of regularly scanning the log and taking necessary action. It can be expected that the operator will start taking a relaxed attitude toward information collection and logging at about the same time that he realizes that his efforts to compile a complete, accurate log are disregarded. Naturally, any logged information must be treated with caution, but it must also be treated with respect. The routines of log entries and perusal can be the basis of a tremendously valuable learning and training process, and the opportunities presented should not be neglected.

4.5 DEFECT AND INCIDENT RECORD

Any defect that occurs during a shift should be recorded. It may be found convenient to write a comment on the current log sheet, but that is not an effective method of forming a history. It is much better to use a form of diary entry in a proper hard-back notebook. By that means the dayworkers—the mechanics and electricians—can peruse the entries on a daily basis and annotate items as and when they receive attention, the operator can review the needed work progress and conduct his shift with regard to the current abnormal or recently normalized conditions, and the superintendent can review the record for repetitious occurrences and initiate corrective procedures. He can also identify patterns and trends of failure or misperformance, thereby modifying existing preventive routines to suit.

The defect and incident record book, if it is properly utilized, can be a most valuable permanent register of all unplanned, unexpected, or unusual events. No page should ever be removed from the defect and incident book. Scrap paper, blackboard, and similar methods of communcation have no permanence, receive scant respect because of their transience, and thus are of negative value and cannot be recommended. The insertion of remarks in the defect and incident record book should not be limited to the operators; all power plant staff should be encouraged to enter comments that will contribute to the welfare or improvement of plant operations.

4.6 LOG ENTRY COLUMNS

A fairly typical log sheet is illustrated in Fig. 4.1. The sample sheet was designed for a 2.5-MW generator engine, and it will be seen that there are many columns in which entries must be made. For the purposes of this discussion, each column has numbers written at its head, and they serve as guides to the maximum figures one would expect to see on this particular engine at full load. The numbers were taken from the original test sheets. Also the columns have been serially numbered to facilitate this discussion.

The subject engine is a 4-cycle unit fitted with two turbochargers and 16 cylinders. It runs at 900 rpm and uses P40 diesel fuel (i.e., pourpoint −40°C).

Column 1, Hour

In ordinary industrial applications, most shifts change at 12:00 a.m., 8:00 a.m., and 4:00 p.m., and thus the time column has three breaks in it to better define the shifts. It is recommended that the oncoming operator read the log immediately upon arriving at the power plant to assume his shift and then go and take his first set of readings. The tour of the machinery necessary to gather the figures must also include a close inspection of all the generating machinery. Never assume that an engine will start and provide optimum performance if it is not inspected or if it has not been used for some time.

Columns 2 to 5, Load and Amperage

Load and amperage are closely linked. Column 2 shows the kilowatt load on the unit, but it is important to see that none of the three phases has a current overload. If there is an imbalance between the phases, it should be brought to the attention of the electrician, but at no time must the average phase amperage be allowed to exceed that marked on the relevant machine's data plate. Taking this a little further:

1. For a 4160-V machine:

$$\frac{\text{Phase-1 amperes} + \text{phase-2 amperes} + \text{phase-3 amperes}}{3} = \text{average amperes}$$

$$\text{Average amperes} \times 7 = \text{kilowatt output}$$

4.4 CHAPTER FOUR

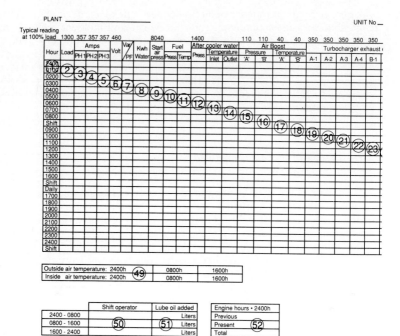

FIGURE 4.1 Sample engine log sheet.

2. For a 2300-V machine:

$$\frac{\text{Phase-1 amperes} + \text{phase-2 amperes} + \text{phase-3 amperes}}{3} = \text{average amperes}$$

$$\text{Average amperes} \times 4 = \text{kilowatt output}$$

3. For a 600-V machine:

$$\frac{\text{Phase-1 amperes} + \text{phase-2 amperes} + \text{phase-3 amperes}}{3} = \text{average amperes}$$

$$\text{Average amperes} \times 1 = \text{kilowatt output}$$

These calculations are only for general guidance, but they are useful for checking the ammeters against the kilowattmeters.

Column 6, Voltage

Voltage is normally under automatic control, but it must be carefully monitored. Manual adjustment is possible, but if frequent adjustments are necessary, consult

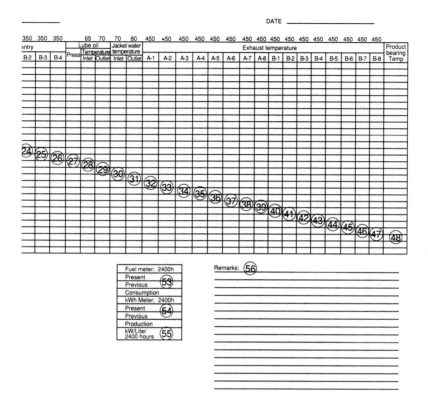

FIGURE 4.1 (*Continued*) Sample engine log sheet.

the electrician. Good voltage control is important because variations can adversely affect consumers' equipment.

Column 7, The Power Factor

The power factor indicates overall efficiency of power usage by the consumer. For example, an all-lighting load, which would use power very efficiently, would result in a power factor of somewhere near 98 percent, but a system in which there were lots of low-efficiency fractional horsepower motors, one would expect a lower power factor, say 85 percent or 0.85.

Column 8, The Kilowatt-Hour

The kilowatt-hour meter shows the kilowatt production, and that eventually represents the gross revenue earned by the generator set. The kilowatt-hour meter reading will be referred to again in this chapter because it is an important factor in the calculation of the engine's efficiency.

Column 9, The Starting Air Pressure

The starting air vessel should be at maximum pressure at all times. The operator must be aware of the routines for tipping the air vessel safety valve and draining any oil or water from the receiver. He must check the compressor oil level and start the compressor to maintain the necessary pressures.

Column 10, Fuel Pressure

"Fuel pressure" is the primary fuel pressure. It may vary if the fuel is too cold, if the primary filters are plugged, or if the level in the supply tanks is too low. If the fuel pressure falls only gradually over many hours or shifts, it may be that the primary fuel pump is becoming worn. That, however, is an unlikely cause of pressure drop.

Column 11, Fuel Temperature

Ideally, the fuel temperature should not be less than 0°C (32°F); preferably, it should be in the 20°C (68°F) range. Some fuel systems are heated. In those cases, the fuel temperature should not be permitted to rise above 65°C (150°F) or the legal limit for the fuel used, whichever is lower.

Column 12, Aftercooler Water Pressure

On many engines the aftercooler water circulation is on a circuit separated from the jacket system, and so the pressure in the aftercooler circuit may differ from that of the main cooling system. Some larger engines also have a pressure gauge at the aftercooler outlet. A comparison of the pressure guage with the inlet gauge will indicate if fouling of the cooler is taking place.

Columns 13 and 14, Aftercooler Water Temperature, Inlet and Outlet

The temperature of the charge air entering the cylinder is important. If the air is being cooled successfully, there will be a difference of several degrees between the water inlet temperature and the outlet temperature.

Columns 15 and 16, Boost Air Pressure A and B

The air boost pressure from both turbochargers should be the same. If the pressures differ, a problem with one of the turbochargers may be developing, the air filters may be getting dirty, or there may be something wrong with one or more inlet valves.

Columns 17 and 18, Boost Air Temperature

The boost air temperature will vary somewhat with the load, but it should reach its permitted maximum only at full load. If the maximum boost air temperature is attained before the engine is at full load, it is probable that either the aftercooler is dirty or the air filters are plugged. Normally, both turbochargers should show similar pressures and temperatures. On the larger engines, on which the turbochargers are equipped with tachometers, it is possible to compare turbocharger rotational speeds as a performance check.

Columns 19 to 26, Turbocharger Exhaust Entry

Multiple exhaust entry turbochargers' temperatures should not vary at full load by ±2 percent of the average. If the turbocharger is permitted to run with greatly varying gas inlet temperatures, there may be a danger of turbo rotor flutter or thermal shock, which will lead to metal fatigue of the blading.

Column 27, Lubricating Oil Pressure

Lubricating oil pressure is the most important reading of all on any engine. During normal operation, the lubricating oil pressure will remain steady all the time. The lubricating oil pump has the capacity to pump excess quantities of oil and thereby maintain the pressure in the system as the engine ages and wears. If the pressure begins to fall, notice must be taken of the oil level in the sump and the dipstick must be carefully examined for indications of fuel dilution or coolant entry. Notice must also be taken of the lubricating oil filter pressures and the pressures and temperatures at the lubricating oil cooler.

    The pressure and temperature gauges in the lubricating systems must be believed, and no chances that will put the crankshaft at risk should be taken. The lubricating oil pressure shutdown device must not be adjusted to suit a falling lubricating oil pressure. If the oil pressure begins to fall, the lubricating oil filter element must be carefully examined for signs of metal debris.

Columns 28 and 29, Lubricating Oil Temperatures, Inlet and Outlet

These temperatures can often tell a story. Besides lubricating all of the moving parts, the lubricating oil acts as a cooling medium on the cylinder walls, the interior of the piston crown, and the piston skirt, and it also carries away excess heat from bearings, crankshafts, and other moving parts. If the oil outlet temperature starts to rise, it will indicate that something in the engine is getting too hot. That will be substantiated by the inlet temperature if it remains normal. If the inlet temperature also goes up, it is possible that the lubricating oil cooler is plugged or dirty.

Columns 30 and 31, Jacket Water Inlet and Outlet Temperature

The jacket water inlet and outlet temperatures will show a difference of 10°C (20°F) once the engine has warmed up regardless of the load. If there is any deviation from the normal in winter conditions, consider overcooling and antifreeze jelling as a cause. Refer to the discussion of cooling systems.

Columns 32 to 47, Exhaust Gas Temperatures at Full Load

At full load, the exhaust gas temperatures should be quite balanced [e.g., within 15°C (30°F) of average] and both banks should average out the same. At less than 25 percent of full load, the temperature readings will be quite ragged, but no attempt at equalization should be made at that time. Further remarks will be found under the heading "Exhaust Temperatures."

Column 48, Pedestal Bearing Temperature

The pedestal bearing temperature should not rise above 75°C (167°F). It is necessary to check the oil level at the same intervals as those for checking the crankcase oil level.

Column 49, Outdoor and Indoor Air Temperatures

The outdoor and indoor temperatures are important components of the daily report. Correlating outdoor temperatures with peak loads is essential to future load forecasting. The indoor termperature reading serves to indicate if the ventilation system or the aspirating air blend and consumption need attention, and although the temperature may vary from day to day, an indoor temperature of 21°C (70°F) should be sought on a year-round basis.

Column 50, The Shift Operator

The shift operator should identify himself in column 50, because it is often necessary to identify the person who manned a particular shift at some time in the past.

Column 52, Engine Hours

The engine hours should always be read off to the nearest whole hour. Great errors can be made when the log data are subsequently processed by other people if fractions or decimal portions of hours are logged. The hour count can be regarded as the real age of the engine, and it is also used to determine when a given overhaul activity will take place. Relating the engine hours to the wear on the various moving parts permits a certain amount of forecasting of future overhaul costs and also determines if parts are being replaced too soon or too late and if they are failing within a warranty period or at too great a frequency.

Column 53, The Fuel Meter

The fuel meter indicates the engine fuel consumption within the shift and over an extended period.

Column 54, The End-of-Day Kilowatt-Hour Reading

This reading shows the total production of electricity by the unit in the past 24-h period. By using the fuel meter reading, column 53, the ratio of kilowatt-hours produced to the liters of fuel consumed can be calculated.

Column 55, Kilowatts per Liter

The kilowatts per liter figure indicates the efficiency and effectiveness of utilization of the machine. A very good performance would be 17 kW/gal (3.73 kW/L) on any unit of 750 kW and larger, and 16 kW (3.52 kW/L) can be expected from a smaller unit.

Column 56, Remarks

The remarks column should contain any item that the shift operator wants to bring to the attention of others. If a request for someone to do something is entered, the person who attends to the request should subsequently initial the remark to indicate that it has received attention. He should also insert notations such that the operator will be aware of what has transpired.

CHAPTER 5
ENGINE FAULT IDENTITY

5.1 FAULT IDENTIFICATION AND TROUBLESHOOTING

A large number of engine breakdowns can be blamed on poor operating practices; thus, improvement in operating skills and knowledge is necessary if there is to be any reduction. It is imperative that the operating personnel in any plant be very familiar with the many minor signs that warn of a developing defect. Much of that familiarization must come from the superintendents and journeymen, who in many cases, have considerable background and experience.

The real secret of trouble avoidance lies in the early recognition of symptoms and the correct diagnosis of cause. Each operator or mechanic should endeavor to expand his problem-identifying abilities by constant study of engine manuals and instruction books. If he is to be effective, the operator must acquire a fault-identifying ability. That ability can only be developed through a process of self-education supplemented by the teaching that other power plant personnel are able to impart.

Checking the instruments for efficacy obviously has to be the first step in any problem-solving routine. In all the instances that will be discussed, it is assumed that the instrumentation showing abnormal conditions is in fact accurate and in good order. It is assumed also that the unit giving trouble is correctly assembled, although the assembly factor must be taken into account if a problem shows up immediately after work has been done on the engine. Individual or detail design defects in new or old installations are not cited in the discussion as causes of engine faults.

Troubleshooting can be defined as an organized activity aimed at identifying and rectifying an incidental defect. It consists of several basic phases:

1. *Recognition and study of symptoms:* To perform the first phase, it will be necessary to consider what warning signs were evident, what that has occurred recently may have induced or fostered the problem, and whether there have already been similar incidents and, if so, what was done or not done then. The most important question of all must be this: Is the engine safe to run if further troubleshooting checks are to be expedited by starting it? If the answer is anything less than a solid affirmative, the engine must not be started.

2. *Identification of the probable causes:* The more time taken to identify the cause of the problem the more accurate the diagnosis will be. The final resolution of the problem will then be the most economical of both money and time. Hasty identification tactics often camouflage the real causes and symptoms. The simple and obvious things must be checked first. Stripping of the engine must not start before the problem has been properly identified. (For example, in the case of high lubricating oil consumption, do not start pulling pistons out to rering them before the lubricating oil cooler and shaft seals have been checked for leakage.)

3. *Planning and following a procedure for the rectification of the defect:* The scope of this activity may be quite small and little planning will be necessary. On the other hand, the job may be very large and complicated. Whatever the magnitude of the job, a plan and a procedure must be evolved so that others can be made aware of how long the equipment will be out of service, an estimate of the cost can be developed, and the necessary parts, manpower and work space can be obtained in good time. Whatever the size of the task to be undertaken, an organized work plan is necessary so that all the people involved know what is expected of them and when the situation will be normalized.

4. *Eliminating the root cause:* Any truly successful troubleshooting procedure will be completed by the elimination of the root cause of the original defect. Elimination is not always wholly practical at the time of repair, but every effort must be made to get the situation rectified at the earliest opportunity.

5. *Reporting the outcome:* At the completion of the troubleshooting exercise, a report should be prepared and circulated so that all those involved learn from the experience and will thus know what to look for next time and how to react to the situation if it starts to develop again. Long-term arrangements for the permanent eradication of the particular and similar problems should be made.

Troubleshooting can be a valuable learning and teaching process. There should be no hesitation in telling other people how the problem was solved, because those others may be able to contribute something that will reduce another problem encountered later. (It is not normal practice to use operating personnel for either repair or overhaul work. Such work is generally done by specialists who are either employed by the power plant owner or provided by the equipment manufacturer or its representative on a contract basis. The subject of overhaul is discussed more fully elsewhere in the book.)

Whatever his ability may be, at some time, the duty operator may need assistance during his shift. At the commencement of every shift he must be aware which relief operator, journeyman, and supervisory staff will be available if a problem develops. In any well-regulated plant there will be a list of the telephone numbers the operator can call if difficulties develop.

5.2 BASIC FAULT INDEX

1. Engine will not turn over.
2. Engine will turn over but not start.
3. Engine stops immediately after starting.
4. Engine stops with rapid deceleration.
5. Engine stops with normal deceleration.
6. Engine stops on signal from a protective device.
7. Engine hunts.

8. Engine vibrates.
9. Engine knocks (mechanical).
10. Engine knocks (fuel).
11. Engine overspeeds.
12. Engine will not attain synchronous speed.
13. Engine will not carry any load after breaker closure.
14. Engine will not carry full load.
15. Engine misfires.
16. Engine coolant overheats.
17. Engine has fuel dilution of lubricating oil.
18. Engine gases.
19. Engine has excessive crankcase pressure.
20. Engine has sludge in sump.
21. Engine lubricating oil level rises.
22. Engine has low lubricating oil pressure.
23. Engine has high lubricating oil pressure.
24. Engine lubricating oil is too hot.
25. Engine uses an excessive amount of lubricating oil.
26. Engine makes black smoke.
27. Engine makes white smoke.
28. Engine makes blue smoke.
29. Engine has high exhaust temperatures.
30. Engine uses too much fuel.
31. Engine makes unusual noises.
32. Load trips off; engine continues to run.
33. Engine will not stop.

5.3 POSSIBLE FAULTS

Symptom 1, Engine Will Not Turn Over

- Cylinder is flooded.
- Piston is seized.
- Shaft bearing is seized.
- Turning gear is engaged.
- Mechanical obstruction.
- Battery is disconnected.
- Jammed starter.
- Battery is disconnected.
- Starting sequence is unable to proceed.

- Low battery voltage.
- Air vessel valve is closed.
- Low starting air pressure.
- Protective devices are not reset.
- Running gear is jammed by metallic debris.

Symptom 2, Engine Will Turn Over But Not Start

- Starting controls are in the wrong positions.
- Low starting air pressure.
- Low battery voltage.
- Fuel is shut off.
- No fuel in tank.
- Water in fuel.
- Air-locked fuel lines.
- Fuel pump rack is not in starting position.
- Fuel pump plunger(s) are seized.
- Fuel filter is dirty.
- Ice in primary fuel line or filter.
- Crushed fuel line.
- Air intake is blocked.
- Shutdown device(s) are not set for starting.
- Governor drive is sheared.
- Lubricating pressure switches are not working correctly.
- Fuel timing is incorrect.
- Air start valve timing is incorrect.
- No compression.
- Defective starter motor.
- Seized injector nozzles.
- Incorrect fuel mode—heavy fuel or light fuel.
- Heavy fuel is too cold.
- Lubricating oil is too cold.
- Air or exhaust valves are in open position.
- Cranking speed is too low.
- Primary fuel pump has failed.
- Cracked injector or injector cap.
- Dirty injector edge filters.
- Grounded magneto. ⎫
- Low fuel gas pressure. ⎬ Gas-fueled engines only
- Faulty gas regulator. ⎭

Symptom 3, Engine Stops Immediately after Starting

- Air in fuel system.
- Fuel starvation.
- Safety shutdown tripping.*
- Primary fuel pump failure (icing).
- Fuel return line is closed.
- Open cylinder head cover (EMD).
- Heavy fuel is cold.
- Crankcase pressure detector.
- Incorrect fuel mode—light fuel or heavy fuel.

Do not make repeated attempts to start until all possible reasons for safety shutdown tripping have been investigated. Severe additional damage may occur.

Symptom 4, Engine Stops with Rapid Deceleration

- Bottom end seizure.
- Main bearing seizure.
- Piston seizure.

(Symptom 4 will probably precede symptom 1.)
  Great caution must be observed before opening the crankcase for examination when symptom 4 occurs. Refer to Sec. 7.3.

Symptom 5, Engine Stops with Normal Deceleration

- Fuel is shut off.
- Fuel supply is exhausted.
- Water in fuel (iced filters).
- Safety shutoff is actuated.
- Governor failure because of linkage detachment.
- Electrical protection trip.
- Air box fire.
- Fuel line is restricted.
- External or internal fuel leaks.
- Fuel pump drive shaft failure.
- Eductor tube is blocked.
- Primary fuel pump failure.

Symptom 6, Engine Stops on Signal from a Protective Device

- Low lubricating oil pressure.

- High jacket water temperature.
- High lubricating oil temperature.
- Low lubricating oil level.
- Low jacket water level.
- High crankcase pressure.
- High vibration level.

Do not attempt restart until the cause of shutdown has been clearly identified and remedied.

Symptom 7, Engine Hunts

- Air in fuel line.
- Plugged fuel filters.
- Governor adjustment is required.
- Governor linkage is loose.
- Voltage regulator defect.
- Low oil level in governor.
- Governor drive gears are too deeply meshed.
- Aeration of lubricating oil in governor.
- Fuel filter is restricted (ice, wax, water).
- Surging load.
- Obstruction in fuel return line.
- Governor compensation is incorrectly set.
- Fuel tank vent is obstructed.
- Exciter defect.
- Flex hoses are distorted.
- Fuel tank is almost empty.

Symptom 8, Engine Vibrates

- Engine-driven fan blade pitch inequality.
- Missing fan blade.
- Loose fan belts.
- Loose engine component(s).
- Uneven firing pressures.
- Stuck injector(s).
- Stuck fuel pump(s).
- Incorrect fuel injection timing.
- Incorrect valve timing.
- Loose coupling.

- Unbalanced turbocharger or generator rotor.
- Phase imbalance.
- Misalignment.
- Loose hold-down bolts.
- One defective turbocharger (V-engine).
- Excessive wear.
- Torsional damper failure.
- Bent connecting rod.
- Defective bearing(s).
- Broken tooth on timing gear.
- Collapsed vibration isolator(s).
- Movement of foundation.
- Obstruction causing direct contact between underframe and floor.

Symptom 9, Engine Knocks (Mechanical)

- Loose valve rocker lever adjusting screw.
- Failed bearing.
- Loose bearing bolts.
- Broken valve.
- Broken valve seat.
- Piston crown loose.
- Sticking valve.
- Loose flywheel coupling bolts.
- Excessive crank float.
- Broken tooth in gear train.
- Loose hold-down bolts.
- Gross valve tappet clearance.
- Bent rocker or pushrod.
- Spalled cam or cam roller.
- Failed injector nozzle.
- Low lubricating oil pressure.
- Low lubricating oil level.
- Air in fuel line.
- Restricted fuel line.
- Broken valve spring.

Do not run an engine with a knock. Do not confuse a fuel knock with a mechanical knock.

Symptom 10, Engine Knocks (Fuel)

- Engine is overloaded.
- Wrong quality of fuel.
- Engine is overheated.
- Misadjusted injectors.
- Unbalanced fuel pump.
- Defective turbocharger.
- Fuel injection timing is too advanced.

Symptom 11, Engine Overspeeds

- Incorrect starting practice.
- Improper use of starting aids.
- Governor load limit is improperly set.
- Fuel rack is in full-fuel position at startup.
- Breaker trips at full load.

Symptom 12, Engine Will Not Attain Synchronous Speed

- Overspeed shutdown is improperly reset.
- Governor load limit is insufficiently advanced.
- Governor oil level is too low.
- Water in fuel.
- Air in fuel system.
- Fuel supply is restricted.
- Fuel rack linkage is jammed.
- Engine timing has slipped.
- Air filters are dirty.
- Exhaust outlet is impeded.
- Fuel pump delivery valves are defective.
- Injectors are defective.
- Primary fuel pump malfunction.
- Valves on fuel supply or return line are not fully open.

Symptom 13, Engine Will Not Carry Any Load after Breaker Closure

- Voltage regulator is defective.
- Boost air pressure is too low.

   Refer also to symptom 10 and symptom 12 causes.

Symptom 14, Engine Will Not Carry Full Load

- Air in fuel supply.
- Rack travel is obstructed.
- Cylinder balance is incorrect.
- Dirty fuel filters.
- Fuel is too cold.
- Dirty air filters.
- Dirty injectors.
- Governor load limit is not set correctly.
- Governor linkage is misadjusted.
- Boost pressure is too low.
- Air inlet temperature is too high.
- Dirty turbocharger.
- Bent pushrod(s).
- Turbocharger seals are leaking.
- Sticking fuel pump relief valve.
- Fuel pumps are worn.
- Poor compression.
- High exhaust back pressure.
- Valve timing is incorrect.
- Injection timing is incorrect.
- Cylinder liners are glazed.
- Internal or external fuel leaks.
- Lubricating oil level is too high.
- Incorrect or defective injectors.
- Fuel pump delivery valves are defective.
- Fuel pump rack settings are mismatched.
- Excessive carbon in cylinder liner or exhaust ports.

Refer also to symptom 10 and 12 causes.

Symptom 15, Engine Misfires

- Engine is short of fuel.
- Water in fuel.
- Fuel is too cold.
- Air leak in fuel suction lines.
- Partly plugged fuel lines.
- Plugged injector nozzles.
- Injector adjustment is incorrect.
- Cracked injector body or cups.

- Poor compression.
- Valve leakage or stem seizure.
- Dirty injector edge filters.
- Broken or collapsed injector springs.
- Injector clamp is loose.
- Engine is mistuned.
- Broken valve spring.
- Broken piston rings.

Symptom 16, Engine Coolant Overheats

- Water pump is not circulating system coolant.
- Fan belts are slipping.
- Thermostat (regulator) is not working.
- Remote fan control is set incorrectly.
- Cooling system is dirty internally.
- Cooling system is dirty externally.
- Obstruction in coolant flow.
- Incorrect coolant mix.
- Valve timing is incorrect.
- Leaking head valves.
- Fuel timing is incorrect.
- Air filters are blocked.
- Poor turbocharger(s) performance.
- Aspirating air temperature is too high.
- Incorrect radiator louver position.
- Coolant jelled in radiator restricting circulation.
- Engine is overloaded.
- Bypass valves in cooling system are not operating properly.

Do not confuse boiling that is due to overheating with gassing (symptom 18).

Symptom 17, Engine Has Fuel Dilution of Lubricating Oil

- Poor compression.
- Dribbling injectors.
- Leaking fuel lines and seals.
- Cracked injector body, nozzle, or nut.
- Long idle or low-load periods.
- Injection is mistimed.
- Wrong injector nozzles or cups.

- Leaking injector cup O-rings.
- Broken or collapsed injector spring.
- Seized injector needle.
- Glazed liners.
- Slipped camshaft coupling.
- Incorrect injection pressure.
- Leaking fuel pump seals.
- Broken or gummed piston rings.

Symptom 18, Engine Gases

- Cracked cylinder head.
- Loose or cracked precup.
- Blown cylinder head gasket.
- Cracked liner.
- Cracked turbo casing.
- Cracked valve cage.
- Stretched or broken cylinder head studs.
- Cracked cylinder liner landing or block counterbore.
- Split or perforated injector tube.
- Incorrectly torqued cylinder head.

Symptom 19, Engine Has Excessive Crankcase Pressure

- Broken, jammed, or worn piston rings.
- Restricted breather.
- Crankcase exhauster is not running.
- Blowby on one or more cylinders.
- Oil level is too high. (Check for flooding.)
- Eductor tube is blocked.
- Defective blower seal.
- Burnt or perforated piston crown.

Symptom 20, Engine Has Sludge in Sump

- Dirty lubricating filters (bypassing).
- Engine is operating too cold.
- Lubricating cooler is improperly regulated.
- Wrong grade or class of lubricating oil.
- Recurrent low-load or idle running periods.

- Intrusion of coolant.
- Excessive blowby.
- Oil change frequency is overextended.
- Sump was not cleaned out at oil change.

Symptom 21, Engine Lubricating Oil Level Rises

- Coolant leaks into the crankcase.
- Fuel oil passes into the crankcase.
- Distorted dipstick or dipstick tube gives incorrect reading.
- Foaming or emulsification of lubricating oil.
- Defective automatic-lubrication leveler.

Symptom 22, Engine Has Low Lubricating Oil Pressure

- Worn bearings.
- Worn lubricating oil pump.
- Lubricating oil pump coupling is defective.
- Lubricating oil level is too low.
- Lubricating oil pressure relief valve is partly open.
- Lubricating oil temperature is too high.
- Lubricating oil is diluted by fuel.
- Lubricating oil filters are plugged.
- Lubricating oil pump is drawing air.
- Lubricating oil suction restriction.
- Wrong grade of lubricating oil.
- Engine is running too hot.
- Lubricating oil thermostats are faulty.
- Oil is diluted by coolant.
- Bearing clearances are incorrect.
- Oil gallery leaks.
- Lubricating oil cooler is dirty.
- Engine is overloaded.

Symptom 23, Engine Has High Lubricating Oil Pressure

- Frozen lubricating oil cooler.
- Incorrect setting of lubricating oil pressure regulating valve.
- Malfunction of thermostatic control valve.

- Incorrect grade of lubricating oil.

Symptom 24, Engine Lubricating Oil Is Too Hot

- Lubricating oil cooler is dirty.
- Lubricating oil cooler thermostat is operating improperly.
- Engine is overloaded.
- Bearings are approaching failure.
- Worn cylinders, piston, and rings (blowby).
- Insufficient coolant circulation in jacket.
- Lubricating oil level is too low.
- Lubricating oil level is too high.
- Coolant is not circulating in lubricating oil cooler.
- Failed water pump.
- Aspirating air is too hot.
- Bad injection timing.
- Inadequate aspirating air pressure.

Symptom 25, Engine Uses an Excessive Amount of Lubricating Oil

- Worn or broken scraper rings.
- Worn valve guides.
- Seal and gasket leaks.
- Perforated lubricating oil cooler.
- Glazed liners.
- Excessive blowby.
- Makeup lubricating oil was added unnecessarily.
- Stop on dipstick is loose.
- Blower or turbocharger seals are defective.
- Wrong grade or class of lubricating oil.
- Carbonized piston rings.

Symptom 26, Engine Makes Black Smoke

- Engine is overloaded.
- Excess fuel or incomplete combustion.
- Aftercooler is dirty.
- Air filter is dirty.
- Defective turbocharger(s).

- Lack of compression.
- Injector timing is late or pressure setting is incorrect.
- Broken valve seat.
- Burnt exhaust valve.
- High exhaust back pressure.
- Incorrect injector nozzles.
- Defective or collapsed fuel pump delivery valve(s).
- Broken piston rings.
- Defective fuel line(s).
- Plugged injector nozzles.
- Incorrect injector cups.
- Cracked injector body or cup.
- Long periods of idle running.
- Broken injector spring.
- Defective turbocharger clutch.

Symptom 27, Engine Makes White Smoke

- Cracked cylinder head.
- Cracked cylinder liner.
- Cracked turbocharger casing (liquid-cooled).
- Cracked valve cage.
- Leaking aftercooler.
- Cracked piston.
- Cracked precombustion chamber.
- Failed grommets (two-piece block).

When the atmospheric temperature is below 0°C (32°F), these symptoms will not necessarily apply.

Symptom 28, Engine Makes Blue Smoke

- Worn or broken piston rings.
- Glazed liners.
- Stuck piston rings.
- Crankcase breather is restricted.
- Worn valve guides.
- Retarded timing.
- Lubricating oil level is too high.
- Piston cooling sprayers are misdirected.
- Cracked piston.

- Turbocharger oil seals defective.
- Oil bath filter oil level is too high.

Exhaust smoke colors are very much a matter of personal interpretation. Also, they vary with atmospheric conditions, and so on.

Symptom 29, Engine Has High Exhaust Temperatures

- Defective turbocharger.
- Burnt or cracked exhaust valves.
- Burnt, loose, or cracked valve seat(s).
- Engine is overloaded.
- Insufficient cooling.
- Poor compression.
- Dirty air filters.
- Dirty aftercooler.
- Valve timing is incorrect.
- Injection timing is incorrect.
- Exhaust back pressure is too great.
- Aspirating air is too hot.
- Tappet setting is incorrect.
- Incorrect fuel rack settings.
- Cylinder liners and heads are badly scaled.

Symptom 30, Engine Uses Too Much Fuel

- Restricted air intake.
- High exhaust back pressure.
- Incorrect injector nozzles.
- Leaking injector nozzles.
- Injection pressure setting is incorrect.
- Generator phase imbalance.
- Valve or injection timing is incorrect.
- Engine is generally worn.
- Turbocharger is not running at required speed.
- Dirty aftercooler.
- Unit is too large for the load applied.
- External or internal fuel leaks.
- Wrong injector cups.
- Cracked injector body or cup.
- Mutilated O-ring on injector cup.

- Engine overloaded.

Symptom 31, Engine Makes Unusual Noises

For fault identification purposes, it is not possible to provide descriptions of unusual noises that are in any way meaningful. However, if in the operator's experience an engine has abnormal sounds, then it must be stopped for inspection without delay. Generally speaking, all unusual engine sounds emanate from components in distress. Never expect to stop a noise by ignoring it.

Symptom 32, Load Trips Off; Engine Continues to Run

- Electrical protection is actuated.

Symptom 33, Engine Will Not Stop

The fuel rack may not be in the minimum fuel position. The clutch on the governor speeder motor may be slipping. Read and reread the notes which follow.

Important Note. If the load cannot be transferred fully from the outgoing engine, DO NOT TRIP THE BREAKER. It is essential that the engine be left loaded while the fuel rack and the governor linkage are inspected for obstruction and the problem is resolved. If the engine is tripped off the line at anything greater than zero load, there is grave danger it will go into uncontrollable overspeed. If there appears to be a problem of this nature, the operator should get assistance at once. DO NOT ACT HASTILY.

5.4 PROTECTIVE TRIPPING OF BREAKERS

Following a breaker trip triggered by a protective device, the plant operator must take note of which indicator flags have fallen. Identification of the flags will assist the operator in determining what caused the breaker to trip. The tripped flags may also provide clues to what must be done to avoid a similar incident. A certain amount of caution must be used in the restoration of service if it is thought that the feeder breaker was tripped by some form of distribution line damage.

5.5 REVIEW OF OUTAGES

A review of the reasons behind the outages incurred over a period of time is most useful and will reveal patterns of performance or reliability. The review, carried out by the superintendent in conjunction with the operators, will furnish ideas for the elimination of repetitious blackout causes. Interruptions in the supply of power to the consumer may be categorized as:

- Operator error

- Equipment breakdown
- Protection deficiency
- Natural causes (wind, snow, or traffic damage to distribution system, and such)

So far as the consumer is concerned, the quality of the power fed to him as an individual is all-important. It makes no difference to him if he is located in the largest city in the world or the most remote desert. If he is served by an electric power supply, he will be dissatisfied when power is interrupted. Further, he will not lend a sympathetic ear to the explanations offered when power outages are repeated.

CHAPTER 6
HOUSEKEEPING

6.1 WIPING-DOWN ROUTINES

All areas of a plant should be wiped down regularly. It is of prime importance that the person delegated to do the wipe-down is first and foremost properly instructed on the safety aspects of cleaning a machine that moves or is likely to move.

The main purpose of wiping down *is not to clean up a dirty machine, it is to ensure that a complete visual inspection of the entire machine is carried out on schedule*. Thus, it is important that the wiper understands what he is looking for or at. He must be able to notice the development of any leaks, loosenesses, or other defects, to report them accurately, and otherwise take suitable corrective action.

A clean plant is a combined product of good maintenance and operating practices. Also, the morale of the staff will be high and the spare parts, tools, other equipment, and facilities will be properly cared for. The clean surfaces will reflect light and give the plant a bright appearance. It will also be found that the safety record of a well kept plant will be above average.

In a dirty plant there will be little success in keeping a good safety or reliability record, morale will be lower, and turnover of labor will be higher. The dirt accumulations will conceal tomorrow's problem and will discourage the mechanic or operator from giving his fullest attention to any work in hand.

Every plant must have a cleaning routine. The work load should, ideally, be shared by a number of persons, each being responsible for a particular area. It will vary from plant to plant, but in every case it is easier to keep a clean plant clean than to clean up a dirty plant. A regularly cleaned plant is a regularly inspected plant. Regular inspections contribute greatly to comprehensive maintenance, which in turn results in improved reliability.

A power plant floor should be swept at least daily by a designated operator as part of an established routine. The mere act of sweeping the floor is an important part of the daily inspection. The aware sweeper will take notice of and act in an appropriate manner when he finds odd pools of oil or water or quantities of soot or other unusual material on the floor. Any of these may well be an indication that something is amiss. (For example, the sweeper should use care when cleaning the floor below a torsional damper and notice evidence of rubber particles in the form of a grayish dust.)

It is highly recommended that all engine wiping down be assigned to the operating staff. The usefulness of wiping down is largely lost if the work is given to a disinterested laborer as just another tedious, meaningless task.

6.2 DRIP TRAYS

A correctly used drip tray serves to monitor a leak condition. In many instances crankshaft seals or pump glands must weep continuously if they are to be effective and a small patch of oil will ordinarily accumulate in the drip tray, but if the drip tray suddenly starts receiving a much-increased flow, corrective action must be taken.

Drip trays do not cure leaks. In certain circumstances, it is impossible to stop all leaks on an engine (e.g., lubricating pump glands, water pump glands, and crankshaft end seals). In these cases, drip trays are permissible. The drip tray must be emptied regularly, and it must never be allowed to fill to the point of overflow. Drip trays must never be used as garbage receptacles, and they must always be regarded as fire hazards. If in a plant where the machinery is anything but brand new, a drip tray is considered for a condition that has arisen, it is almost certain that the condition can be corrected by fixing the leak and thereby eliminating the need for a drip tray.

Servicing drip trays is the operator's duty, and the operator must log any change in the pattern of catchment. Drip trays must not be sanded or loaded with oil-absorbing material such as vermiculite because what was a moderate fire hazard will thereby be converted to a major fire hazard.

6.3 CLEANING CONCRETE FLOORS

Concrete floors are best left always unpainted because the high pressures imposed on the floor during the movement of oil barrels and spare parts crush the paint structure and cause peeling and flaking. The continuous washing of a painted floor requires the frequent usage of detergents or other solvents which also have a decaying effect on the paint. A painted floor that has become damaged over a period of time cannot be properly swept clean, and successive coats of paint only accentuate the problem.

Plain unpainted floors can be kept clean quite easily through the use of water and detergents, with occasional use of solvent to remove the denser stains. Much cheaper than solvents and usually more effective is trisodium phosphate. The crystals are spread on the soiled area and covered with rags or cardboard sheets which are kept damp. After several days, the oil stain will be gone. If oil, coolant, or other liquid repeatedly falls onto the same place, there is a clear need for a permanently corrective action.

6.4 SPILLS AND CLEANUP

Various absorbent media can be used for soaking up oil spills. At one time, sand and sawdust were the common media, but in more affluent times, vermiculite has become popular and certain other materials of improved absorbency are available. Peat moss also is in common use.

None of these materials are recommended for general use; for specific jobs, any of them are suitable within limits. To soak up accidental liquid spills from rough surfaces, any absorbent product will do, provided both the spill and the sponge material are totally removed. All too often, the spent absorbent material is left lying in a sodden mass that becomes dirtier by the day and presents a wonderful fire opportunity. When cleanup does not take place, the spill-absorbent mix begins to get tramped around the plant, and then a major cleaning operation has to be launched.

The granular, absorbent materials are quite mobile; they lodge in floor cracks, cable troughs, under engine support frames, in doorways, under mats, or any other place where cleaning is difficult and where the material is a fire hazard. The answer to the problem lies in the conscientious handling of fluids by all plant personnel and cleaning up spills as they occur. Peat moss should be used only for the cleanup of large scale out-of-door fuel spills. Under no circumstances should it be used indoors.

Ideally, absorbent material should be kept in the storeroom and be obtained only with the special permission of the plant superintendent. He will want to hear, after the materials has been used, that the spill and the surrounding area have been thoroughly cleaned up and returned to normal.

6.5 WIPING RAGS

Great savings in costs can be achieved through a little care in the use of wiping rags. New rags are usually supplied in quite large pieces which are simply not safe to use because they can become caught so easily in moving machinery. The storekeeper issuing the rags or the person using them should cut the rags down to pieces approximately 12 in square. A rag of that size is perfectly adequate for wiping purposes, and when soiled is more easily washed for reuse. A rag should not be considered used until it is totally oil or dirt laden.

All rags should be recovered, washed, and reused. Rags in storage, whether new or used, must be kept dry because there is a considerable risk of spontaneous combustion from massed damp or wet rags. New supplies should be checked for dampness before being stored. Used rags must be kept in closed containers before recycling, and the containers must not resemble a garbage can in any way. Never put oily rags in a garbage can. Rags must not be left in drip trays, nor must they be used to dam oil leaks.

The use of cotton waste in any diesel engined plant must be prohibited. Cotton waste, once used, cannot be recycled. It is a tremendous fire hazard because it is much more supportive of flame than ordinary rags. It is also very susceptible to spontaneous combustion when damp. Detached threads of waste will always be found on rotating shafts, valve spindles, far blades, or anywhere else from which removal can be difficult.

For any cleaning purposes, rags or paper towels are superior to cotton waste. Rags can be washed many times over, and paper woven towels also can be washed more than once. On a cost basis, if rags cost $1.00, cotton waste will cost $1.50, paper towels will cost $3.60, and paper roll will cost $4.40. When the rags are recycled, the economic advantage is profound. The good operator will take these numbers into account and be much more selective in the disposal of spent wiping materials. Most of the dirty wiping materials will go into the washing machine again and again. The provision of a simple washing machine with a wringer attachment will result in substantial savings.

6.6 SELECTION OF COST-EFFECTIVE CLEANING AGENTS

The tremendous range of cleaning materials in solid, crystal, powder, and liquid form is a source of constant confusion to the potential user. It is difficult to decide which product is safer or more suitable than the next, and the asking price of any product is no indication of superiority or effectiveness. The circumstances of the intended use can often have a marked effect on the efficiency of the selected material.

The purchaser must make a solid effort to find cleaners that are truly the safest and most cost-effective that can be obtained. Any serious potential supplier will provide samples of his product for experimental use. An appropriate cleaner having been procured, the personnel using it must be fully instructed on its proper use, and it must be seen to that they use the material frequently, not wastefully. The purchase of ready-mixed water-based cleaners is not advisable. It is far cheaper to purchase water-dilutable crystals or powder.

Cleaning materials are always composed of some sort of solvent. The solvent agent selected must be able to dissolve the foreign material that acts as a binder of the included solids. It naturally follows that the solvent chosen must not in any way attack the parent metal, fabric, insulation, or coating of the object being cleaned. The composition of any cleaning material considered for use should be reviewed from a personnel health and safety aspect before it is purchased. The product should not be dermatitic; its vapors and fumes cannot be harmful to internal organs; nor must it be harmful to the eyes. The vapors must not be explosive when exposed to an electric spark or naked light.

If a new product is being tried for the first time, it is advisable to apply a small quantity to various test surfaces such as aluminum, paint, resin, plastic molded or bonded components (i.e., polyester, polycarbonate, polyethylene, and polystyrene), brass, rubber, and gasket and O-ring materials. Some solvents may have an undesirable reaction, and pretesting is better than repairing damage.

Before any attempt is made to use a cleaning agent, the instructions for use and the attendant cautionary notes must be understood by the employee and the employer. Read all product instructions and ensure that first-aid materials commensurate with the cleaning solvents in use are available. Understand the first-aid instructions and remember that no cleanser is 100 percent safe. Even common bar soap has its traditional hazard.

6.7 CLEANING MATERIALS

It is clear that numerous factors are to be considered in the selection, storage, and use of cleaning solutions, materials, and compounds. It is suggested that lasting purchasing standards be evolved for the individual plant so that the dangers of multiple product mixing are lessened and at the same time training and information needs are reduced to a minimum.

All powder and crystal cleaning agents must be stored in dry areas. Liquid solvents and acids must be kept cool and away from sunlight. The storage rules for any combustibles must be observed, and all storage areas must be adequately ventilated. Powders and crystals may become caked in storage and require reduction before use. Crumbling must not be done with bare hands, and the creation of dust when breaking down lumps of the material must be avoided. Simple powdered detergent has to be handled with reasonable care. The rough handling of bulk quantities will generate a dust which can have an asphyxiating effect if inhaled.

Physical contact with all cleaning agents should be minimal. The user must take notice of softening of his fingernails, slickness, burning, or itching of his skin, prickling or tearfulness of his eyes, nasal itch, throat irritation, or any other physical discomfort.

When opening any container of cleaning material, care must be taken to gradually relieve any internal pressure that may have developed through ambient temperatures of the plant or environment, gassing, aging, and so on. The container in which the cleaning agent is to be carried while in use must be proof against the effects of the cleaner. (Zinc-coated buckets, plastic buckets, and aluminum pans may be at risk.) If there is a reaction between the cleaning agent and the ready-use container, the container may melt, rot or otherwise decay and in some cases evolve noxious fumes.

6.8 PERSONAL PROTECTION AND SAFETY

During the use of evaporative solvents or sprayed cleaning agents, the density of the atmosphere may increase without its being noticed by the user. It is, therefore, important that all areas where cleaning is carried out have sufficient ventilation so that no personal ill effects develop. To protect the respiratory organs, cotton wool masks will prevent the inhalation of dust, but if noxious fumes result from the cleaning solvent or its effects, a catalytic respiratory known to trap the specific fume must be used. Such a respirator is required, for example, where trichlorethylene-based solvents are used.

In all cases in which intensive cleaning operations are being conducted, the personnel must wear eye protection. Form-fitting flexible plastic goggles are inexpensive and are highly recommended. Neoprene or polyvinyl gloves and aprons, close-fitting head gear with hard hat superimposed, and liquid-proof boots or shoes are pretty much standard requirements. The skin must be protected from cleaning compounds containing chromium. If chromium-containing cleaning agents are used for spray cleaning, an approved mask must be worn by the user.

Employer-provided protective equipment shall be used in preference to all other. The use of provided protective clothing or equipment shall be the responsibility of the employee, but the provision and suitability of all protective equipment remains the responsibility of the employer.

6.9 MECHANICAL APPARATUS, CLEANING

Whether the cleaning agents are supplied in liquid, powder, or crystal form, caution must be exercised when mixing any of them because some chemicals used in cleaning materials can occasionally have a violent reaction when brought into contact with other chemical products.

When mixing alkaline cleaners, add the cleaner to water—*do not add water to the cleaner*. Violent reactions can be expected if improper mixing methods are used. Do not add large quantities of fresh alkaline material to already prepared hot mixes. Trickle in the new supply premixed in cold water to avoid violent disturbances.

When acidic materials are used, agitation by air may induce rapid corrosion. Agitation should be by means of stirring or thermal circulation. If acidic materials are being employed in a cleaning operation, there is a likelihood that gases will be generated. Adequate ventilation is required.

The following are miscellaneous effects and uses of cleaning materials.

- Alkaline solutions used on aluminum will generate highly explosive hydrogen gas.
- Do not use acidic solutions on any cyanide-hardened parts. Acidic solutions must not be mixed with any cyanide-based solutions. Toxic cyanide gas may be generated.
- Do not use acidic solutions on any chlorine-based materials. Toxic chlorine gas may be generated.
- Acid-based cleaners may etch metal surfaces. The highly polished surfaces of camshafts and fuel injection equipment are especially susceptible to damage of this kind.
- Whenever possible, in the interests of reasonable economy, scrape mechanical parts to remove bulk dirt before going into a chemical cleaning process.
- The wires and slings used in acid-cleaning operations should be regularly examined for corrosion wear. Some solvent cleaners may affect nylon slings and sisal or polypropylene ropes.

Before any spent cleaning solution may be poured away, all municipal or other environmental regulations regarding sewers, lagoons, and effluent discharges must be consulted and followed. Arrange to recover solvent drainage whenever possible; pour it into a settling can and use it again after the solids have dropped out.

Using an air gun tends to be wasteful unless the solvent employed is either very inexpensive or is being drained down into a catchment tray and recovered. Whenever possible, the most effective method of solvent application is with a bristle brush or paintbrush. The solvent should be worked into the dirt-covered surfaces, and then an air blast should be used to clear the loosened debris after the binders have been dissolved. It is, however, grossly inefficient and shockingly wasteful to use an air gun and solvent blast just to get dirt away from a surface. If such a method must be employed, perhaps it is better to use kerosene or fuel oil. At any rate, the fluid material used for such a cleaning method must be absolutely the cheapest possible.

Compressed-air nozzles, pressurized spray guns, or lances must not be pointed at any person at any time. Compressed air must not be blown on either a person or his clothing. Never use compressed air or cleaning solvents to clean any part of the body. Use only specifically approved hard, soft, or liquid soap for personal washing.

6.10 ELECTRICAL APPARATUS, CLEANING

- Fluids used for cleaning electrical components should have a dielectric strength of 25,000 V or greater. Cleaning materials for electrical apparatus should embody a nontarnish agent.

- Preclean electrical apparatus by air blast to reduce solvent expenditure to the smallest amount. Air, steam, water, or solvent lancing pressures must not exceed 35 psi.
- Detergent solutions used to clean heat exchangers, oil coolers, and similar equipment can suspend only a certain amount of solids. Therefore, the circulation system must be fitted with a filter so that the entrained solids can be trapped and the full effect of the detergent action can be sustained.
- Do not permit welding or cutting to take place near a cleaning operation; there may be some risk of vapor or fume ignition. Do not weld or cut a container that contains or has contained a cleaning solution.
- Before a cleaning operation commences in the vicinity of electrical apparatus, be sure the current is switched off. Apply warning tags. Most cleaning solutions, especially alkali-based ones, are highly conductive. Check with the electrical supervisor or journeyman in order to avoid short circuits or personnel injury.
- Cleaning materials, particularly those for electrical equipment, are best used under the supervision of a journeyman of appropriate certification.
- Cleaning solvents containing trichlorethylene or carbon tetrachloride must not be exposed to heated surfaces, sparking connections, or naked flames because highly toxic fumes will be generated.
- Whenever a rotating machine is cleaned internally and the use of sprayed cleaning solvents is involved, all of the machine's bearings must be stripped and the lubricating medium must be renewed.

6.11 TEMPERATURE OF CLEANING FLUIDS

The temperature of any fluids used for spray, lancing, or pressure wand cleaning methods should not exceed 50°C (120°F) if there is any possibility of the fluid's coming into contact with the person. This temperature may seem to be very low, but any temperature, hot or cold, that causes intense discomfort is rated as damaging from a medical point of view. Temperatures above and below those that cause discomfort cannot be accurately judged by touch or feel, and they may therefore be positively dangerous.

High temperatures may be resorted to when the object to be cleaned is placed in a circulating or agitating closed bath, but there must be a cautionary attitude toward any fumes that may be generated.

6.12 DISPOSAL OF SPENT CLEANING MEDIA

One of the main points that must be debated when selecting any cleaning solvent is that of disposal when spent. After most cleaning operations, there are quantities of fluid that must be dispersed in an acceptable manner. The product container often carries instruction for its use; sometimes the instruction advises on the disposal of the spent material. Such advice should be followed whenever practical.

Dirty water and ordinary household-type detergent cleaning powders may be poured into a sewage system, but any acidic, caustic, or vaporizing cleaner may have spectacular, if not disastrous, effects upon a sewer. A concrete or ceramic sewer pipe may not be affected, but a plastic sewer pipe will be at risk. Cleaners that are of a biocidal nature or contain a petroleum base cannot be passed into any stream, river, or lake.

When any disposal terminates in the natural environment, expert advice must be sought and accepted. The burning of small percentages of oil-laden solvents mixed with the main fuel in a boiler is permissible, but no attempt can be made to burn off any wastes that may contain carbon tetrachloride or trichlorethylene because toxic gases will be evolved. No solvent wastes whatsoever shall be mixed into a diesel fuel supply.

6.13 COIR MATS

Coir, or coconut, mats should be placed at each power plant entrance, at the entrance to the control room, and the entrance to the office. These mats will gather a great deal of oily mess that otherwise would be walked all over the floor areas and increase the housekeeping work load. The mats can be washed regularly in solvent (Varsol, for example), dried and returned to service many times over.

Generally, the mats are left on the floor until they are totally impregnated with oil and dirt, and then they are thrown away. If they are subject to a regular routine, they will assist in keeping the everlasting floor-cleaning problem under control. The best solution to the problem is not to use coir mats at all, but instead use self-cleaning metal mats.

6.14 MOPS, BRUSHES, AND PAILS

When not in use, the housekeeping tools (i.e., mops, pails, and brushes) should be returned to the same parking place every time. None of the cleaning equipment should be carelessly left in any walkway or passage where, in the event of a power outage, it would be a tripping hazard.

6.15 ELECTRIC LIGHT CLEANING

Cleaning of all electric lights is the reponsibility of the electrician. Cleaning of windows, washrooms, and lunchrooms is the responsibility of the operator(s) or other designated personnel.

6.16 CLEANING FANS

Cooling and ventilating fans in diesel power plants have to be cleaned quite frequently. It is quite common for the fan blades to become coated with an oil film mixed with dust, insects, and pollen that will cause fan imbalance, vibration, and heavier starting loads. Before the cleaning work commences, the power to the fans must be switched off and locked out in a formal manner. A safe work platform must be placed in a position that will allow the cleaning and inspecting personnel to perform the task safely.

Most important of all, the fan itself must be fastened or secured so that it cannot rotate. Changes in the atmospheric pressure inside or outside the power plant building can cause an idle fan to rotate quite suddenly and quickly. Serious injury of an unaware person can result. Atmospheric changes can occur when another engine is started, a boiler is lit, the main doors of the plant are thrown open, or a gust of wind blows over the plant.

6.17 PAINT AND PAINTING

A clean plant is usually a well-maintained plant in total; cleaning and painting are very much part of the upkeep routine of a well-run facility. In almost every plant, except for the largest of the diesel generating plants, the business of painting falls upon the station operators or operators to be, and some care must be taken to get the painting work done without neglecting the operating routine. The ongoing and very necessary wipe-down routines must not be put aside in favor of or because of new-paint applications. A well-cleaned plant rarely needs new paint—least of all the floor.

A few diesel power plant owners have their own color schemes, but the majority of the world's diesel generation, pumping, or propulsion plants share a common color coding and pipe identification scheme:

Color	Identity
Green	Raw water
Medium blue	Treated water (including ethylene glycol-water mix)
Light blue	Potable water
Yellow	Lubricating oil
Brown	Fuel oil
Red	Fire-fighting water
White	Compressed air
Black	Electrical conduit

The use of black paint for anything but electrical conduit (which most owners prefer to leave unpainted anyway) is very much discouraged. The color serves no useful purpose and adds nothing to the power plant decor.

Every power plant employee must be cautious in his regard for the various paint hues as applied to the pipe systems in any power plant complex. He must never assume that a pipe of a particular color carries the same medium as it did in his previous workplace. The variations from established codes are commonplace, and the indiscriminate use of color codes applied in other industries can render pipe content recognition a rather hazardous business. The employee is recommended to enquire very thoroughly into the color and medium relationships in any given plant.

Better than any pipe color coding method are the adhesive labels that state the medium—lubricating oil, fuel, jacket water—carried in a particular pipe. Used in conjunction with directional arrows, these labels are particularly effective as training aids.

It is recommended that all of the structural steel work, the main machinery, the auxiliaries, the ancillary equipment, and the floors (if the floors must be painted at all), be painted in light shades of gray, stone, or buff. Light colors, blues excepted, will reflect light and give the plant interior a bright aspect. Any leaks, cracks, or other defects that develop will show up more readily against a well-painted and frequently cleaned background.

Flexible connections, safety valves, hoses, rubber boots, chains, cables, adjustable threaded items, pressure gauge lines, and plastic fittings must not be painted. Individual components such as governors, safety shutdown devices, instrument, and breaker panels can usually be left in the original factory paint for many years because they often have special enamel finishes that readily wipe clean. Do not paint any fuel injection equipment whatsoever.

Never paint any data plate that shows serial numbers, performance figures, specific instructions, dimensions, rotation direction, and part numbers. Similarly, do not paint any coil springs, because their original color usually indicates their performance grade.

Study the instructions on the can before using the paint. Today's paints are technically advanced, and many of them have peculiarities appropriate to their intended use. Take note of the correct thinners, the curing time, the surface preparation requirements, and the suitability. Incorrect use or selection can turn a simple paint job into a cleanup nightmare if the directions are not followed. Spray painting is good for larger surfaces provided that any masking necessary is applied and that the spraying is not carried out near a running engine. Spray painting in the vicinity of a running engine can be hazardous:

1. The paint fumes may fuel the engine, particularly if the paint being used has a cellulose or methyl ethyl ketone base.
2. The air filters of the running engine will plug up very quickly as they pick up the atomized paint from the atmosphere.
3. The danger of flash from the hot exhaust is always present.

Except for minor touch-up work, the use of aerosol paint cans cannot be justified owing to the extreme cost.

Never attempt to apply paint to a surface that is above 30°C (85°F) or below 5°C (40°F) or to apply paint to a piece of equipment that is in motion. Applying paint to oily, dirty, flaking, or an otherwise improperly prepared surface is just a waste of time, money, and effort, and the future effect is demoralizing.

Paint must be stored carefully and should not be exposed to temperatures below 0°C (32°F) or above 20°C (68°F).

6.18 ENGINE PAINT

The paint used on an engine ideally is the one recommended by the engine manufacturer. Should a paint be acquired from local sources for use on a diesel generator, it must meet several basic requirements. It must be proof against the effects of contact with fuel oil, lubricating oil, and coolant, and it must be capable of withstanding the heat of engine operating temperatures. It must not discolor under normal cleaning routines, nor must it flake, crack, or chalk with temperature fluctuations. The cured finish must be a durable, impact-resistant skin with an expansion coefficient similar to that of the basic engine material.

The paint should be based upon an acceptable solvent that is readily available, be compatible with the paints already applied, and require no special skill in its application. The surface preparation needs should be minimal and without resort to machine wire-brushing or sandblasting. Neither spray paints nor epoxy type paints can be recommended, because they have severe economic and application drawbacks.

6.19 PAINT COVERAGE

For the purposes of estimating the quantities of paint required for various maintenance jobs around the power plant, the coverage of oil-based machine paint can be taken as:

- 600.0 ft^2/imp gal
- 500.0 ft^2/US gal
- 160.0 ft^2/L
- 14.5 m^2/L

These coverage figures will be about right for any metal or other previously painted surface. In the case of concrete or wooden surfaces that have not been painted before, the coverage rate, for the first coat only, will be from 25 to 50 percent less than that stated above.

Heat-resistant paint as applied to mufflers, exhaust pipes, and other high-temperature surfaces will go approximately 800 ft/US gal or 16 m/L.

6.20 PAINTING ENGINES WITH EXPOSED FLYWHEELS

Many engines have a number of timing marks stamped on the rim of the flywheel. Frequently, these marks get painted over when the rest of the engine is painted. It is suggested that before any painting activity commences, all of the flywheel marks should be located, thoroughly cleaned with a wire brush (DO NOT USE EMERY CLOTH!), and then masked while the painting is completed.

It is very important that all timing marks be easily discernible and very legible. The station mechanic should review the marks and satisfy himself that all the marks actually stamped on the wheel have been found, cleaned, and are ready for masking in preparation for paint application. At no time should the flywheel marks be obliterated, altered, or otherwise damaged, nor should any other marks be added.

6.21 CRANKCASE PAINT

Many engines are painted internally. The paint is applied to the crankcase, pan, timing gear casings, and other surfaces exposed to the flushing action of lubricating oil to bind any particles of scale, core sand, machine cuttings, or other debris originating in the engine manufacturing processes.

It is not normal to ever have to repair or touch up the engine's internal paint surfaces during the life of the unit. Occasionally, however, the paint will be damaged by the action of a leaking coolant medium or by the introduction of an inappropriate lubricating oil.

Crankcase paint will soften and slime under the action of ethylene glycol, but the state of the oil on the dipstick will indicate something is amiss long before the engine is opened and the paint is found to be soft. Lubricating oil analysis will also advise of the presence of coolant in the lubricating oil system. Crankcase paint will decay more rapidly under the action of propylene glycol. Propylene glycol is not in common use in engine cooling systems, but it will be found where waste heat from the engine cooling system is utilized to temper a potable water facility. Such a facility is often connected some time after the original cooling system was installed, and it will therefore be necessary to obtain OEM advice on the effect of the new coolant on the old crankcase paint.

Synthetic lubricating oil may have an adverse effect on crankcase paints, and again the OEM should be consulted before any conversion from the standard lubricating oil is made. Hydraulic oils and automotive transmission oils have a very pronounced stripping effect on crankcase paints. If paint problems are sudden and dramatic, the possibility of wrong oil being introduced to the engine cannot be ignored.

If paint is softened to a slime or is semidetached from the metal surfaces, the engine must be thoroughly scrubbed internally so that all loose material is washed down into the base of the engine, from which it can be removed. Should the paint appear to be blistered or flaking, then all of the paint that has lost its adherence must be scraped off and removed from the crankcase. At each oil change the lubricating oil filter elements must be examined for evidence of crankcase paint breakdown. If it is found necessary to repaint any internal area of the engine, the only paint that can be considered is that supplied by the OEM. The instructions for its application must be followed very carefully.

6.22 WELDING ON PAINTED SURFACES

Although it is not anticipated that there will ever be much need for welding operations in an established diesel power plant, a certain caution is required when welding does take place. Alkyd and polyurethane paints release toxic fumes when burned, and it is therefore necessary to remove such paints from surfaces that will be heated to welding temperatures. In any event, the work area must be fully ventilated and individuals not directly concerned with the welding operation should stay away from it.

6.23 PREPARATION OF SURFACE FOR PAINTING

The use of shot- or sandblasting techniques for surface preparation prior to painting within the power plant cannot be accepted. Although it may be convenient to take metal components outside the plant to sandblast or shotblast them clean, considerable care must be taken to see that any machined surface is protected from pitting or other metal loss. Sand- or shotblasting inside the power plant will lead to metallic dust entering switchgear, electronic equipment and generator windings. Air filters, turbocharger inlets, pump glands, and seals will be contaminated by minute hard particles.

A stiff wire brush is quite adequate for most surface-cleaning purposes, followed by washing with a detergent. Machine wire-brushing will have the same deleterious effect as sandblasting. Individual engine components removed from the engine during overhaul can be cleaned in an acid bath, and perhaps sacrifice all of their paint in the process. Caution is required to ensure that none of the nonferrous metals—copper, brass, or aluminum—are put in the bath, where they will corrode very quickly.

Most power plants tend to get painted too frequently, especially in the easy-to-reach places. Layer after layer of paint builds up without any benefit, and the surface strength of the painted area becomes very weak. Crushing and crumbling of the layered paint then lead to a very rough surface which is difficult to keep clean. For the most part it is better to concentrate on better housekeeping with regular cleaning and forget about paint.

6.24 TEXTURED OR NONSLIP PAINT

The use of textured paint cannot be recommended in any operational or administrative power plant application. Nonslip paint can be used successfully only where foot traffic is nominal. The use of either textured or nonslip paint makes plant housekeeping duties extremely difficult, and it is almost impossible to maintain such painted surfaces in a clean state. Slippery surfaces are best treated by eliminating the cause of slippage: preventing ice formation and avoiding spilled oil and coolant leaks.

6.25 FLOOR PAINTING

Many power plants have their floors painted, and they look very nice when the paint is fresh and new. Unfortunately, painting engine room floors creates another work load that has no real benefit. Soon after it has been applied, the paint begins to chip or flake because of the point loads imposed by barret rims, engine components, dropped tools, and so on. Some paints are susceptible to the decaying effects of fuel oil, and ethylene glycol will cause most floor paints to soften. A painted floor with any liquid lying upon it is very slippery.

Eventually when the floor paint has been damaged extensively, the application of new paint is considered. All the fluids that have got onto the floor from time to time will have penetrated the concrete to a certain degree and will prevent any new paint from obtaining a proper key. Furthermore, the remaining old paint will be somewhat loose and will have to be scraped off. If it is not scraped clean, the floor will assume a rough appearance when the new paint is slapped on.

All cleaning solvents, including soap powder and Varsol and other degreasers, are ruinous to floor paint. Plain unpainted floors can be kept clean quite easily through the use of water and common detergent with the occasional use of solvent to remove stains. Much cheaper and usually most effective is trisodium phosphate.

If oil, water, fuel, or coolant repeatedly falls onto the same place, there is a clear need for permanently corrective action. A clean but stained concrete floor is easy to look after and represents a minimum work load. A painted floor will be an ongoing work load and will not look good for long.

6.26 CRACKING OF CONCRETE FLOORS

Hairline cracking of concrete floors is commonplace and is not cause for concern. Major cracking or deterioration of the floor, such as the following, must be reported so that engineered remedies can be proposed and activated.

- Some cracks open. (Cracks exceeding 100 mm (0.375 in shall be reported to the owner.)
- The concrete surface assumes various levels on each side of a crack.
- The floor shows evidence of subsidence or heaving.

The first step in the information feedback is to read the floor levels so they can be compared with the original construction signatures (Sec. 43.3). That will give some indication of the extent and severity of the floor movement that has taken place. A copy of the readings should be sent to the company headquarters for reference.

Depending upon the season, it may be some time before a permanent repair can be made. In that case, the plant staff shall take it upon themselves to do whatever is needed to keep the generating machinery in an operable state until permanent repair can be made. Chocking, realignment, and adjustment of flexible connections, are some of the elementary things that can be done to start with.

6.27 EXIT DOORS

In all plants and associated buildings, certain doors are designated as EXIT. Illuminated signs over these doors are intended to identify escape routes for personnel in the event of fire, power outage, or other severe disturbance. The doors are usually fitted with panic bars so that a person can leave the building in haste.

In plants subject to heavy snowfall or snowdrifting, the operator must as part of his safety duties ensure that the designated EXIT doors are clear to open at any time. If there are no climatically induced dangerous conditions, the operator must see that no oil barrels, vehicles, trash cans, or whatever obstruct the doors at any time.

CHAPTER 7
CRANKCASE

7.1 CRANKCASE PRESSURE

Crankcase pressure is almost always developed by combustion gases that leak past the piston rings. In an engine fitted with new liners, pistons, and rings, the crankcase pressure will be minimal; indeed in those engines on which a crankcase exhauster is fitted there may be a small negative pressure. As the engine wears, a certain amount of combustion gases will leak past the piston rings and the crankcase pressure will begin to rise.

All engines are fitted with some means of relieving the pressure. In some instances, it will be an exhauster as mentioned above; in most engines a simple gauze oil trap and pipe venting to atmosphere is utilized; and of late, some engines are fitted with an apparatus that extracts most of the oil from the vented fumes. The oil passes to a collector and the remaining fumes are induced into the engine air-intake manifold.

It is most desirable that the engine crankcase fumes be kept under control. If the crankcase pressure rises too high, the introduction of makeup lubricating oil will become difficult, the crankshaft seals will commence leaking, and the breather discharge will slobber badly. Lubricating oil consumption will rise.

In a normal situation, an engine may run many thousands of hours without developing excessive crankcase pressures. In fact, most engines will go from major overhaul to major overhaul without developing any abnormal crankcase pressures. The cylinder liners, the pistons, and the piston rings will all wear down to their replacement dimensions without any undue difficulties.

However, if the engine, at some point in its life, is subjected to a great deal of light-load operation, say, 50 percent of capacity or less, then trouble may be expected. Under light-load running, liner glazing will begin to develop, particles of carbon derived from incompletely burned fuel will become embedded in the walls of the liners, the piston rings will begin to ride over the particles, and eventually a glass-like surface will evolve. As the conditions continue to develop, the liner and piston ring concentricities will no longer conform to each other, the sealing capabilities of the piston ring(s) will deteriorate, and combustion gases will begin to leak past the piston into the crankcase. This condition is known as *blowby*. Liner glazing and subsequent blowby may also be induced by running an engine at incorrect coolant or lubricating oil temperatures for prolonged periods.

Another cause of blowby is one or more broken scraper rings. After a scraper ring breaks, it fails to sweep the lubricating oil from the cylinder walls. The oil film on the liner wall comes into contact with the compression rings and overheats to the point of carbonization. Gradually the compression ring grooves become packed with carbon-thickened lubricating oil, and the rings jamb in their grooves and cease to function as seals against the combustion gases.

Combustion gas leakage will take place in an engine that has scored liners. The liner scoring may be caused by the rubbing of broken rings on the liner walls, a breakdown of the piston and liner lubricating system, or intermittent gross overheating coupled with partial seizure of one or more pistons.

Undue crankcase pressures may develop if the valve guides are badly worn. Pressurized exhaust gases will pass up the guides, into the rocker covers, and thence through the oil drains and into the crankcase. Worn guides on the air valves will permit pressurized aspirating air to pass up the guides and enter the crankcase in a similar manner.

If the coolant temperature in the cylinder jacket and cylinder head is too low, incomplete combustion may take place and create the conditions outlined in the preceding paragraph. If the lubricating oil temperatures are too low, the oil sprayed onto the liner walls will not be so effectively swept by the scraper rings. The remaining oil will come into contact with the piston sealing rings (which operate at a much higher temperature than the scraper ring) and be semicarbonized. As the carbon builds up, the piston rings become unable to conform to the cylinder liner and blowby commences. This condition is known as *ring gumming*.

The source of excessive crankcase pressure can be determined by examining both the breather and the lubricating oil filler cap. If the breather is functioning correctly, the pressure at the breather outlet will be the same as that at the lubricating oil filler cap. If the breather pressure is lower, then there is a probability of an obstruction somewhere in the system. In winter conditions, ice derived from crankcase vapor commonly causes back-pressure buildup at the breather discharge. Operators must make a point of checking the breather outlet at 8-h intervals when the outside temperature is 0°C (32°F) or lower.

If the lubricating oil filter elements begin to display abnormal amounts of carbon at recommended element change intervals, it may safely be assumed that the piston rings are not performing properly. It should be noted that none of the materials collected by the lubricating oil filter elements come from lubricating oil itself under ordinary operating circumstances; the material is carbon formed by burned or semiburned fuel.

A plugged breather pipe or dirty breather outlet filter will cause excess pressure to build up in the crankcase. If the breather outlet is restricted in any way, there is a tendency for both crankcase seals to leak simultaneously. If the crankshaft seals should start leaking, it is quite improper to install drip trays and believe that the problem is under control. The leakage can be tolerated for only the briefest period, after which remedial work must be carried out. Changing the seals in such cases is costly and incorrect.

In many engines that run with excessive crankcase pressures, it will be found quite difficult to obtain an accurate dipstick reading of the lubricating oil level. A gross fuel or coolant invasion of the crankcase can cause a sudden buildup of crankcase pressure and although these events are rare, they are not unknown. Fast-growing or suddenly occurring crankcase pressures are caused by:

1. Major piston or cylinder liner damage, which will also be quite audible
2. Broken piston rings, possibly audible
3. A frozen breather—a sudden change in ambient temperatures

If the breather has previously operated satisfactorily, the ambient temperature has not changed greatly, and the breather suddenly starts freezing over at the discharge end, review the coolant and lubricating oil levels. An accelerated condensation output may indicate an internal coolant leak.

7.2 CRANKCASE BREATHERS AND EXHAUSTERS

The purpose of the breather is to permit the engine crankcase interior to operate at about atmospheric pressures. If an excessive crankcase pressure builds up through the leakage of combustion gases past the piston rings, lubricating oil will begin to leak past both crankshaft seals. It will be difficult to introduce makeup oil because of the fumes blowing back out of the oil-fill hole, and an undesirable density of atmosphere will prevail in the crankcase. Excessive crankcase pressure may influence false lubricating oil levels, and escaping fumes will contribute to early air filter plugging.

Several breather arrangements have been developed to allow the crankcase fumes to vent successfully. The simplest is to take a pipe from the engine through the power plant wall and just let the fumes blow. In the winter, however, this method must be monitored by the operator because the exiting vapors tend to freeze immediately on contact with the atmosphere, and the ice closes off the end of the pipe. Unless there is an arrangement for catching the oil drops that are blown out of the line, the power plant exterior can assume a most unsightly appearance. Also, the method is environmentally unacceptable.

Another method in common use is to bring the breather pipe off the engine down to the bottom of a drum that is filled with steel wool and/or sheet-metal baffles. In the base of the drum is a drain cock for the removal of precipitated oil. This is a simple method, and it is inexpensive, but it involves a certain amount of housekeeping activity.

Yet another and more sophisticated method has the crankcase fumes drawn off the breather, passed through a filter, and introduced to the aspirating air system. In this manner the engine consumes its own fumes. The oil that collects in the filter housing can be drained back to the crankcase by a suitable connection.

Whatever type of breathers or vents is fitted, the plant operator must observe the performance and take note of any change. The gauze element normally fitted at the crankcase outlet should have a pressure not normally exceeding 1 in of water. If the engine has a crankcase exhauster fan, a vacuum of -0.5 in of water can be expected.

There are some engines, usually in the 1-MW and larger range, that have motor driven crankcase exhausters. These perform the same function as a common breather with the added advantage that they give the crankcase a negative pressure. That helps to maintain a much cleaner engine and power plant, but the motors and fans need more maintenance and attention than when less sophisticated types of breathers are used.

7.3 CRANKCASE OVERPRESSURE

Crankcase overpressures are caused by blocked crankcase breathers, excessive piston blowby, or overheated bearings. In the first two cases the pressure increase will be quite gradual and will be evidenced by leakage at the crankshaft seals. Some difficulty will be found in adding makeup lubricating oil against the blow-back at the oil filter opening. In the third case the pressure will build up very quickly indeed and will vent itself either by way of the crankcase pressure relief valves (if fitted) or possibly by bursting the crankcase doors. The cause of such incidents, which are fortunately quite rare, are overheated bearings igniting the normal oil-laden atmosphere in the crankcase. In some instances there may be a fair amount of fuel vapor that allows ignition to occur at a lower temperature. Usually when the crankcase doors are burst open, there is a double explosion, the second being much more violent than the first.

The first explosion is fueled by the oil-oxygen mixture present in the crankcase immediately before ignition. This explosion lifts off the doors, and a large quantity of air is able to rush into the open crankcase and provide the necessary oxygen for a second, larger explosion. It is to prevent the second and more dangerous explosion that many engines are fitted with crankcase pressure relief valves (Fig. 7.1). At the first explosion, the valve opens, vents the built-up pressure, and then closes instantly, thereby shutting off the fresh supply of oxygen necessary for the secondary explosion. The operator, without any delay what-

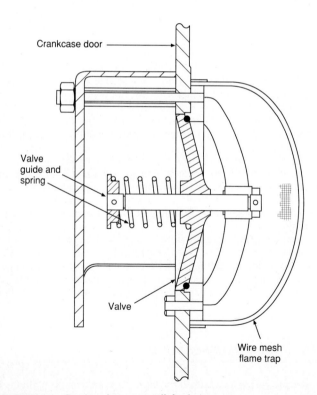

FIGURE 7.1 Crankcase pressure relief valve.

ever, must stop the engine. Under no circumstances, however, should he open the crankcase doors until the engine has cooled down.

When it is safe to remove the crankcase doors, the engine's interior can be examined, but it is recommended that nothing be dismantled before a journeyman mechanic is able to attend. If the crankcase pressure relief valves open, it is quite probable that the low oil pressure switch does not work properly, but it will not be the main reason for the explosion.

7.4 CRANKCASE NEGATIVE PRESSURE

Crankshaft seal wear may occur on an engine that normally runs with a negative crankcase pressure if the induced negative pressure is too great. When the exhauster sucks too much oil vapor out of the crankcase, the oil mist upon which the crankshaft seal relies for lubrication no longer exists and wear commences. When it is possible to regulate the negative pressure, every effort must be made to maintain a level of 0.5 lb or −5 kPa.

In an EMD engine, in which a negative crankcase pressure is always maintained, a coolant leak into the crankcase will lead to condensation forming on the lubricating oil separator screen at the eductor. When the eductor becomes unable to evacuate the crankcase atmosphere, a shutdown will automatically occur.

Regular inspection of the crankcase pressure regulating and/or modulating devices and a continuous awareness of the pressures prevailing are a standard part of any engine operator's duties and must not be neglected.

7.5 CRANKCASE DOORS

Crankcase doors, particularly those on the larger engines, often have oil leaks. Almost without exception, the cause of the leaks is either overtightening of the fasteners or incorrect gasket application. When a leak becomes visible, the natural reaction is to retighten the fasteners, but all too often this action compounds the problem. Under the pressure of the capscrews or nuts, the door is pressed into the gasket material and, in section, the door assumes a ripple profile that allows even more leakage to occur (Fig. 7.2). The correct torque applied to all of the fasteners in conjunction with the correct type and thickness of gasket material will always provide an adequate seal. However, deformed doors may be present on engines that have been in service for some time. They should be taken off, inspected for flatness, and straightened out as necessary.

When gasket compound is applied, it is better to apply it only to the gasket and crankcase door interfaces. It should not be allowed to get onto the gasket and crankcase interface because it will result in the destruction of the gasket the first time the crankcase door is taken off. Also, the gasket debris left adhering to the crankcase has to be scraped off, and some fragments will always fall into the crankcase.

On crankcase doors fitted with handwheel-type dogs, as on EMD engines, the dogs should not be any more than hand tight. It should not be necessary to use a tool of any kind to either install or remove the door.

Most engine manuals quote the correct torque value that should be applied to

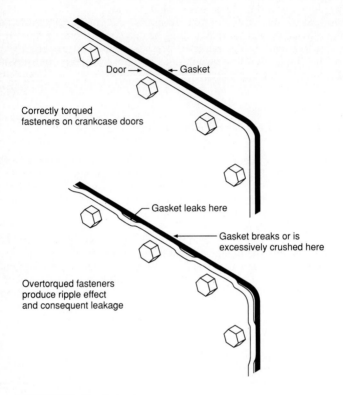

FIGURE 7.2 Crankcase door.

crankcase door fasteners. Usually those numbers are surprisingly low—not much more than *finger tight*, in fact.

7.6 CRANKCASE AND SUMP CLEANING

The crankcase and the sump (otherwise known as the pan) of an engine is a fairly well sealed chamber. When the engine is running, the atmosphere within the crankcase and sump is one of a very agitated, hot, lubricating oil vapor. As the engine runs, combustion gas and carbon leak past the piston rings in small quantities, lubricating oil that touches hot spots carbonizes, and in isolated cases, coolant may enter the crankcase.

Most of the solid and semisolid debris is swept up in the lubricating oil stream and is retained in the lubricating oil filters. Eventually, the quality of the lubricating oil deteriorates, the engine is stopped, and the used lubricating oil is drained out and replaced by fresh, clean oil. The filter elements are also replaced or cleaned at this time, and the engine is ready to go back to work. In spite of the oil change and new filters, deposits of sludge, carbon, and other debris remain in the engine, principally in the sump.

At some point in the upkeep schedule the interior of the engine must be thoroughly cleaned out. If it is not, the capacity of the sump becomes reduced and the life of any new oil is shortened. It is not uncommon to find an engine with 10 percent or more of its lubricating oil volume occupied by sludge after 3000 or 4000 h of running without cleanout. Under such circumstances, the life of the engine will be curtailed by an escalated wear rate. Because a smaller volume of lubricating oil will be available for circulation, the lubricating oil will deteriorate at a faster rate, the lubricating oil filter elements will plug up sooner than they should, and unfiltered oil will bypass the filters.

The person directed to do the cleanout job must be instructed to report anything unusual found in the sump. He must look for nuts, washers, locks, split pins, pieces of gasket material, broken piston rings, flashes of bearing metal, or anything else that is apparently foreign.

On the larger engines it is often expedient for the cleaner to get inside the crankcase to do the cleaning work effectively. Before he enters the engine, both he and the person in charge of the work must satisfy themselves that the turning gear is disengaged and its power shut off. No other work should take place on the engine while the cleaner is in the crankcase. A second person must be stationed at the crankcase door during the whole time the cleaner is inside the engine, and that person is responsible for the cleaner's safety.

7.7 CRANKCASE INSPECTION

Part of the regular maintenance routine should include a crankcase inspection. This can be done very conveniently when the lubricating oil is being changed. The pan has to be cleaned out at this time, and all of the solid debris removed from the pan should be examined. In the normal course of events, the debris will be a black mud or sludge composed entirely of carbon. The inspector must look carefully for any fragments of metal, rubber, gasket material, or other non-oil products. The origin of any metal found must be determined. After the engine's interior has been thoroughly cleaned, each liner and piston may be scrutinized by means of a lamp and a mirror. All of the main and bottom end fasteners must be tap-tested and each split pin must be seen to be whole.

The skirt of each liner can be examined for evidence of cracking or coolant streaking indicative of a lower seal leak. With the piston at TDC, the liners are searched for signs of scoring, scuffing, and glazing. The piston is inspected again for evidence of cracking in the vicinity of the skirt or wrist pin boss. It may be possible to sight the camshafts, some of the timing gears, lubricating oil jets, lubricating oil suction strainers, or other internal components. The connecting rods should be examined for any sign of bending, fretting, or cracking.

Crankcase inspections can usually be carried out quite conveniently on any engine having inspection doors or plates on the crankcase sides. In small engines the pan of which has to be dropped for such an examination, the inspection would be scheduled to occur when the pan is removed for cleaning. On larger engines with high capital values, is recommended that crankcase inspections take place whenever the unit is stopped for servicing at, say, 1000-h intervals.

On engines of 1 MW capacity or greater the internal inspection should also include a crank deflection reading, a tap test of all of the hold-down bolts and a tap test of the crankshaft balance weight bolts.

CHAPTER 8
CRANKSHAFT AND TORSIONAL VIBRATION DAMPER

8.1 CRANKSHAFT DEFLECTIONS

Deflection can be defined as a dimensional alteration of a component that is brought about by cyclic forces; thus, a valve spring can be said to deflect proportionately to the valve travel. Similarly, the distance between the webs of a crankshaft varies micrometrically during each revolution owing to the application and relief of the combustion forces imposed upon and transmitted by the connecting rod. Also, if the alignment of the crankshaft is incorrect, if the crankshaft bearings are worn, or if the engine bed is distorted, abnormal crankshaft deflection measurements will occur with each revolution. Any deflection of the crankshaft will represent some flexing of the metal which, if excessive, will result in metal fatigue and subsequent crankshaft breakage.

Besides worn bearings and misalignment, factors such as failing foundations, cracked engine bases, unevenly loaded vibration isolators, driven-load misalignment, or unbalanced cylinder firing pressures may also contribute to wayward crankshaft deflections.

The engine operator can play an important role in limiting crankshaft deflection by ensuring that the engine runs at design temperatures. Undue "hogging" of the engine will occur if the cylinder heads and jackets run at the proper temperatures while the lubricating oil is run at abnormally low temperature; "sagging" will occur if the cylinder heads and jacket run cold while the lubricating oil runs hot. The effects of improper temperature care are generally more pronounced on larger multicylinder engines.

Before commencing any realignment procedures or making an alignment check, it is advisable to refer to the relevant workbook and OEM manuals to determine if the previous readings were, or should be, taken hot or cold. Ideally the workbook will show sets of deflection readings that were taken both hot and cold (Figs. 8.1 and 8.2).

8.2 CRANKSHAFT SEALS

Although most crankshaft seals are trouble-free and present few maintenance problems, leakage of seals can be expected at some point during the life of the

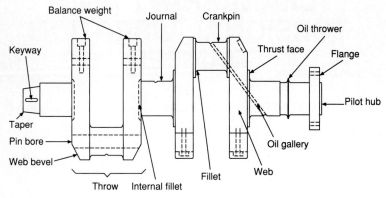

FIGURE 8.1 Crankshaft Terminology.

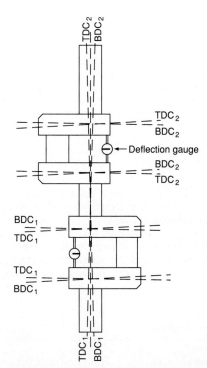

FIGURE 8.2 Crankshaft Deflection.

engine. Remember that in the case of surface-contacting seals, some oil must pass through the seal in order to lubricate it. Leakage of seals is brought about by:

- Wear
- Crankcase pressure
- Fit
- Sticking
- Overload

It is important that the cause for the leakage be identified correctly in each case so that remedial work is effective.

It is not uncommon for new engine seals to leak, particularly in engines that have been in storage for some time. In those cases, it is recommended that the seals be stripped and that both they and the shaft seal surface be thoroughly cleaned of all gum, rust and dirt. Shaft corrosion in the vicinity of the seals may occur when the engine is stored or is in a prolonged standby state. Obviously, if common rusting is taking place, weatherproofing is necessary. Corrosion can take place if an engine is shut down for a considerable period of time with old lubricating oil in its crankcase. An engine that is to be laid up should always have its oil changed and then be run for a few minutes before being left.

Excessive crankcase pressure will cause seal leakage, and abnormal crankcase pressures resulting from piston blowby will cause lubricating oil to pass the shaft seals. Before seal leakage is really apparent, however, difficulties will have been met in getting a clean dipstick reading. Crankcase pressure can easily build up and have an effect on the seals if the crankcase breather becomes plugged. Breather outlets should be inspected at least daily, particularly during cold weather, to prevent ice buildup. If both seals start leaking more or less simultaneously, then a breather obstruction can be considered as the most likely cause.

Seal wear may show up as grooving of the shaft in the way of the seal. Such grooving can be remedied by fitting an OEM sleeve specifically designed for the purpose. Many engines today are supplied with wear sleeves that may be changed as part of the overhaul routine. Grooving and subsequent leakage is frequently associated with engine operation in a dirty environment. Worn sealing surfaces will be found on engines that have run a very long time or that have received improper care. A certain amount of wear can always be expected but dirty lubricating oil, misalignment, and incorrect seal loading, contribute to accelerated wear rates.

Seal gumming may be attributed to antifreeze in the lubricating oil or too long a period between oil changes. If the former is suspected, immediate action must be taken before all the bearings are affected. Sometimes a seal is incorrectly installed. (It is not unknown for a complete seal or parts thereof to be assembled the wrong way around.) Care must be taken when installing new seals to ensure that all the marks are read and properly positioned.

The operator must consider the environmental conditions within the power plant building when shaft seal leakage becomes apparent. In drawing its aspirating air from within the building, the engine may be creating a partial vacuum and thereby increasing the pressure differential between the engine internal atmosphere and the plant interior atmosphere. In such conditions seal leakage will become pronounced. Merely opening the power plant door for a period of time is sufficient to prove the point. It is quite likely that both seals will leak in the circumstances described. Leaving the door open is not, of course, the solution to the problem!

Seals will always leak when the level of the crankcase lubricating oil is too high. Although the correct level should always be maintained, note must be taken of deviation. High-speed engines (1800 rpm and above) have the ability to churn their lubricating oil up into a semisuspended atmosphere, and it then becomes exceedingly difficult to get a clear dipstick reading. When that occurs, it is not unknown for the operator to add an "insurance" quart, which further compounds the problem. If a high level is noted, a check must be made to see there is neither a fuel nor a coolant intrusion into the lubricating system.

Some engines are designed with seal drains. If the drain holes get plugged, the seal spills oil out of the engine. It is therefore important that every engine be examined for such drain passages at the time of overhaul and the holes be cleaned if necessary.

8.3 CRANKSHAFT DAMAGE THROUGH ELECTRICAL DISCHARGE

When an electric current passes through a crankshaft, damage to the journals and main bearings results. The presence of stray currents traveling through the crankshaft and discharging to the main bearings or vice versa is not readily detected, but close visual examination during overhaul of the journals may reveal some random streaking of the bearing surfaces. The streaks will not necessarily be in line with the axis of shaft rotation.

Pitting of the crank journal creates an abrasive surface which will lead to an accelerated rate of bearing wear. Generally speaking, the wear will be approximately the same on each of the main bearings, but one or more individual bearings may show more pronounced wear.

In cases of electrical discharge wear of crankshaft bearing surfaces, it is unlikely that any damage to the crankpins or bottom end bearings will occur. Lubricating oil analysis will provide notice of metal loss. Lead, aluminum, steel, copper or other bearing metals can be expected to be present, but in the earlier stages of the difficulty no metallic debris will be visible in the lubricating oil filters.

Electrical discharge damage to crankshafts and their bearings is caused by ineffective grounding of the electrical systems attached to the engine or by improper welding practice when making repairs or modifications to the unit. If it is found necessary to do any electric arc welding on the engine-generator installation, it is imperative that the ground cable clamp of the welding equipment be firmly located as close as possible to the point where the welding will take place. If it is placed some distance away from the point of welding, there is grave risk that total destruction of one or more main bearings will take place.

Many engine installations are wired with substandard grounding methods for the attached electrical equipment; the grounding does not provide known, controlled, and insulated paths through which stray current may pass. In all cases when stray currents cause crankshaft damage, a deficient ground system can be cited.

To test the effectiveness of the normally installed ground strap, the engine should be run at low idle speed with all the electrically powered accessories switched on. Use a voltmeter to measure the voltage potential between the engine block or its mountings and the grounded battery terminals. The voltage will be less than 0.5 V if the grounding arrangement is adequate.

8.4 TORSIONAL VIBRATION DAMPERS

Torsional vibration is developed in a crankshaft by the imposition of varying, intermittent and cyclic forces along the crankshaft's length. The combustion forces developed in the cylinder are transmitted to the crankshaft, where they induce the initial deflection XB and, because of the shaft's inertial reaction, XA. The distance AB is known as the *amplitude* (Fig. 8.3).

At a given speed of rotation the amplitude will be at its maximum. This is the critical speed. The term *critical speed* is also used to define speeds at which the induced stresses exceed the safe limits of the material(s). The critical speed of an engine may be above or below the engine's normal operating speed, and some engines may have more than one critical speed band. When the critical speed of

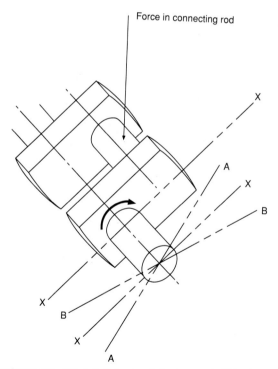

FIGURE 8.3 Crankshaft torsional displacement. X is the static centerline of the crankshaft; XA is the degree of twist imparted to the shaft; it is caused by the shaft's own inertia during a firing stroke. XB is the reaction to XA. AB is the torsional angular displacement.

the engine is below the normal operating speed, the engine will be appropriately marked or plated.

The purpose of a torsional vibration damper is to reduce the amplitude of torsional vibration in an engine crankshaft. To perform its function, a torsional vibration damper incorporates a designed structure of energy-absorbing materials. That material may be either metal in the form of a spring, a silicone-based semifluid of low viscosity, or rubber. The action of the torsional damper is such that the energy-absorbing and converting materials are greatly stressed owing to their dampening or restrictive qualities. In time, they deteriorate through fatigue. At that point, the torsional damper becomes ineffective and must be recognized as such before the crankshaft suffers damage.

Three sorts of torsional vibration dampers are used in *internal combustion* (IC) engines:

1. A heavy metal ring mounted on a rubber bushing.
2. A heavy metal ring set in a casing loaded with a highly viscous fluid, usually a silicone.
3. The free-pendulum type.

The third type is complex and expensive and is rarely used on generator-set engines. Dampers of the first type are constructed in the two ways illustrated in Figs. 8.4 and 8.5. In each illustration, sound and failed dampers are shown. The second type of damper, which depends on a viscous fluid, is illustrated in Fig. 8.6 which is discussed in Sec. 8.8.

Still another version of a torsional damper employs a bonded asbestos lining instead of an encased silicone or rubber transmission medium. The asbestos lining is attached to the control disk by means of spring-loaded fasteners. At a given tension on the fasteners, a small movement of the mass ring relative to the disk will be possible. This type of damper is generally fitted within the crankcase, where it is cooled by lubricating oil.

Torsional vibration dampers in good working order contribute to the reduction of gear noise. They do not, however, affect the balance of an engine. Thus, they will not reduce linear vibration caused by imbalance of moving parts, but a failed torsional damper may contribute to an increased linear vibration.

A rubber element type of torsional damper will have a radial movement of the mass ring relative to the crankshaft of not more than half a degree. The mass ring in a viscous type damper can tolerate a greater amount of movement, perhaps as much as one degree. The mass ring on a torsional damper rotates at a more constant speed than that of the crankshaft. The crankshaft is subject to pulsative forces which occur in continuous acceleration and deceleration of the individual crank throws and thereby create the phenomenon of torsional vibration.

8.5 RUBBER-BONDED TORSIONAL DAMPER INSPECTION

Evidence of a failing torsional damper may not be immediately visible except in the case of a rubber-bonded damper. At approximately 1000-h intervals, rubber-bonded torsional dampers should be examined for evidence of failure. If inspections have been neglected, the most usual sign of something amiss will be an excessive vibration when the engine is running.

With the engine stopped, the damper must be examined for accumulation of

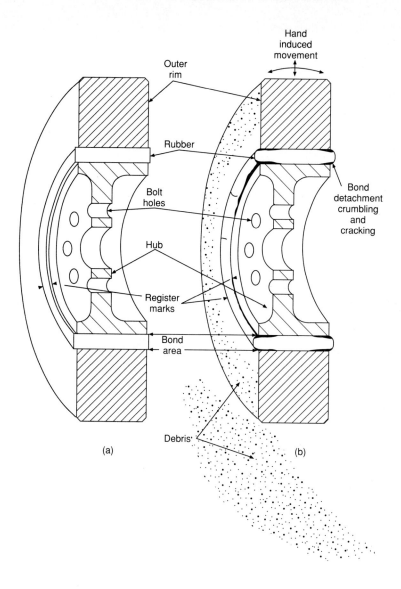

FIGURE 8.4 Typical rubber-bonded torsional vibration damper, type 1. (*a*) Sound; (*b*) failed.

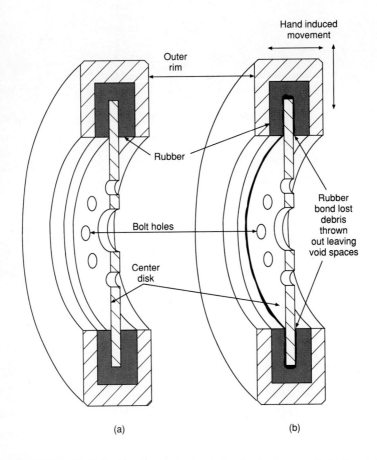

FIGURE 8.5 Typical rubber-bonded torsional vibration damper, type 2. (*a*) Sound; (*b*) failed.

what appears to be carbon on the radial surfaces of the damper (Figs. 8.4 and 8.5). This "carbon" may be pulverized rubber thrown off by the elastic core of the damper. The damper can be tested for breakdown by applying a light leverage in both forward and backward mode on the damper outer ring. The measured movement can then be compared with the OEM specifications in the respective manual. Also, radial leverage applied to lift the damper rim will show if there is any untoward movement away from the concentric. If the rim is loose, the entire damper must be replaced.

Rubber-bonded dampers fail readily if they are badly ventilated or if they are continuously exposed to fuel oil or lubricating oil. A failing or failed rubber-bonded damper will have produced a quantity of rubber dust which will accumulate on the floor in a radial area projected from the damper. There may also be signs of cracking, flaking, or other breakdown of the rubber component of the damper. Some rubber-bonded dampers have register marks on the metal surface

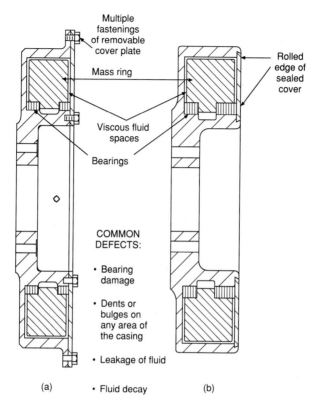

FIGURE 8.6 Typical fluid-type torsional vibration damper.

each side of the rubber. If the marks are out of register, the damper can be regarded as defective because the metal-to-rubber bond has failed.

8.6 RUBBER-BONDED TORSIONAL VIBRATION DAMPER CONDITION IDENTIFICATION

Damper Is Sound

The outer ring of the damper is clean; the rubber component is intact and presents a smooth outward appearance; the hub is clean; and the bolt holes are bright [Figs. 8.4(*a*) and 8.5(*a*)]

Damper Has Failed

The outer ring of the damper is coated with a carbon-like black dust; the rubber is distended, cracked, or crumbling. There may be some clearance between the rubber and the ring or hub. The bolt holes may have a reddish-brown color that indicates some fretting. The peripheral areas surrounding the damper are sprinkled with a gray-black dust. The outer ring may feel loose. Overheating the rubber element will have overcured the rubber. The rubber has become brittle and lost its elasticity; the bond between the rubber and the metal has begun to separate [Figs. 8.4(*b*) and 8.5(*b*)].

8.7 RUBBER-BONDED TORSIONAL DAMPER SERVICE LIFE

The service life of a rubber-bonded torsional damper may, in good operational conditions and environments, exceed 30,000 h. It is necessary to change the damper only when the rubber element begins to break down. Regular inspection of the damper will therefore be necessary, and it is recommended that it be done at each oil change.

8.8 VISCOUS FLUID LIFE IN TORSIONAL VIBRATION DAMPERS

The service life of a sealed viscous damper should not be permitted to exceed 30,000 engine hours. Because the fluid in such dampers cannot be sample-tested, it is recommended that changeout of viscous-type torsional dampers be included in the periodic major overhaul schedule.

Torsional vibration dampers that are fitted to larger engines are often provided with a sampling plug that allows a sample of the fluid to be drawn off for laboratory testing by the OEM. A sample should be taken at approximately 5000-h intervals. The efficiency of the fluid will be reported to the engine owner by the OEM. Comparison of successive reports will show the decline in efficiency and offer some prediction of when the damper must be removed for servicing.

Viscous torsional dampers can be checked in situ by observing the temperature buildup in the damper and comparing it with the temperature of the engine crankcase. The temperature of the damper will increase after start-up at a faster rate than that of the crankcase, and it will continue to do so until all the normal operating temperatures have been reached.

Commencing with a cold engine, a start is made and the engine is run until the crankcase just begins to feel warm. The engine is then stopped and the temperature of the damper is felt. It should be considerably hotter than the crankcase, in which case the damper can be considered to be in good working order. If the damper is cold or at the same just-warm temperature as the crankcase, it can be judged to be defective.

A sealed-case type of damper that is found to have a dented or leaking case must be removed from service without delay. In almost all configurations, the torsional damper is protected in part by a wire or sheet-metal guard, but when an overhaul takes place, the guard is often removed. The case of the viscous damper is quite thin, and it can be readily dented by tools falling on it, by crane-suspended loads contacting it, or by the most frequent cause of all: prybar damage when thrust clearances are checked.

Most viscous dampers are recoverable, and they should be sent away for rebuilding provided the OEM approves. Before being shipped, the damper can be further tested by standing it on edge on a flat surface. If the damper rolls until it has found its center of gravity, it is probable that the viscous filling has solidified. On the other hand, if the damper does not roll no matter what point of its rim it stands on, the viscous fluid is still in good condition and the damper is not defective. In a reasonable environment, a viscous damper can be expected to survive for better than 30,000 h, but testing frequencies must increase with age. On large engines the viscous damper is usually fitted with a sampling plug, and the OEM provides sample vials. Samples are taken at approximately 5000-h intervals; they are sent to the damper manufacturer for evaluation of the life remaining in the fluid.

A mistuned engine will exhaust a damper's life much sooner than an engine that is in good tune. It is also thought that the silicone fluid of a viscous damper will decay prematurely if it is exposed to subzero temperatures for prolonged periods of time. If it is necessary at any time to remove the viscous damper from the engine, it must be handled with care. It must not be dropped or otherwise mistreated, because the sheet-metal shell is quite susceptible to damage.

It is of the utmost importance that at the completion of an overhaul or other engine work activity and before the engine is started for test running, the engine is barred over while a final close visual inspection of the damper casing is carried out. Most crankshaft breakages can be attributed to torsional damper failure, and therefore it is not permissible to operate an engine when it is known that the torsional vibration damper is defective.

8.9 TORSIONAL VIBRATION DAMPER VENTILATION

A rubber-bonded torsional vibration damper generates heat when it is working properly. The heat must be continuously dissipated to the surrounding atmosphere without hindrance if the damper is to remain serviceable. If the damper gets hot, the rubber bond will overcure or harden and render the damper inoperative. It is therefore necessary to see that the damper had adequate ventilation. External dampers should not be painted, nor should any wire guard over the externally mounted damper be replaced by a sheet-metal guard.

CHAPTER 9
BEARINGS

9.1 PURPOSE OF BEARINGS

Broadly speaking, the purpose of a bearing is to enable one component of a machine to rotate within another under conditions of minimum friction and wear. All bearings in diesel generator sets are designed very carefully and are thus able to provide a long and predictable service if the machine is correctly operated and maintained. If a bearing does fail, however, the consequential damage is always costly to repair, particularly so if the crankshaft, camshaft, or turbocharger is affected. Almost all bearing damage can be minimized by vigilance, awareness, and prompt action on the part of the operator.

Many of today's engine manufacturers produce excellent bearing condition guidebooks. Of particular interest is the Caterpillar publication *Engine Bearings and Crankshafts* (SEBD 0531), and it is highly recommended to all diesel power plant personnel.

9.2 WEAR IN ENGINE BEARINGS

Bearings are changed out when normal wear or damage renders them unsuitable for further service. Normal wear can be taken to be the gradual metal loss that takes place over a period of time and is acceptable to the owner and the manufacturer and is within planned overhaul schedules. Abnormal wear or damage can be categorized as that which forces unprogrammed attention.

Within the engine operating industries, it is estimated that an incidence rate of premature bearing failure is caused by dirt or foreign matter in 45 percent of cases, followed by misassembly in 13 percent, misalignment in 12 percent, lack of lubrication in 10 percent, overload in 9 percent, corrosion in 4 percent, and other causes in the remaining 7 percent.

"Dirt or foreign matter" can be any material likely to be adrift in an engine. Particles of iron, steel, or aluminum, wood splinters, cloth, plastic, jointing compound, honing grit, carbon, coolant, and coolant additives or any other material of a nonlubricating nature that circulates one or more times through the bearings falls into this classification.

The abrasive action of any of these materials will break the lubrication film and permit surface-to-surface contact of the bearings. The metal will lose designed clearances and sustained oil pressure will be reduced, with resultant loss of oil film retention.

9.3 CAUSES OF DIRT IN AN ENGINE

Dirt in the engine lubrication system results from improper engine upkeep. Some of these improprieties are:

1. Running the engine too long between oil changes.
2. Running too long between oil filter changes. [(1) and (2) are not necessarily changed at the same time.]
3. Lax or improper servicing of breathers.
4. Perforated, badly installed, or absent air filter elements.
5. Improper care of fresh oil supplies.
6. Careless handling of makeup oil.
7. Dirt entering and remaining in the engine during repair or overhaul work.
8. Metallic debris from failing engine components.

(8) may commonly be detected from scheduled oil analysis or lubrication filter elements. The case for thorough cleanout of the sump and crankcase at oil change cannot be overemphasized.

9.4 CONDITIONS CONTRIBUTING TO BEARING FAILURES

Foreign Particles on the Back of the Bearing Shell

A localized area of wear will be seen on the bearing surface. Also, evidence of foreign particles may be visible on the bearing shell back directly behind the area of surface wear. Foreign particles between the bearing and its housing prevent the bearing back from being in full contact with the housing bore. As a result, the transfer of heat away from the bearing surface is not uniform, and the result is localized heating of the bearing surface which reduces the life of the bearing. It will then be necessary to install new bearings while being careful to follow proper cleaning procedures before installation. It may also be necessary to grind and polish journal surfaces to OEM specifications.

Review the changing of engine oil and oil filters at the recommended intervals. Ascertain that the air filters and crankcase breather are serviced at the proper intervals.

Misassembly of Bearings

Misassembly of bearings will occur when substandard work practices are permitted. At all times the OEM manual must be consulted during assembly if problems related to excessive crush, insufficient crush, mislocation of bearing cap, overtorquing, fillet riding, or housing or journal dimensional variations are to be avoided.

Each bearing half insert is slightly oversized when the two halves are placed in the bore. They project very slightly above the lands (as little as 0.002 to 0.005 in depending on the shell diameter). When the bearing bolts are torqued up, the shell halves are pressed together. This force fit is known as bearing crush, and it provides complete contact between the back of the bearing shell and the surface of the housing bore. The degree of crush also assists in keeping the shell from turning in the housing. Lock tabs formed into the bearing shell halves at the parting line serve to locate the shell in its proper position in the housing bore. It will be seen, therefore, that the correct installations of the shells, together with accurate torquing of the bearing cap bolts, is imperative if premature bearing problems are to be avoided.

Loose Bearing Cap Fasteners

Housing Bores. If the back of a bearing shell has a shiny appearance, it is possible that the shell has been loose in its housing. Fretting of the bearing back has occurred. Fretting takes place when there is intermittent contact or slipping between two metallic surfaces. On steel, evidence of fretting shows up as debris formed at the interface, and it is accompanied by pitting of the metal. The debris, which is iron oxide, is usually red-colored, but it may be dark brown or black. Fretting damage may occur on any metal. Fretting conditions can exist where there are high contact pressures, oxygen at the contacting surfaces, no means of replenishing lubricant at the contaminating surfaces, and there is a minute degree of oscillating or reciprocating motion.

Insufficient torque applied to bearing cap bolts will tend to induce fretting. Overtorquing of bearing cap bolts will lead to the distortion of bearing housings and overstraining of the bearing bolts.

Engine overheating is rarely the cause of bearing breakdown, but excessively high lubricating oil temperatures will certainly shorten bearing life. An overheated bearing may be caused by insufficient crush, and failed bearings should be examined with that in mind.

In bearings damaged by overspeeding, there will have been a breakdown of the oil film, but the damage in the shell will be predominant in the upper half of main bearings and in the lower half of bottom end bearings. Recognition of this feature will assist in the determination of cause following engine breakdown.

Excessive Crush

Excessive crush will be identified by extreme wear areas adjacent to the parting faces. Bearing shells subjected to excessive crush will not be suitable for service because they will be distorted, out-of-round, and unable to accommodate a viable oil film. The engine will probably be too tight to turn over. The condition should not be mistaken for dirt behind the bearing shell, but that possibility must be investigated.

Insufficient Crush

Insufficient crush can be identified by the appearance of highly polished areas on the bearing shell back or on the edge of the parting line. Insufficient crush will result in the shell having inadequate contact with the bearing housing, thereby inhibiting heat transfer from the bearing shell. The bearing surfaces will quickly break down in the circumstances and damage to the bearing housing may be done. Corrective action to be taken would include disposal of the bearing, checking the torque wrench, and measuring the bearing housing. Review the length of the bolts and the holes in which they are applied to eliminate possibility of bolts bottoming in blind holes.

Shifted Bearing Cap

Excessive wear areas will be seen near the parting lines on opposite sides of the upper and lower bearing shells. The bearing cap that has shifted will cause one side of each bearing half to be pushed against the journal at the parting line. The resulting metal-to-metal contact and excessive pressure cause bearing surface scoring and/or fatigue.

Mislocation of Bearing Cap

Mislocation of the bearing cap frequently occurs when the match marks are not taken into account or the caps are mismatched. The indicator of mislocation is excessive wear on the parting lines of the top and bottom shell opposite to each other. The engine will be tight to turn, and the bearing will not be suitable for reuse.

Mislocation is often brought about by using a socket with an overlarge outside diameter to tighten the bearing cap. The socket crowds the cap and forces it away from its proper position. Caps must be placed in their proper positions and checked to see they are not reversed. Stretched cap pockets can also contribute to the problem.

In these events the corrective action would be to check the journal surfaces for undue wear. Polish the shaft if necessary and install new bearings of correct size (beware of oversizes). Apply correct torquing procedure and values. Replace all bolts or screws suspected of stretch or thread damage. Measure main and bottom end bolts against OEM data. Gauge and discard any overlength fasteners.

Out-of-Round Crankshaft Diameters

If the crank journals or pins are out of round or parallel, the marks on the bearing shells will give an indication of the problem. If after micrometer measurement the shaft bearing surfaces are found to be beyond the OEM limits, the shaft must be polished or reground. Uneven wear on the crank bearings may be caused by misalignment, incorrect thrust clearance, or bent connecting rods.

In general, if a bearing has been damaged because of an out-of-shape journal, an uneven wear pattern is visible on the bearing surface. These wear areas can be in any one of three patterns. An out-of-shape journal imposes an uneven load on the bearing surface, thereby increasing the heat generated and accelerating bearing wear. An out-of-shape journal also affects the bearing's oil clearance and makes the clearance insufficient in some areas and excessive in others, thereby upsetting the proper functioning of the lubrication system.

An hourglass or barrel-shaped journal is always the result of improper machining. If the journal is tapered, there are two probable causes:

1. Uneven wear of the journal during operation (misaligned rod)
2. Improper machining of the journal at some previous time

Regrinding and polishing the crankshaft to specifications is the best way to remedy out-of-shape journal conditions. Install new oversize bearings in accordance with proper installation procedures.

Out-of-Round Bearing Housings

An out-of-round bore may be encountered on a connecting rod, particularly in an engine that has been overspeeded. The main bearing housings may also be out of round following bearing overheating. In either case, reboring or replacement will be necessary. During the inspection, areas of overheating may be fully determined. Out-of-roundness of bearing caps is commonly caused by overtorquing the bearing bolts.

The bearing shells will have an appearance of localized excessive wear areas visible near the parting line on both sides of top and bottom shells. Oil clearance near the parting line is decreased to such an extent that the metal-to-metal contact between bearing and journal takes place and results in areas of above-normal wear or bearing material fatigue. Also, improper seating between the bearing back and the housing bore may be present. This can impede proper heat transfer and cause localized heating of the bearing surface and increased fatigue. The condition will be brought about by overtorquing or undertorquing of housing bolts. Check the roundness of bearing bores before installing the new bearings. If they are found to be out-of-round, recondition the bearing housings or replace the connecting rod. Tighten bearing housing bolts to the proper torque by using an accurate torque wrench.

Fillet Ride

When fillet ride damages a bearing, areas of excessive wear are visible on the extreme edges of the bearing surfaces. Fillet ride will show up as pressure marks on the extreme edge of the bearing shell where the shell was riding on the fillet but not on the journal. This type of damage can readily destroy the crankshaft.

If the radius of the fillet at the corner where the journal blends into the crank is larger than specified, it is possible for the edge of the engine bearing to make metal-to-metal contact and ride on this oversize fillet. Fillet ride results if excessive fillet radii are left at the edges of the journal at the time of crankshaft regrinding. The situations will be corrected by regrinding the crankshaft while paying particular attention to specified fillet radii. The minimum radius shall not exceed specifications, since this can weaken the crankshaft at its most critical point. Check the crankshaft end clearance and thrust washer thickness.

Surface Fatigue

Surface fatigue will appear as small, irregular areas of surface material missing from the bearing lining. Heavy, pulsating loads imposed on the bearing by the engine cause the bearing surface to crack owing to metal fatigue. The fatigue cracks widen and deepen perpendicular to the bond line. Close to the bond line, fatigue cracks turn and run parallel to the bond line. They eventually join and cause pieces of the bearing face metal to flake out. Bearing surface fatigue is the result of abnormally high bearing loading. Improper assembly and improper reworking of main bearing, block line bores, and crankshafts can cause excessive localized loading and hence fatigue.

Check block line bores and reassemble correctly. Replace all other bearings in the same subsystem. Check out overspeed shutdowns. Remedial steps include an examination of bearings, renewal of the bearings if necessary, and polishing the crankpins and journals. Also:

1. Grind to OEM specifications if surfaces are beyond polishing.
2. Enforce correct lubricating oil and lubricating oil filter element change intervals.
3. Clean or replace air filter elements and breathers at recommended intervals.
4. Review and revise clean lubricating oil supply and service.
5. Ensure cleanest possible operation, maintenance, and overhaul practice.

Lack of Lubrication

Oil temperatures or pressures may vary in any engine. It is true to say that the first revolutions of many engines upon starting are made with zero oil pressure in the bearing. However, the shaft is supported on an oil film that has a strength momentarily equivalent to the normal operating oil pressure. When the engine starts, the lubricating pump builds the oil pressure that is necessary to maintain the film strength required to separate the shaft from the bearing. In effect, the shaft does and must float on oil.

Film penetration will cause the shaft metal to contact the bearing metal, at which time wear will take place. That wear during normal engine operations is continuous but barely perceptible even after many thousands of hours of operation. Discontinuity in the maintenance of the oil film will eventually cause such wear that the bearings must be replaced. If for some reason the bearings wear out prematurely, the problem is usually classed as lack of lubrication, the common causes of which are:

- Low engine oil level
- Plugged oil pump screen
- Failed lubricating oil pump
- Break or restriction in the pressure oil supply line (oil gallery)
- Failed pressure-regulating valve
- Degraded lubrication (i.e., diluted or contaminated)
- Disturbance of the lubricating oil pump pressure relief valve

When investigating the engine after a crankshaft bearing failure, the state of the other bearings in the assembly will provide some clues. (For example, a single bearing failure in the middle of the engine is unlikely to have developed because the lubricating oil supply to the system failed.)

9.5 BEARING DIFFICULTIES

Bearing Running Too Hot, I

If it is suspected that an engine bearing(s) is running too hot, the engine must be stopped forthwith. The engine must then be allowed to cool down before any attempt is made to open the crankcase for inspection. Upon opening the crankcase, each of the bearings can be checked for heat, movement, and color. There is a possibility that the hottest bearing cap will look drier than the others. Metal flash between the crankcase and the bearing cap will indicate a totally failed bearing.

While waiting for the engine to cool down, the opportunity to open the lubricating oil filters is present. Remove the lubricating filter elements, cut them open, and study them for signs of metal debris. It should be noted that it is not possible to effectively clean small tube-type lubricating oil coolers of metal fragments, and thus they must be changed out. Refer to the OEM recommendation.

Bearing Running Too Hot, II

Antifriction bearings that run too hot will quickly fail because they generally contain their own supply of lubricant which is effective only under normal operating temperature conditions. Antifriction bearings are not recoverable and must be discarded when suspect. In every case of failure in an antifriction bearing, there must be a close examination of both shaft and housing, because they frequently suffer some damage subsequent to antifriction bearing failure. The bearing housing must be carefully checked for enlargement of diameter and radial cracking, and the shaft may have developed heat colors or be scored. The fit and alignment of the assembly must be most accurately checked after the new parts have been installed.

Spark Erosion of Engine Bearings

Cases of bearing failure caused by electric spark erosion are not uncommon. The spark may be caused by:

1. Alternator stray currents
2. Electric starter ground currents
3. Improper electric arc welding practices

Case 1 will be readily detected through regular lubricating oil sampling and analysis, but examination of the lubricating oil filters will not show any visible metal debris (at least not in the early stage of the problem). Case 2 will take a greater span of time to develop because the current leak will occur only during the starting sequence. A ground current that would affect the bearings badly and with easy recognition would, almost certainly, have an earlier and more pronounced effect on the starter itself. Case 3 can inflict immediate and serious damage on main and bottom end bearings.

In no case should welding to an engine generator underframe assembly be permitted without observing the utmost caution. In any event, when electric arc welding is to be considered, the technical advice of the owner must be obtained. If the owner's advice is not forthcoming, the manufacturer's representaive must be contacted. The welding work should be closely supervised and all technical advice should be closely followed.

If welding has taken place on an engine and there is any reason to think that technical advice was not obtained or applied, it will be necessary to examine all of the main and bottom end bearings for damage. Spark erosion damage on bearings is characterized by deep pitting, and the walls of the pit will be near 90° to the plane of the bearing surface. There will be little sign of wiping of the remaining bearing surfaces, and the pitting may have occurred on the top or bottom shell, on the bearing surface, or on the thrust surface. Once spark erosion of the bearing surface has commenced, total bearing failure through loss of lubrication pressure can be expected to occur eventually.

9.6 BEARINGS IN STORAGE

New and unused bearings in storage must be kept wrapped and fully protected from dirt, bruising, scratching, corrosion, stack stressing or any other form of damage (Fig. 9.1). Ideally, they should be handled only by persons well acquainted with the value of each bearing, the ease with which bearings can be destroyed, and the consequences of fitting substandard bearings in any engine. New bearings are cheaper than crankshafts; if there is any doubt about the quality of a bearing that is to be installed, the bearing must not be used.

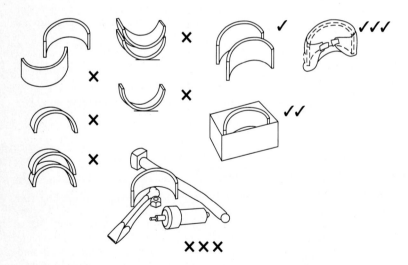

FIGURE 9.1 Bearings in storage.

Antifriction bearings are made to fine tolerances and are very predictable in their performance. Their life is determined by the fatigue strength of the materials from which they are made, the quality and quantity of lubrication, and the nature of the service in which they operate.

In the field of diesel generation, antifriction bearings will be found in the larger turbochargers, governors, virtually all electric motors, machine tools, all types of pumps, fan hubs, and other auxiliary equipment. In total numbers within any one plant, antifriction bearings exceed those of plain bearings by a great majority.

Causes of antifriction bearing failure are dirt, lack of lubrication, incorrect grease, and fatigue. Dirt enters the bearing through sloppy lubrication practices, by running the equipment without its covers, shields, and guards properly installed, or by lax attention during installation. Lack of lubrication frequently occurs following bearing changeout. All too often a prelubricated sealed bearing, upon nearing the end of its reliable life term, is scheduled for changeouts. In ignorance, it is replaced by a standard bearing that requires regular lubrication, and because the bearing it replaces needed little or no attention, it receives none either.

9.7 TURBOCHARGER BEARINGS

The bearings in turbochargers rotate at speeds up to 25,000 rpm and are subjected to elevated temperatures. Because of the circumstances of their operation turbocharger bearings are highly stressed and their service life should not be extended beyond 8000 h. After 8000 h their reliability is questionable and failure of the turbocharger caused by bearing failure can only be at a horrible cost. It is recommended that the turbocharger bearings be charged out at 8000 h intervals or exchange turbochargers fitted.

9.8 ANTIFRICTION BEARINGS

The term *antifriction bearing* is generic for all ball, roller, or needle bearings that rely on a rolling metal element rather than a hydrodynamic or hydrostatic fluid film to carry an imposed load without wear and with minimum friction. Although the basic principles of design do not vary greatly from one application to another, there are many variations in the materials used and the dimensions of the bearing components.

The life of an antifriction bearing of either ball or roller configuration depends upon the amount of use and load imposed on the bearing. Many ball and roller bearings are similar in appearance, but not all look-alike bearings have the same capabilities. Thus a "near enough" or "will fit" approach to replacements is not acceptable.

When a bearing fails, the maintenance record should be checked to see when the bearing was installed. The date may suggest that the bearing was in use for too long a period, was improperly selected, or was incorrectly installed. It is not safe to copy the number marked on the failed bearing and obtain a similar replacement. The OEM manuals should be consulted for the correct bearing number, and if the desired information is not found the OEM representative should be contacted. Crossover numbers may be used when the quoted number can be relied on.

Ball and roller bearings will be damaged by excessively tight fits, incorrect axial alignment, and overloading or distortion of the bearing housing or the shaft. Evenly spaced indentations or depressions in the raceway of the bearing suggest an improper mounting method such as beating the bearing with a hammer. The same condition may occur if when the housing is offered to the bearing, one or the other is tilted out of line.

The term *Brinelling* or *false Brinelling* is used to describe a particular form of antifriction bearing damage. This condition occurs when an antifriction bearing is subjected to static overload, shock, or vibration while motionless. A bearing thus damaged shows depressions formed in the races by the balls or rollers. It is a rare condition which may result from transportation or transmitted vibration. (A Brinell test is used to determine the hardness of a metal in which a steel ball 1 cm in diameter is pressed onto the tested material under a standard force. The area of indentation is then measured and compared with the load. The result is expressed as the Brinell hardness rating. Brinnel testing is done in a laboratory.)

Many antifriction bearing failures have their origins in corrosion, and some of the conditions that can cause corrosion may be avoidable. If corrosive action occurs in a bearing while the bearing is in service, it is highly probable that other parts of the machine also will be corroded. Acid fumes and damp or saline atmospheric conditions lead to such effects. Bad storage conditions or incorrect lubricants may provide a corrosive condition that is limited to the bearing only.

9.9 GREASING OF ANTIFRICTION BEARINGS

Unsealed bearings will require a shot of grease every 2000 h, approximately. Many of the troubles attendant on greasable bearings are avoided when prelubricated sealed bearings are used as replacements. Antifriction bearings must be serviced with the correct type of grease. Grease types and grades used in any one bearing should not be changed without the manufacturer's advice. If the type of grease is to be changed for any reason, the bearing must be thoroughly cleaned before the new grease is applied. Antifriction bearings that have been cleaned must not be spun by hand until they have been relubricated.

If when the bearing is dismantled, the grease has a stiff or caked appearance or has changed in color, it is an indication of bearing failure. The grease will have a smell of burnt petroleum and will have lost its lubricating qualities. In some cases the grease may have a varnish-like brittleness, and a bearing with grease in that condition must be discarded.

Usually the first indication of bearing lubrication failure will be a rapid rise in operating temperature. At the same times the bearing will begin to give off a whirring or whistling noise. If the bearing is allowed to continue running, the temperature will continue to rise, the bearing hardness will be lost, and total failure will ensue. A brown or bluish discoloration of the same or all of the bearing parts will be indicative of excessively high temperature operation. These failure symptoms will prevail for only a few minutes and will probably be undetected by the operator.

9.10 ELECTRIC ARCING EFFECT ON ANTIFRICTION BEARINGS

When an electric current passes through an antifriction bearing, it is interrupted at the contact surfaces between the races and the ball and arcing results (Fig. 9.2). This phenomenon produces high localized temperatures. Spark erosion takes place on both the race and the ball. Eventually a condition known as *fluting* occurs. As it develops, noise and vibration result.

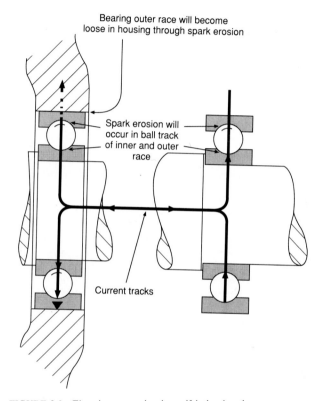

FIGURE 9.2 Electric arc erosion in antifriction bearings.

The causes of the current may be static electricity derived from belt drives and similar attachments or from electrical leakage past faulty insulation. The current may be of very low value and be difficult to detect. If the current is of higher amperage, such as that which can be expected from a partial short circuit, the ball track of the bearing will assume a rough, granular appearance.

Heavy currents passing through the bearing may produce a welding effect: there will be a migration of metal from the race to the ball. The noise and vibration of the bearing will then be marked. There is some thought that similar spark erosion may be responsible for eventual looseness of the outer race in the bearing housing of generators and other rotating electric machines. None of the arcing effects are visible on an operating machine, and any suspect bearing can only be replaced, because the interior cannot be examined without destroying the bearing. It is suggested that whenever there has been trouble in the motor or generator windings or a failure of the exciter, an examination of the antifriction bearings is appropriate.

9.11 BEARING INSPECTION

If the bearings are visible, the person responsible for routine inspection should examine the bearings. If there is evidence of grease running down the housing from the bearing, it can be assumed that the bearing has been or is being overheated. The situation must be rectified before total failure of the bearing occurs. In normal operation, the grease will always remain in the bearing. Check the type of grease being used against the OEM requirements.

9.12 REUSE OF ANTIFRICTION BEARINGS

An antifriction bearing taken out of service must be destroyed and sent away from the site. The reuse of an old antifriction bearing is absolutely unacceptable.

9.13 STORAGE OF ANTIFRICTION BEARINGS

Any antifriction bearings in storage must be kept in clean, dry conditions and should not be removed from their wrappings until they are installed. If there is any doubt about the cleanliness of a bearing, the bearing should not be used until it has been thoroughly flushed out and repacked with the correct grade and type of clean grease.

9.14 INSTALLATION OF ANTIFRICTION BEARINGS

When an antifriction bearing is being installed, the application force must always be exerted against the race being mounted. If the bearing is to be placed upon a shaft, the force or pressure must be on the inner race. If the bearing is being set in a housing, the force must be on the outer race (Fig. 9.3). Generally speaking, the race with the tightest fit is the one which will rotate when in service.

The force imposed upon an antifriction bearing race must be applied evenly and steadily. Misapplied force will damage the bearing. The races will be indented, and although the damage may not be visible, failure of the bearing will be inevitable.

9.15 VERIFICATION OF UNDERSIZE BEARINGS

Shafts that have been damaged in some way may be put back into service after their bearing surfaces have been refinished if the shaft damage was not too ex-

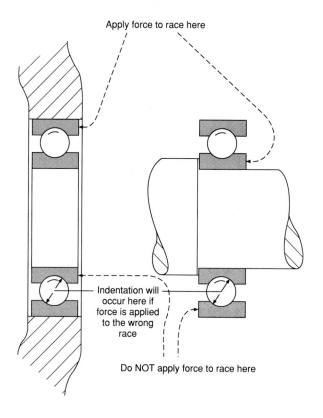

FIGURE 9.3 Installation of an antifriction bearing.

tensive. Depending on the nature and extent of the damage, crankshafts and camshafts are commonly repaired by metal spraying, grinding to original size, and then polishing; by grinding to a given undersize and then polishing; or by polishing alone.

To accommodate repaired shafts, it is quite normal to obtain and fit undersized bearings each having a dimensional match for each shaft bearing surface. A crankshaft that has been rehabilitated may have anything between one and all bearing surfaces undersize and all matched by appropriate undersize bearings.

Few engines that have been fitted with undersize bearings carry any plain notice of the variance; there may or may not be a written record. No matter what information is available, the mechanic reinstalling the bearings after an inspection or other work must routinely verify the bearing sizes and their shaft counterparts. All bearing clearances must be checked by the use of a plastic-type compressable gauging filament or similar material and the readings recorded in the engine workbook.

CHAPTER 10
CYLINDER LINER

10.1 CYLINDER LINER PITTING

Cavitation erosion, commonly known as *pitting*, may occur on the coolant side of a cylinder liner. During the combustion process, the pressure built up in the cylinder liner induces flexing of the cylinder liner wall. This motion creates the continuous development of a very large number of extremely small vapor bubbles on the coolant side of the liner wall. The life of each bubble is almost infinitesimally short, but the total chemistry of the bubbles acts adversely upon the metal of the liner wall.

The bubbles collapse or implode, and chemically altered metal then departs from the liner surface in microscopically small quantities. This process of erosion will eventually result in visible pitting. The pits will occur on the wetted areas of the liner, and severe pitting may be found in the liner seal ring grooves. Behind the O-rings, the pitting is accelerated by the sideways movement of the liner within the cylinder block.

It is not generally possible to detect pitting as it takes place, and the effect of pitting will be noticed only when coolant starts to invade the lubricating oil system or when the engine is stripped down for overhaul. To combat liner wear of this type, it is normal practice to include corrosion inhibitors in the antifreeze solution. Also, corrosion protection supplements are available for engines that operate without frost protection in their cooling systems. Additives are available to replace spent inhibitors, and test kits can be provided so that an operator can monitor the condition of his engine's coolant on a regular basis. The anticorrosive agents carried in the cooling medium also play an important part in the protection of lubricating oil coolers, aftercoolers, and radiator cores, which are particularly susceptible to corrosive attack.

10.2 CYLINDER LINER SCORING

Cylinder liner scoring is a form of damage found on the inner surface of a cylinder liner. The surface may be only lightly scratched and readily restorable by honing. A very few deeper score marks may be relieved by careful stoning, but deep multiple score marks will render the liner unserviceable. It must then be replaced. No remedial work should be attempted on a chrome-plated cylinder liner that exhibits any sort of bore surface damage.

Whenever a cylinder liner is found to be scored, the origin of the condition must be determined before the engine is restarted, particularly after a replacement liner has been fitted. Cylinder liner scoring is not usually associated with piston seizure. Cylinder scoring may be brought about by:

- Using one or more chrome-faced piston rings in a chrome-plated liner.
- Running with one or more broken piston rings.
- Using injector nozzles with an incorrect spray pattern.
- Supplying the cylinder with overheated aspirating air.
- Using incorrect or improperly installed scraper rings.
- Using ether for starting.
- Running the engine with perforated air filters, leaking air intake ducts, or no air filters at all.
- Running for prolonged periods on unduly light loads.
- Using dirty fuel.
- Providing insufficient piston cooling.
- Overheating the engine.

When a cylinder liner becomes scored, the gastight seal formed between the cylinder liner wall and the piston crown by the piston rings is penetrated. Gases escape from the combustion chamber and pass into the crankcase, where they raise the atmospheric pressure, contaminate the lubricating oil, and lead to accelerated wear of the moving parts. The lubrication film is blasted off the lower cylinder walls during the firing stroke. That results in increased liner wear rates and overheating of the piston skirts. On the compression stroke, the cylinder atmosphere will not attain the designed firing pressure. The cylinder will thus fail to carry its share of the load, and the engine will suffer a capacity loss.

10.3 CYLINDER LINER SCUFFING

Whereas cylinder liner scoring gives the appearance of a few or multiple grooves over the length of the piston stroke, another condition known as cylinder liner scuffing will sometimes be found toward the top or combustion end of the cyliner liner. This scuffed surface will be intermittently rough and without any pattern. Metal will appear to have been torn away from the cylinder liner surface. The condition can be attributed to the generation of excessive amounts of carbon in the combustion spaces, ineffective piston cooling, or defective fuel injection.

10.4 GLAZED LINERS

Glazed liners are almost always found on engines that run extensively at idle speeds or at very low loads. Poor engine loading can also cause undue piston ring gumming and a heavy buildup of carbon on the ports and valves. Carbon cutting of the piston crowns will be evident. As the engine continues to run under the same light load, the compression rings will cease to function and unburned fuel will pass between the piston and the cylinder liner. When the engine is later required to carry its full rated load, the conditions that have developed in the cylinders will be such that the engine cannot meet the demand. (See the discussion of fuel dilution of lubricating oil.) The liner surface can be restored by honing if there has been no scoring of the liner walls.

10.5 FLOODING OF CYLINDERS

Cylinder flooding will occur only when the engine is at rest. The cylinder can be filled with either coolant or lubricating oil. The coolant entry will be by way of a cracked head, liner, precup, or valve cage. Also coolant has been known to leak past the injector tube.

If an engine is to stand for a considerable period of time, a routine of barring the engine by hand should be instituted so as to give early warning of a hydraulic condition developing. Some engines which are idle most of the time get frequent and automatic lubricating oil priming. That delivers lubricating oil to the valve gear, and over a period of time it is possible to load a cylinder with lubricating oil. This is a rare circumstance, but it is one of which station personnel should be aware.

If it is suspected that one or more cylinders are flooded with fuel, lubricating oil, or coolant, there must be no attempt to start the engine. If the engine is turned by the standard air- or electric-starting method and one or more cylinders is either partly or fully flooded, there is a grave danger of the crankshaft or connecting rod being bent by the hydraulic locking of a flooded cylinder (Fig. 10.1).

No attempt to turn the engine should be made until that cylinder has been properly vented. Only then may the engine be rotated, and then by hand. Using the starting system to jolt the engine will probably result in damage to the engine. There is no good reason for wanting to start and run an engine with a suspected cylinder flood.

10.6 CHROME-PLATED LINER VERIFICATION

It is economical to recover the worn liners of certain engines for reuse by chrome-plating the bore back to the original size. Problems may arise after recovered liners have been in service for some time, however. Although almost every engine is fitted with a piston ring pack the top ring of which (the fire ring) is chrome-plated, it is not possible to run a chrome-plated ring in a chrome-plated liner. If a chrome-plated liner is used, the top ring must be a plain cast-iron one.

The trouble comes in the correct identity of a used chrome-plated liner. The true surface of the liner can be proved by the use of a solution of copper sulfate hydrate. Dissolve copper sulfate crystals in the ratio of one ounce of crystals to one pint of water. The crystals are available in drug stores, dispensaries, and veterinarian supply stores.

If the crystals do not readily dissolve, add a few drops of liquid detergent. A piece of rag or paper towel is then soaked in the copper sulfate solution. When it is applied to a clean area at either end of the liner, it will instantly identify chrome-plated or cast-iron surfaces. An iron surface will assume a copper-red appearance, but a chrome surface will not change color (Fig. 10.2).

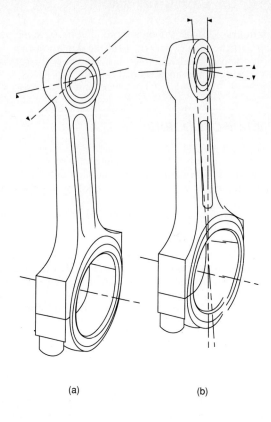

FIGURE 10.1 Connecting rod hydraulic lock damage. (a) Twisted connecting rod; (b) bent connecting rod.

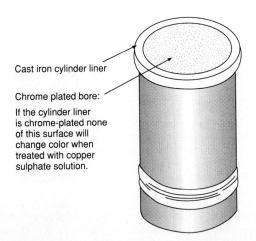

Cast iron cylinder liner

Chrome plated bore:

If the cylinder liner is chrome-plated none of this surface will change color when treated with copper sulphate solution.

FIGURE 10.2 Cylinder liner chrome-plating test. A piece of rag or paper towel soaked in copper sulfate solution and applied to the unchromed areas will cause the iron to assume a copper-red color.

10.7 CYLINDER LINERS IN STORAGE

All cylinder liners should be stored standing in vertical positions with their top ends upward. Liners stored in a laid down position have been known to assume a degree of ovality. The liners should stand on wood laths and be covered by plastic bags; garbage bags are ideal for the larger sizes. Any nonstandard liners should be clearly marked (i.e., oversize dimensions, chrome-plated, etc.) so that any anomalous liner installed in an engine is clearly identified and recorded.

CHAPTER 11
PISTONS

11.1 PISTON TEMPERATURES

Temperatures of 540°C (1000°F) can occur in the crowns in cast-iron pistons, and in aluminum pistons temperature of 260 to 360°C (500 to 680°F) can be expected. Because the melting temperature of iron is about 1500°C (2800°F) and that of aluminum is 650°C (1200°F), the risk of piston failure on engines fitted with aluminum pistons is greater if thermal overloading takes place. One of the main causes of thermal overloading is the introduction of aspirating air into the engine at elevated temperatures. Dirty aftercoolers, improperly controlled aftercooler coolant, and inadequate piston cooling are some of the conditions that may lead to the overheating of one or more pistons or all of the pistons in one bank. Individual pistons may become overheated if the piston cooling oil supply is obstructed or deflected.

The minimum effect of running the pistons at too high a temperature will be short piston ring life. The maximum effect of running the pistons too hot will be either piston and liner seizure or piston separation, particularly in the case of an engine fitted with aluminum pistons. If the pistons run at too low a temperature, cylinder liner glazing, gummed piston rings, high crankcase pressures, and plugged and bypassing lubricating oil filters are conditions that can be expected.

11.2 PISTON COOLING

Almost without exception, diesel engine pistons are cooled by the passage of lubricating oil over the interior heated surfaces. In smaller engines, lubricating oil is ejected from the top of each connecting rod, and it provides a form of splash cooling to the underside of the piston crown and the wall behind the piston rings. In larger engines, a jet of cooling oil is projected into the piston interior from a rail set in the lower part of the crankcase.

In the very large engines, the piston cooling oil is generally collected at the bottom end bearing. The oil travels through a rifle bore the length of the connecting rod; at the top end, the oil passes into the wrist pin and then into a chamber in the piston crown. Having accumulated a quantity of heat, the oil then falls from the piston into a catchment tray and then onto the lubricating oil cooler, where the excess heat is dissipated. The largest engines are so arranged that the oil returning from each individual piston passes over a thermometer, which permits the operator to monitor each piston temperature in comparison with the original specifications and other pistons on the same engine.

11.3 PISTON CROWN SCUFFING, AND SCRATCHING, OR SCORING

The scuffing, scratching, and scoring that take place on the piston crown surface above the piston rings can often be attributed to low-load operation over prolonged periods. Carbon formations brought about by incomplete combustion processes accumulate on the cylinder liner above the surfaces swept by the piston rings to a point at which the piston crown is contacted and carbon cutting of the piston then commences. A certain amount of carbon cutting over the life of the engine is quite normal, and the pistons can be reused after cleaning during overhaul. If a piston has been badly scored, however, it will need replacement.

Scuffing will also take place when the piston crown becomes overheated and, in expanding beyond its design limits, contacts the cylinder liner at some point in its travel. Usually the area of maximum contact will be at the top of the piston stroke. In most cases, some piston metal will adhere to the cylinder liner.

Leaking fuel injection nozzles or incorrect fuel timing will commonly cause piston crown overheating, but high aspirating air temperatures or jacket water circulation problems may also be causes. If the crown damage is limited to only one or two cylinders, faulty injector nozzles are the prime suspect. Pitting of the piston crown combustion surfaces also can be taken as evidence that injector nozzles or the fuel timing are at fault.

11.4 PISTON SKIRT SCUFFING

Scuffing and scoring of the piston skirt, the area of the piston below the piston rings, is usually caused by some inadequacy in the jacket cooling system. Also, scuffing of the piston skirt can be brought about by defective lubrication, the failure of the piston-cooling jets, repetitive cold starting, or the use of ether as a starting fuel.

11.5 PISTON SEIZURE

An engine at rest with a seized piston will have demonstrated several unusual characteristics immediately before it last stopped. The engine would probably have been stopped by either the high water temperature shutdown or the low lubricating oil pressure shutdown switches. The engine may have been overloaded; the aspirating air may have been very hot; the engine sounds would have been unusual; and the engine would have come to rest rather abruptly. In circumstances of suspected seizure of any engine component, it is quite unwise to open any part of the engine until the engine has cooled down, because there is always a possibility of explosion upon venting an overheated engine.

11.6 PISTON SEPARATION

Piston separation occurs most often on engines fitted with aluminum-alloy pistons. The usual failure points for the separation of the crown are the lowest piston ring groove and the wrist pin bore. Generally, the piston will fail at the point of maximum heat concentration or the smallest cross-sectional area in the piston configuration.

Piston separation is brought about by overheating the pistons to the point at which the metal is weakened and will not withstand the alternate tensile and compressive loads. Piston overheating is most frequently brought about by the admission of insufficiently cooled aspirating air to the cylinder or by running the engine in an overload condition. The aftercooler should reduce the temperature of the air leaving the turbocharger to about 52°C (125°F) as a general rule, but some engines require even lower temperatures for effective operation.

If it is not possible to reduce the air temperatures to the recommended level, it is quite probable that:

1. The air filter is dirty.
2. The aftercooler is dirty.
3. The engine is overloaded.

Piston separation can cause catastrophic damage as one or more connecting rods is released from the failed piston(s). It is imperative that the engine be stopped immediately. The engine suspected of having a separated piston can be barred round easily after it has stopped; it will be necessary to look up into the bore of each cylinder to ascertain which piston skirts still have their crowns attached. One or more crowns may have been left at the top of the stroke. Examine the valves and pushrods, several of which may be bent and display excessive clearance.

11.7 PURPOSE OF PISTON RINGS

The purpose of an engine's piston ring is to provide a seal between the piston and the cylinder liner that prevents the escape of combustion gases past the piston on the power stroke. On the compression stroke, the piston ring again prevents the aspirating air from being lost between the piston and the cylinder and thereby permits the development of compression pressures. Piston rings also serve to control the amount and dwell of the lubricating oil on the cylinder liner and are often fitted simply as a seal in hydraulic applications. In all cases the sealing principle is the same.

Primary sealing contact is made between the piston ring face and the cylinder bore when the ring is installed. The contact is maintained in service by the inherent tension of the piston ring. Sometimes a spring is used to supplement the tension of the ring. Secondary contact occurs when the gas or fluid pressures developed in the cylinder cause leakage around the sides and the back of the piston ring. The flow of gas or fluid through the minute clearances causes a differential pressure to develop across the ring, and the unbalanced forces cause the ring to press firmly against the low-pressure side of the groove and provide a secondary seal.

Internal-combustion engine piston rings perform two major roles: They seal off combustion gases from the crankcase, and they control the flow of lubricating oil in the cylinders. Generally, the upper rings are referred to as *compression rings* and the lower rings are referred to as *oil control* or *scraper rings*. Compression rings are subject to very high temperatures, stresses, impacts, corrosion, and abrasion. Many compression rings are chrome-plated, molydenum-plasma-coated, or carbide-coated in an effort to reduce wear of both the ring and the cylinder, but in all cases the cylinder material and the piston ring material have to be compatible.

Oil control rings do not suffer the same thermal stresses that are imposed upon the compression rings. All oil control rings must scrape lubricating oil off the cylinder wall. At the same time, however, they must leave a sufficient amount of oil behind to provide a lubricating film for the compression rings to ride on during their next stroke. There is a great variety in the design of oil control rings. Some rings, known as *conformable rings*, have their natural tension supplemented by a spring; others have great variations in their profile design.

Depending on the engine design, many of these piston rings have a "witness" feature as a guide to the piston ring wear. It may be a minute chamfer on the ring face that contacts the cylinder liner or concentric grooves machined into the contact face of the ring. As wear takes place during the life of the engine, the edge of the chamfer wears off gradually and the amount of wear that has occurred is visible when the rings are inspected. Similarly, when the piston rings have grooved contact faces, wear will eventually reduce the grooves and indicate the degree of wear that has taken place. On some two-cycle engines, notably the EMD marque, it is possible to view the piston rings through the scavenge ports without having to dismantle the engine.

It may be found within a particular plant that two basically similar engines, both from the same manufacturer, but one in the in-line and the other in the V-configuration, will have identical running gear (i.e., pistons, connecting rods, cylinder heads, and liners). The oil control rings will probably be different, however, and the plant personnel will have to exercise some caution when fitting replacement parts.

11.8 CHROME-PLATED PISTON RINGS

The top piston ring is frequently chrome-plated. Chrome-plated piston rings must not be fitted to a piston that is to run in a chrome-plated liner. If there is any doubt whatever, test both the ring and the liner by using the procedure outlined in the topic "Chrome-plated Liner Verification," Sec. 10.6. Use of a chromed piston in a chromed liner will lead to the rapid failure of both.

Less common are chrome-plated second and third compression, or scraper, rings. However, such plated rings are occasionally supplied, and it is suggested that all piston rings about to be installed in a chrome-plated cylinder liner be thoroughly checked by the previously described copper sulfate method.

11.9 SCRAPER RINGS

Scraper rings, also known as oil control rings, will wear rapidly under the influence of dirty lubricating oil. Some ring wear is quite normal over the life of the engine, but if there is occasion to change out the fire or compression rings, the scraper rings also should be changed. Worn scraper rings can be blamed for excessive lubricating oil consumption, but far more often the failure of scraper rings to control the amount of oil on the cylinder walls will be due to the scraper rings being stuck or broken. It is not unknown for scraper rings to be installed upside down. The installer must examine new rings for positioning marks, which are etched on the top or bottom lands.

11.10 STICKING RINGS

There are three classes of stuck rings. A *cold-stuck ring* is one that lies stuck in the piston ring groove when the engine is not operating. Carbon deposits will be the normal cause of the sticking. After starting and warm-up, the ring will become free and function normally. The entire face of a cold-stuck ring will have a normal bright appearance which indicates the ring is sealing properly. Cold-stuck rings should not have a detrimental effect on engine performance, and their replacement should not be necessary.

A hot-stuck ring sticks in the piston ring groove during operation. The ring face will be black owing to blowby, and it may be marked by improper lubrication. Any hot-stuck compression ring has to be changed out. Some liner scoring may be evident. A heat-set, collapsed, or broken ring will have lost all sealing capability. The face of the ring will be black from blowby and the ring may be broken in several places. Obviously, the ring must be changed at the earliest opportunity before liner scoring occurs.

CHAPTER 12
CYLINDER HEAD

12.1 CYLINDER HEAD CRACKING

Cylinder head cracking is invariably caused by a sustained, intermittent, or momentary absence of coolant from the cylinder head. Evidence of a cracked cylinder head on a running engine may be some or all of the following:

1. White smoke at exhaust stack
2. Slowing of turbocharger
3. Appearance of boiling at the header tank
4. Exhaust temperature excursions
5. Loss of power

In the case of (2), the turbocharger may make unusual noise; it may also stop completely or vibrate badly.

It is important that an engine with a cracked cylinder head be stopped as soon as possible after detection so that further damage is minimized or avoided. Other cylinder heads may be affected by the same force that cracked the first cylinder head; the turbocharger powered by the exhaust gases coming from the cylinder with the cracked head may be fouled by scale; the scale which precipitates from the leaking coolant will cause turbo rotor imbalance and attendant vibration; and the lubrication system of the engine may be contaminated with coolant. In the remedial process following failure, these factors must be taken into account.

The operator may encounter a situation in which an engine overheats and either shuts itself down or is deliberately and abruptly shut down by the operator because of an apparent overheating problem. In such a case, it is very prudent to bar the engine over by hand after it has stood still for some time and cooled down. It is not uncommon for one or more cylinder heads to crack through an overheating incident but fail to be recognized before the engine is shut down in the ordinary manner. In such cases, the first indication of trouble will appear when an attempt is next made to start the engine. The engine will not be able to turn over at all, or it will make only part of a turn because one or more cylinders has flooded with coolant.

Depending on the position of the piston in a flooded liner when the engine is started, either the cylinder head studs will be grossly overloaded when the piston moves up on a compression stroke or quantities of coolant will be passed through the exhaust valves and enter the turbocharger and possibly another cylinder with an open exhaust valve. In any event considerable damage to the engine can or may result. Besides the overstressed cylinder head studs, each of the connecting rods in each flooded cylinder may have been bent to a degree that is discernible just by looking at it, but most degrees of distortion can be determined only by proper measurement techniques.

Other possible damage that must be looked for includes burst gaskets, cylinder head gasket rupture, stretched or broken cylinder head studs, and bent connecting rods. It is recommended that any engine that has either been shut down manually or automatically because of an overheating problem or has stood waiting to go to work for any great length of time should be barred over at least one complete turn before a start is attempted.

It is commonly supposed that either loss of coolant or overheating will cause cylinder heads to crack, but that belief is a little inaccurate. When a running engine loses its coolant circulation for any reason, the cylinder heads will become overheated. If the engine is then shut down and coolant still does not flow through the cylinder heads before the engine has cooled down to starting temperatures, it is quite probable that no damage to the cylinder heads will ensue. Conversely, if the coolant is reintroduced to an overheated cylinder head, be it running or stopped, the cylinder head will be subjected to an abrupt and uneven loss of heat, known as *thermal shock*, and cracking may then result.

A cylinder head that is not accurately torqued down may permit combustion gases to pass the seal between the cylinder head and the cylinder liner. The gas will enter the cooling spaces and create voids. After passing through the cooling system, the gases will eventually vent in the header tank. If the gas leak is small, the header tank will demonstrate a certain turbulence. If the gas volume is great, the header tank will give the appearance of boiling. Because of the coolant displacement caused by the combustion gases, coolant will overflow from the header tank.

A similar effect can be caused by leaking precombustion chambers. If the leakage is more than a few minute bubbles visible in the coolant entering the header tank, there is a real danger of coolant flow impediment in the engine. Voids will develop in the cooling spaces, and hot spots in the cylinder heads liners or jacket may occur. The high coolant temperature sensors will not operate if they are not immersed in the coolant and will not provide reliable protection. When movement of the gas bubbles is such that the coolant body follows and any hot spots are abruptly cooled, there is considerable risk that metal cracking through thermal shock will take place.

The combustion face of a cylinder head is exposed to the highest temperatures prevailing in the engine, and a great deal of heat must be dissipated to the coolant. To enable this dissipation to occur, the metal thickness of the firing face is kept to a minimum. That contributes to the inherent probability the cylinder head will crack under conditions of unusual and extreme thermal stress.

Frozen or blocked suction lines from the radiators may culminate in engine overheating, but the coolant pump will still be able to maintain an almost normal jacket coolant pressure even though there is no circulation. Eventually the engine will overheat and the jacket water temperature switch will shut the engine down without any real damage being incurred. Some engine installations are equipped with a visible flow indicator in the coolant discharge line and thus provide good warning of poor circulation.

If an engine has no jacket water temperature protection at all or has protection that does not work properly, it will continue to run and will heat to the point at which the coolant boils. The next stage in this particular event will be for one or more pistons to seize in the liner. Shortly after the engine has come to a forced stop most if not all of the cylinder heads will crack.

Overheating can be caused by worn or loose coolant pump impellers, valves in the cooling circuit being in the wrong position, overloading the engine, overheated aspirating air, incorrectly set or slipped timing, poor turbocharger output, scaled cooling spaces, or dirty air filters.

12.2 EJECTION OF COOLANT FROM HEADER, SURGE, OR EXPANSION TANKS

For the purposes of this discussion, the term *header tank* will be used exclusively. The ejection of coolant from the header tank will be caused by one of the three conditions described in the following subsections.

Overheating

If there is a defect in the engine's cooling system or in the heat rejection system, the coolant within the engine can reach boiling temperatures. In that case the high jacket water temperature switch should shut the engine down. After the engine has come to rest, the power plant personnel can examine it, determine the cause of overheating, and make the necessary repair. If the high jacket water temperature switch is malfunctioning, it may happen that the coolant will boil and will cause the header tank to overflow through either expansion of the coolant or severe turbulence of the coolant.

Overfilling

The engine cooling system and its attached heat rejection system have the capacity to hold a certain volume of coolant. When the engine is at rest and at a normal standby temperature, the whole system except the header tank will be filled with coolant. The header tank is provided to accommodate any coolant displaced from the system by expansion of the coolant as it rises to standard operating temperature with the engine running. If the header tank is filled with coolant when the engine is at rest, overflow at operating temperature is inevitable.

Gassing

Gassing occurs when combustion gases leak into the coolant. Leakage takes place through cracks which develop in the cylinder head, cylinder liners, or precombustion chambers and also when the cylinder head is not fully seated on the cylinder head to cylinder liner gasket. Improper seating may be caused by uneven torquing or breakage of the cylinder head fasteners. The extremely fast expansion of the combustion gases in the coolant will give the coolant the appearance of boiling. Much coolant will be displaced with the result that substantial quantities will be ejected from the header tank. Radiator caps will vent; surge or header tanks will probably overflow. The condition will be more marked at high load than at low load.

Combustion gas leakage may be caused by cracked cylinder heads, cracked liners, leaking cylinder head gaskets, cracked precombustion chambers, or leaking precombustion chamber seals. When gassing takes place immediately after an overhaul or repair work centered around a particular cylinder, it is probable that the cylinder liner or cylinder head is improperly seated and the copper seal rings or gaskets are not uniformly crushed.

An overheated engine will continue to boil after it has stopped; an engine that has been gassing will show normal temperatures immediately after the load is removed and before stopping, and turbulence in the header tank will be much diminished. When the engine is gassing, it is highly unlikely that the high jacket water temperature switch will sense any temperature rise. Therefore, it will not shut the engine down nor will the jacket cooling temperature gauge show any increase in temperature.

No useful purpose will be served by draining coolant in an attempt to stop overflow. If air appears to be present in the coolant in either at the engine in the header tank or in the radiator, it must be believed that the "air" is in fact combustion gases. It is highly improbable that it is air entrained in the cooling system through the coolant pump glands. In a gassing event, the operator will have no alternative but to take the engine off load and initiate diagnostic procedures.

Gassing may be proved by attaching a plastic hose to a vent cock mounted on the coolant outlet manifold and terminating in a bucket at floor level. With the engine running on full load and the vent cock open, bubbles will be seen coming out of the hose if there are combustion gas leaks. On smaller engines the hose may be fitted to the radiator overflow pipe. On engines with individual fuel pumps the faulty cylinder can be identified more positively by pushing the fuel rack into the full-fuel position, one cylinder at a time, while observing the coolant and bubble flow. Two people are necessary for this test.

Running an engine with evidence of gassing may lead to the multiple cracking of cylinder heads. It is possible for the leaking combustion gasses to pressurize the engine coolant spaces to a degree that would prevent the entry of further coolant to the other cylinder jacket and/or cylinder heads.

12.3 CYLINDER HEAD COVERS

In many instances, it is possible and seemingly convenient to operate an engine when the cylinder head covers are not in position or are open. Similarly, some engines have detachable covers for the fuel injection pumps and their control gear. Many engines are equipped with hinged valve covers that can be opened while the engine is running. When hinged covers are provided, the operator should take advantage of the feature once per shift so that he can inspect the valves and valve-operating mechanism. That will ensure that the fuel injection pipes, valve cage cooling pipes, and other connections are not leaking. At the same time, the operator can check to see that all the valves are rotating correctly.

Many operators get in the habit of leaving hinged valve covers or fuel pump covers open when the engine is operating. Two main difficulties can be expected if that practice is followed.

CYLINDER HEAD

1. The oil mist that is supposed to condense on the fuel pump racks and fuel pump layshaft and otherwise provide lubrication to many minor parts will not be available to do so, and the general wear rates of the exposed parts will be accelerated.
2. The oil mist rising off the engine will condense on the power plant walls and ceiling, the ventilating fans, and other cool structural surfaces and make the housekeeping chores more difficult and widespread. Also, much of the oil vapor will be drawn into the engine's air filter, the radiators, and the alternator. Upkeep personnel will be reluctant to work in such a plant because of the oily film that covers everything.

12.4 CYLINDER HEAD STUD OVERLOAD

When cylinder heads are installed on an engine, it is necessary to torque the nuts on the cylinder head studs in an even and balanced manner. The correct sequence of tightening and the appropriate torque valves are stated in the shop manual, and they must not be ignored. Failing to observe the OEM recommendations will lead to warped cylinder heads, uneven crush of the sealing ring between the cylinder heads and the liners with resultant leakage, uneven loading of the cylinder head gasket which may lead to subsequent blowout, and overloading of some cylinder head studs. An unevenly crushed head-to-liner seal cannot be saved; it must be replaced.

Should any movement be initiated by the forces of combustion, there is danger that fatigue cracking will take place in either the liner or cylinder housing. Movement of the lower end of the liner will induce pitting of the lower seal ring groove and will also permit the development of coolant leaks.

Cylinder head studs may also be overloaded by turning an engine over on a flooded cylinder. One or more cylinder head studs may break under the conditions noted. Obviously, the broken stud has to be replaced, and it is recommended that all the other studs holding the same cylinder head be replaced at the same time because they too will probably have been stressed beyond their material limits. Any studs discarded in this manner must be removed from the site or destroyed so they are not inadvertently used again.

12.5 LOOSE CYLINDER HEAD STUDS

All the cylinder head studs must be tightened down fully before the cylinder head is put in position. If the studs are not tight, there may be debilitating consequences. Although the cylinder head nuts may have been torqued down to their correct value, subsequent settlement of the cylinder head gasket and the action of combustion forces may cause one or more of the cylinder head studs to loosen up.

When studs in the cylinder block are loose, it is possible that there will be movement of the cylinder head and perhaps the cylinder liner. Evidence of cylinder liner movement may be seen on the cylinder block to cylinder liner landing. Normally, both faces of the landing should be clean: free of rust, carbon, or other contaminants. If the condition has existed for a long time, pitting of the nose ring grooves in the lower part of the liner can be expected.

12.6 CYLINDER FIRING PRESSURE INDICATORS

All low-speed and a large proportion of medium-speed engines have their individual cylinder heads fitted with stems and cocks to which cylinder pressure gauges can be attached for routine test purposes. While the engine is running at a steady load, which should not be less than 50 percent of the rated capacity of the unit, the cylinder pressure gauge is connected to each cylinder in turn. The pressure being developed in each cylinder is noted, and the steadiness of the load is constantly checked. When all of the cylinder pressure readings have been obtained, they are compared. If all is well, cylinder pressure readings should be within 2 percent of the mean pressure of all of the cylinders. Any engine that is running with cylinder imbalances will be noisy and will have an undue level of vibration and a reduced fuel efficiency.

Causes of Incorrect Cylinder Pressures in Individual Cylinders

- Defective injector nozzle
- Weakened or broken injector spring
- Incorrect fuel injection pressure
- Defective fuel pump delivery valve
- Damage in valve-operating mechanism
- Displaced sectional-type camshaft coupling
- Burnt or bent valve
- Burnt or cracked valve seat
- Broken piston ring(s)*
- Scored, worn, or glazed cylinder liner*
- Cracked or perforated piston*

Causes of Incorrect Cylinder Pressures in Several Cylinders

- Water in the fuel
- Dirty fuel filters

Causes of Incorrect Cylinder Pressures All on One Bank of a V-Engine

- Dirty air filter

---

*In these instances the crankcase pressures will be above normal.

- Slow turbocharger
- Dirty aftercooler

Whenever a deficient cylinder pressure is observed, the original test log shall be consulted for the correct pressure figures. Any rectification work necessary to restore the cylinders, cylinder heads, or fuel equipment shall be carried out only by a certified journeyman mechanic.

12.7 INDICATOR COCKS AND CYLINDER PRESSURE GAUGE

Before the cylinder pressure gauge is attached to the cylinder head stem, the cock should be opened momentarily to evacuate any carbon or other solids lodged in the stem. The gauge is then attached, the cock is again opened, and the cylinder pressure is read. The cock is closed, and the gauge is removed. The gauge will have become very hot, and it must be handled only with asbestos gloves.

It is common practice to slack off the pressure gauge coupling nut with a few smart hammer blows on the coupling nut lugs, but that has a very damaging effect on the instrument and shortens instrument life considerably. In the metal case that is supplied with the gauge is also a special key that fits the coupling nut lugs and this tool must always be used to tighten or loosen the coupling nut of the gauge. There is no reason why the cylinder pressure gauge supplied with an engine should not last for the life of the engine if it is given a proper degree of professional care.

CHAPTER 13
CYLINDER HEAD VALVES

13.1 VALVE CLEARANCES

When the engine is at rest and at a standing temperature of approximately 21 to 26°C (70 to 80°F), there will be a clearance between the top of the valve stem and the valve rocker lever. That clearance, generally known as the *tappet clearance*, must be set to the dimensions given in the OEM manual, and it should be verified at about 1000-h intervals on all engines. Because of the valve seat and valve face wear that normally takes place when the engine is running, the original clearance will gradually change.

The clearance is carefully calculated so that when the engine attains its full operating temperature, the valves can be operated without any lost motion and at the same time give a perfect seal. Without any lost motion in the valve actuation there will be no abnormal noise, the timing of the exhaust and aspiration portions of the cycle will be exact, and the conbustion process will take place most efficiently.

It is important that the valve or tappet clearance be set when the engine's temperature has settled to an at-rest state. If the clearances are checked and adjusted when the engine has just been stopped, it is probable that any adjustments made will be false. In subsequent running, the valve clearances will then be incorrect. When the clearance is habitually incorrect, valve pounding can be expected. The valve face will become stepped; the valve seats will be radiused; and the sealing ability will be lost. Any engine that is kept standing with the block heater passing too much heat will not have a correctly adjusted valve clearance.

Some engines are so arranged that two valves are operated from one rocker lever by means of a bridge piece. In extreme cases of valve misadjustments on these engines, breakage of the bridge piece guide is to be expected. When the valve clearance is too closely set, the valves will not be able to close properly at full operating temperatures. In that case, valve seat and face burning will occur; exhaust temperatures will be abnormally high; and the turbocharger, if fitted, will be endangered. Any valve that leaks will be exposed to temperature excursions, and that may result in cracking of the valve head or cracking and loosening of the valve seat. Valve and valve seat dimensions are shown in Figs. 13.1 and 13.2.

13.2 VALVE CLEARANCE SETTING METHODS

How valve clearance adjustments are made on engines with four valve heads very much depends on the arrangement of rocker levers, articulated rocker le-

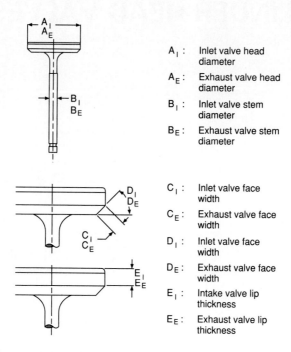

FIGURE 13.1 Valve dimensions.

vers, crossheads or bridge pieces, and tappets. No one method of valve adjustment is applicable to all engines, but in every case the OEM manual must be consulted and followed exactly.

The methods applied to one particular model of engine should not be transferred to a different engine just by force of a habit previously acquired. On two-valve head configurations, the clearance-setting requirements are usually quite straightforward. Nevertheless, care should be exercised.

13.3 EFFECT OF HIGH-SULFUR FUELS ON VALVES

The corrosive effects of high-sulfur fuels can lead to certain types of valve failure. Exhaust gases passing between the valve guide and the valve stem become cooled, and the partial condensation of the gases produces sulfuric acid, which attacks the bore of the valve guide and results in fast wear rates. In the initial stages of the attack the guide bores will assume a barrel-like interior profile that will gradually develop to a point at which the bore is no longer concentric and the true alignment of the valve with its seat is lost. At about this time serious problems of valve bending, seat stepping, leakage, and the like can be expected.

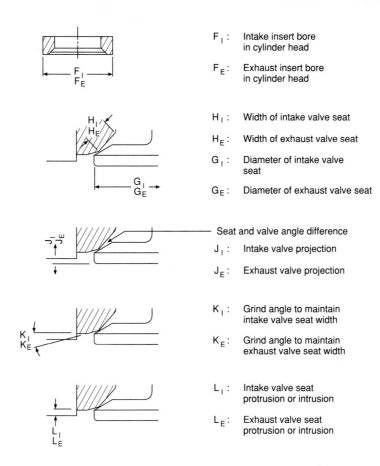

FIGURE 13.2 Valve seat dimensions (subject to specifications).

13.4 EFFECTS OF FREQUENT COLD STARTS ON VALVES

In engines subjected to frequent cold starts followed by brief running periods there may be some effect on the valves. Quantities of soft carbon that will build up on the valves and in the cylinder heads will impede the proper operation of the valve mechanisms. In the case of two-cycle engines, loss of power owing to port plugging will be noticed. The carbon deposits will occur because, until the engine is thoroughly warmed up, there is a tendency for the exhaust gases to condense before they leave the engine. Unburned fuel and lubricating oil will be precipitated on surfaces normally exposed to elevated temperatures. It is a good practice to allow the engine to warm up to the jacket operating temperature before applying the load. By this means, extraordinary carbon deposits can be avoided.

13.5 VALVE GUIDES

Valve guide wear is quite normal to a certain extent: As long as the fuel grade and the mode of operation remain unchanged, the rate of wear of the guides is somewhat predictable. However, if an increase in the fuel sulfur content occurs, a change in the wear rates in the engine can be expected and the wear rate change will initially be more noticeable in the valve guides than in any other part of the engine (Fig. 13.3). Fuels for use in very cold weather contain much less sulfur than those fuels used in hot climates.

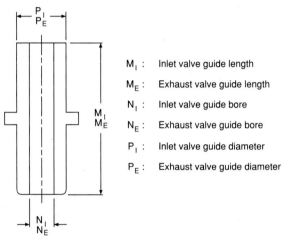

M_I :   Inlet valve guide length
M_E :   Exhaust valve guide length
N_I :   Inlet valve guide bore
N_E :   Exhaust valve guide bore
P_I :   Inlet valve guide diameter
P_E :   Exhaust valve guide diameter

FIGURE 13.3   Valve guide dimensions.

13.6 NIMONIC EXHAUST VALVES

Certain high-performance diesel engines have exhaust valves made from Nimonic alloys. These alloys are better able to withstand the thermal and mechanical stresses imposed upon the valves than standard valve materials. Like any other exhaust valve, the valves need servicing, but because of the softness of the Nimonic alloy it is not usually necessary to grind them to renew the valve-to-seat seal. Simple hand lapping of the valve on a freshly cut seat will suffice to provide a perfect gas seal.

Nimonic alloy valves are expensive; they cost up to 10 times as much as ordinary valves, and simple economics demands that they be made to last as long as possible. Grinding Nimonic valves will generally result in the unwarranted and wasteful removal of metal, which will shorten the overall life of the valve. There is, therefore, a need for care and attention in the servicing of this type of valve.

13.7 REPLACEMENT AND REHABILITATED VALVES

Inlet and exhaust valves are commonly composed of more than one material; the stem is made of one steel alloy and the head is made of another. Some areas of the valve may also be plated. Often the valve sealing face is a welded-on alloy such as stellite, which offers a superior resistance to the effects of exhaust gas scouring, high temperature, and impact. The mating valve seats are of a material that differs from that of the parent cylinder head. The match of the valve head and the seat insert material is very carefully selected at the time of design, and it must not be upset. A great deal of caution must be given to the selection of the rebuilding shop that may be asked to rehabilitate worn valves, because inadequate or incorrect rework practices can have devastating consequences if a rebuilt valve fails in service.

Similarly, all too often jobber valves are not the correct material specified by the OEM, and thus problems of fit, fatigue, and compatibility may develop. Jobber valves can be used with successful results if the OEM specifications are met in all respects and the jobber is willing to warrant his product.

13.8 RECOGNIZING INDICATIONS OF VALVE FAILURE

Valve failures will announce themselves in one or more of several ways. If the engine is running, unusual noises may be heard, particularly in the vicinity of the affected cylinder. The engine may lose power. In the case of an exhaust valve failure, the exhaust temperature of a particular cylinder may climb well above the prevailing average. Against a night sky, sparks may be seen in the exhaust plume.

The kilowatt meter may demonstrate a rhythmic flutter, and the normal vibration readings for the load being carried will be disturbed. There may be a loss in boost pressure, or the turbocharger may vibrate badly. If an air valve has failed, the air inlet manifold may become overheated adjacent to the affected cylinder.

Valve problems can be divided into two basic groups. In the first are the situations in which all of the inlet and/or exhaust valves are part of the problem; in the second group are those in which only individual valves are involved. When all the valves of a kind are seen to be in trouble, it is probable that the root of the problem is one of lubrication temperature timing or something similar. If only one valve has failed, the incident is probably related to misalignment, misadjustment, or fatigue. In the first instance, however, the failure may be only the initial sign of a more complex difficulty. It is, therefore, important that the true cause of any valve failure be thoroughly investigated so the problem is rectified without further or subsequent damage being done to the engine.

13.9 LEAKING AIR INLET VALVES

On some engines, a failing air inlet valve will be evidenced on some engines by a localized overheating of the aspirating air manifold. The OEM manuals carry much dimensional information. It should be noted that, in many cases, the dimensions of exhaust valves, exhaust valve seats, and exhaust valve guides will differ from their inlet counterparts. There may also be specified differences between one engine and another of the same model for valve seat angles, depending on the application, speed of the engine, the fuel used, and other considerations. Departure from the OEM specifications can have a serious effect on the valves. Multiple air-inlet valve problems can be expected to occur after an overhaul or other major work.

13.10 DAMAGE ASSOCIATED WITH VALVE FAILURE

When a valve failure takes place, damage may be sustained by other components, and it will be necessary to fully inspect the engine with that in mind. If the head of the valve has broken up in any way, there is a strong likelihood of fragments from the valve being carried in the exhaust gas stream through the exhaust manifold and into the turbocharger with resulting impact damage to the turbocharger blading. It is also probable that there will be some damage to the related piston, other valves serving the affected cylinder, and the valve-operating mechanism.

When a valve has contacted the piston, the piston crown may be peened, the valve guide may have suffered excessive sideways loading, and the valve and valve seat sealing faces may be degraded. The rocker lever fasteners may be strained or broken, and the valve bridge and its stem may be bent or broken. When valve stem seizure has occurred, piston crown contact may be expected. There is a probability that the pushrods will be bent and perhaps the rocker levers or rocker shaft distorted.

Although an exhaust or inlet valve is only a small component relative to the rest of the engine, its failure can lead to the most destructive damage to the engine. It is therefore important to remove the engine from service and stop it immediately when a valve failure is indicated.

On many engines it is possible to hinge back the valve covers while the engine is running in order to inspect the action of the valves and their operating mechanism. When that is possible, it is suggested that such an inspection be made one or more times per shift. However, on engines that operate at high speeds (i.e., above 1200 rpm), on those on which multiple fasteners are used, or on those that have positive crankcase pressure shutdowns, the removal of valve covers while the engine is running is impossible.

13.11 INLET AND EXHAUST VALVE REHABILITATION

Before commencing any rehabilitation work on cylinder head valves, the mechanic must review the specifications laid down in the OEM manual for the valve and valve seat geometry. Any alteration in the width of contact faces or any change in the interference angles will have prejudicial effects on the engine's performance, and the engine's exhaust emissions will be degraded. Usually there is a half to a full degree of difference between the angles of the valve face and valve seat.

13.12 VALVE STEM SEIZURE

Seizure of the valve stem is more common in inlet valves than in exhaust valves because inlet valves are more susceptible to carbon formations. Running the engine for prolonged periods with a low load or at too low an operating temperature will contribute greatly to carbon buildup on the inlet valve stem. If a valve failure can be attributed to excessive carbon formation developing on a valve stem, the coolant regulator should be checked for correct operating temperature.

The method and standards of engine operation should also be reviewed. The use of an incorrect grade of lubricating oil can lead to abnormal amounts of hard carbon deposits, particularly on the valves and it is probable that all of the valves in the engine will then require attention.

13.13 BENT VALVES

The most common cause of bent valves is piston contact. Valves can also be bent by debris lodged between the valve and the seat. A bent valve is commonly identified by a substantial change in the normal tappet clearance. There will also be an exhaust temperature excursion, and the valve may cease to rotate.

Caution must be exercised in the fitting of new or rebuilt valves, new or rebuilt valve cages, and cylinder heads that have replacement seats inserted. If there is a departure from original design dimensions in the rebuilding process, it is possible for the valve and valve cage or the valve and cylinder head assembly to give insufficient clearance for the valve to operate without contacting the piston at some point in the engine's cycle.

Many manuals dealing with engines fitted with valve rotators recommend a tap test to check that the rotators are operative. This test is valid in the shop when the cylinder head is properly blocked and there is ample clearance below the valves. Considerable caution is necessary if this test is done on a standing engine. It will be necessary to bar the engine around as each set of head valves is tested so the piston relative to the valves to be tested is in the lower half of its stroke. Failure to do so can result in bent valves.

13.14 BURNED EXHAUST VALVES

Exhaust valves may burn under the heat and flow of escaping combustion gas. This usually occurs when the exhaust valve has lost its sealing ability through wear, overdue, or incorrect adjustment of the tappet clearance, failure of the valve to rotate properly, or overheating of the valve induced by a thermal overload of the cylinder. It is not uncommon for the corresponding seat or seat insert in the cylinder head to suffer similar and matching damage.

Indications of distressed exhaust valves and seats will be seen in elevated exhaust temperatures on one or more cylinders, higher than normal turbocharger gas inlet temperatures, glowing exhaust manifolds and turbocharger casings, sparks being emitted from the exhaust stack, and high coolant outlet temperatures on one or more cylinder heads.

13.15 LARGE-ENGINE VALVE STEM LUBRICATION

Under certain conditions, excessive amounts of lubricating oil being passed to the exhaust valve guide and stem can initiate valve stem seizure. In larger engines an arrangement is sometimes made for the direct lubrication of the exhaust valve stems. A regulating valve rations the amount of oil supplied relative to the load being carried by the engine. At full load the regulating valve senses the boost pressure and permits a full flow of valve stem lubrication. As the load drops, the oil flow is decreased and is virtually stopped before the engine comes to rest.

The adjustment of the valve is quite critical: The valve must provide sufficient oil for the lubrication needs under various loads, but if overlubrication takes place and excessive oil remains on the valve stem when the engine has stopped, the normal residual heat in the valve will carbonize the oil, alter the design clearances, and provide the conditions for seizure following the next startup. It is very important that the regulating valve be adjusted only by a journeyman mechanic, who can do such work only in consultation with the relevant manual.

13.16 VALVE PROBLEMS CAUSED BY OVERSPEED

Damage to the valves or valve-operating mechanism following an engine overspeed incident can be expected, and it would be quite imprudent to restart and run an overspeeded engine without carrying out a careful inspection of the engine. Besides the examination of the alternator windings, the lubricating oil filter elements, the main bearing bolts, and the bottom end bolts, it will be necessary to make a careful inspection of the valve operating mechanism (Fig. 13.4). In the first instance, a check of all the valve clearances will reveal any distortion of either the pushrods or the rockers or any bending of the valves. The valve springs, keepers, and cotters must be carefully scrutinized.

The turbocharger rotor will have been damaged if there has been any metal detachment from the valve head. Manual spinning of the rotor will usually be sufficient to show if there is any rotor and blade damage.

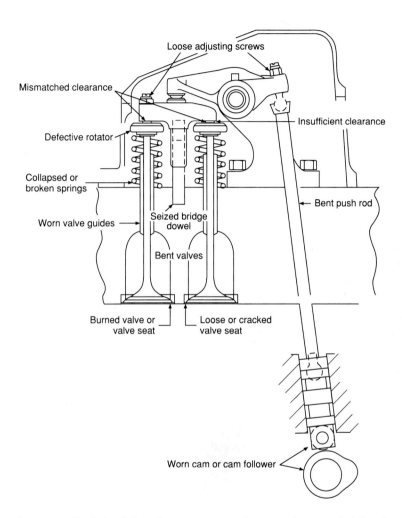

FIGURE 13.4 Typical cylinder valve arrangement and common damage or fault locations.

CHAPTER 14
CAMSHAFTS AND ENGINE TIMING

14.1 CAMSHAFT PURPOSE

The purpose of the camshaft is to convert a rotary motion into a reciprocating motion that will actuate the exhaust and inlet valves. The camshaft can also be utilized to drive the individual fuel pumps. The disposition and profile of the camshaft must be in an extremely accurate relationship to the crank rotation. Any mistiming of any valve or fuel injector will impair the power and efficiency of the engine. The camshaft in a four-cycle engine turns at half of the crankshaft rotational speed, but in a two-cycle engine the camshaft and crankshaft speeds are the same.

14.2 CAMSHAFT PROBLEMS

Camshaft problems are fairly uncommon, and failures are usually confined to individual cams or coupling slippage in sectional camshafts. Single-cam damage in the form of surface spalling or chipping is usually the result of overloads being imposed by or upon one or more components of the valve-operating mechanism. A typical overload situation would be caused by valve and piston contact with resultant excessive contact pressure between the cam and the cam roller.

Misalignment of fuel pump tappet blocks will contribute to point contact between the cam roller and the cam. This condition is often evidenced by breakdown of the surfaces at the edges of the cam lobes or the cam roller. In the early stages, the spalling will show as the detachment of localized flake-like chips, and the debris may be visible in the lubricating oil filter elements. As the spalling condition develops, the cam and cam roller contacting surface will break down entirely and the rate of metal loss will accelerate. It is probable that complete failure will take place before there is any warning advice from the lubricating oil analysis reports.

On engines fitted with individual exhaust temperature gauges, the failure of a fuel pump cam will be accompanied by low exhaust temperature of the corresponding cylinder, and it is almost certain that there will be some audible warning in the vicinity of the fuel pump. On other engines, a marked increase in valve mechanism noise, or valve clatter, may indicate some sort of cam problem. It is possible that individual valves will stop visibly rotating, and an unexplained increase in tappet clearance may also point to a cam or cam follower failure.

14.3 CAM INSPECTION

A cam does not necessarily have the same profile on both the approach and the departure sides. Some attention must be given to the correct positioning of replacement cams on a built-up camshaft, particularly in the case of cams with less pronounced degrees of asymmetry.

Cams are case-hardened, and they should be examined occasionally for signs of surface cracking. At the same time, it is expected that whatever fastening devices are used to position the cams on the shaft, they will be examined for indications of looseness. On sectional camshafts it will also be necessary to inspect the couplings between the camshaft sections. The muff couplings, the fasteners, the splines, and the keys may disclose signs of fretting or other wear which will demand attention.

14.4 CAM PROFILE AND POSITION

The profile of the individual cams is of critical importance to an engine's performance. Any looseness of a cam on the camshaft or loss of metal in either the cam or the cam roller will have a very adverse performance effect on the related cylinder(s).

14.5 CAM ROLLERS

On certain larger engines it is possible to inspect the motion of the camshaft and cam rollers while the engine is in operation. If that can be done, the inspection should be included in the shift operator's tour. Each cam roller should be seen to be rotating correctly, and the corresponding cam should be sighted. The inspector should look for any evidence of roller skidding or wobble, and he should also look for any signs of flaking, spalling, or flats developing on the mating surfaces of the cam and the roller. If any degradation is noticed, the engine must be removed from service so repair can be effected.

14.6 ENGINE TIMING

When any work is done on the timing gear train, the timing chain, the camshaft, one or more fuel pumps, or the fuel pump drive and its couplings, the engine fuel timing must be checked before any attempt is made to restart the engine. Gross mistiming errors are probably detected more easily than minor timing errors, but the results of a gross error can be catastrophic. The person who works on the fuel system or its timing is failing in his duty if he neglects to thoroughly check the fuel injection timing at the end of the work exercise (Fig. 14.1).

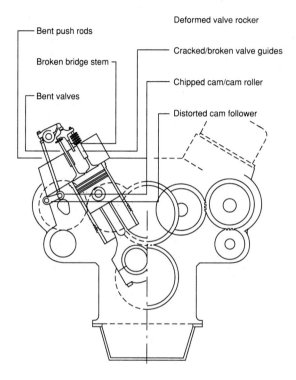

FIGURE 14.1 Failed timing gear damage.

On engines fitted with sectional camshafts the timing must be checked at both ends of the engine. It is possible for the camshaft section couplings or their keys to work loose. When that happens, the sections of the camshaft remote from the drive end will lag and the timing of the affected cylinders will be late.

On larger engines the individual cams are often fastened to the camshaft by means of tapered cotter pins. Although these pins are firmly driven home and finished flush with the cam profile, it is not unknown for them to work loose. When mistiming of only one valve or fuel pump is suspected, it is a good idea to inspect the cotter pins for fit.

14.7 TIMING REFERENCE MARKS

Virtually all engines have timing marks stamped upon their flywheels. On large engines there will be marks referring to each cylinder, but on smaller units only the marks pertaining to the No. 1 cylinder will be provided. Whichever marks are applied, none should ever be defaced or altered in any way, nor may any new marks be added. Other marks necessary to the setting or adjustment of the engine's timing may be found on the timing train gear wheels or sprockets, and in some cases there will be register marks on the timing gear casings. In engines fitted with sectional camshaft(s), additional register marks will be found at the free end of the camshaft.

14.8 INJECTION TIMING

The commencement of injection will be late and the cessation of injection early if the injection pressure is too high. If the injection pressure is too low, the commencement of injection will be early and the cessation late. Improper timing of fuel injection will impair the engine's fuel efficiency and its ability to carry full load. The injection timing will be late if there is excessive wear in the timing gears the fuel pump cam(s), or the fuel pump tappet or if the fuel pump hold-down bolts are loose or the fuel pump coupling is loose. The engine will not develop full power if the timing is late.

When the fuel injection timing on an in-line engine is badly out of phase with the rest of the engine, the engine will probably not start. On a V-engine it is possible in some cases for one cylinder bank to be in time with the engine but the other bank to be out of time. In that event the engine will probably start and run without load, but very roughly. There would be some possibility of fuel washing of the liners with the worst timing.

14.9 TIMING AND DRIVE GEAR APPEARANCES

Timing gear teeth that have run for some time in proper mesh with an adequate supply of lubrication and while carrying their designed load will have a smooth polished appearance. No heat discoloration marks will be visible. (Fig. 14.2). Ra-

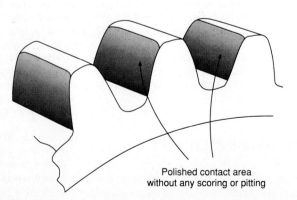

FIGURE 14.2 Gear wheel in good condition.

dial scratch marks on all of the gear teeth will suggest an abrasive action by particles in the lubricating oil over a prolonged period.

Pitting, often observed about halfway up the height and right across the gear teeth, is indicative of inadequate lubrication (Fig. 14.3). Overload or slight misalignment of the driven gear(s) may also contribute to this defect. Spalling and flaking, not to be confused with minor pitting, occur in conditions of severe misalignment or overload (Fig. 14-4).

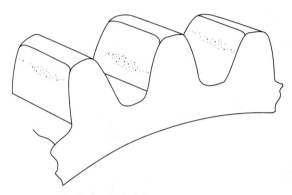

FIGURE 14.3 Gear wheel pitting.

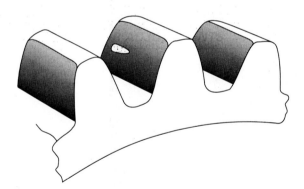

FIGURE 14.4 Gear wheel spalling and flaking.

Point overloading causes the case to be hardened or heat-treated surfaces to crack and metal flakes to detach from the main body of the gear teeth. As soon as it is apparent that such metal loss is taking place, immediate remedial action will be necessary. Spalling or flaking will not necessarily be observed on all the teeth of a gear wheel in distress. Metallic debris may be picked up in the lubricating oil filter elements, and it is possible that the lubricating oil sample will show high iron or nickel readings and provide the first clue that something is amiss.

Scuffing or undercutting will take place when matching gear wheels are improperly meshed. There may be too much or too little clearance between the teeth of the driven and the driver gears, or there may have been a gross lack of lubrication. This condition is highlighted by significant losses of metal in a fairly even pattern from the contacting surface of both driver and driven gears. The tooth profile will be distorted on one side of each tooth, and the metal loss will be evident in the lubricating oil report sample. (Fig. 14.5).

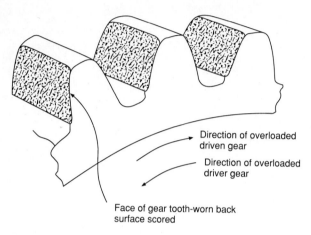

FIGURE 14.5 Gear wheel scuffing and undercutting.

Cracking of gear teeth occurs through the imposition of extreme overloads, as when a foreign object passes between the gears. Such an event is quite extraordinary, but if at any time a broken or cracked tooth is discovered, the entire gear train must be subjected to a detailed search for associated damage. Tooth cracking usually occurs at the root, and it is most visible when the sides of the gear wheel are examined (Fig. 14.6).

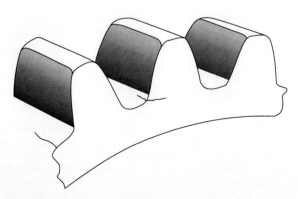

FIGURE 14.6 Gear wheel cracking.

When an engine is assembled at the factory, marks are applied to the timing gears so that the gears can subsequently be dismantled and reassembled without any retiming difficulties. The marks must never be changed or obliterated, and if any new gears are installed, it will be necessary to see to it that the timing has been correctly reset before the engine is started.

14.10 TIMING CHAIN ADJUSTMENT

Few engines that are built today utilize a chain drive for the camshaft drive and timing, but many generator set engines that are chain-driven and operate at speeds of up to 900 rpm are still in service. Timing chains cause few difficulties, but noisy running can be expected if insufficient tension is applied to the chain. Conversely, the chain will stretch unduly if it is overtensioned. All chain tension settings and adjustments must be in strict accordance with the engine manufacturer's specification. The manufacturer will quote a length to which the chain may stretch before it must be renewed. Checking the dimension will be one of the activities within a major overhaul.

CHAPTER 15
FUEL INJECTION EQUIPMENT

15.1 OVERHAUL AND REPAIR

The in-plant overhaul and general repair of fuel injection equipment should be approved only by the owner who can satisfy himself that he has the proper tools and test equipment and the required special skills. The alternative, when *complete* facilities are not available, is to have the work done by qualified specialists. In most cases, that can prove to be the most economical method, and it is certainly the most reliable one.

The fuel injection system of a diesel engine is essentially a hydraulic system which introduces and atomizes fuel into the engine cylinder at a specific point in the engine cycle to produce combustion at an exact moment. The fuel injection system is subjected to severe service conditions. At the same time, the system is required to meter precise and consistent quantities of fuel at injection pressures, which can approach 15,000 psi, through nozzle spray holes that in some cases are no larger than 0.010 in (0.025 mm) in diameter.

The nozzle tip, located in the combustion chamber, must operate efficiently and withstand a wide range of temperatures possibly reaching over 800°C (1472°F). Special materials and precise methods of manufacture are essential to meet these conditions. Limits and fits on fuel injection equipment critical components involve measurements in millionths of an inch; they are the reasons why fuel injection equipment requires special resources for development, manufacture, and service.

Single-cylinder pump maintenance can be comparatively simple and rewarding if it is confined to (1) detail aspects of unit replacement, (2) maintenance of control system and linkage, (3) cleaning and protection of exposed parts (e.g., control rods and pointers), and (4) prevention of random damage by careless handling. Particular emphasis is placed on the need to observe manufacturer's recommended torque figures for delivery valve holders and fuel pump hold-down bolts, and so on. Nevertheless, replacement of an element, control sleeve, or control rod can disturb the balance between cylinders on an engine, and it is therefore generally advantageous to have any job requiring recalibration performed by specialists.

Critical components (e.g., elements and delivery valves) can rarely be field-serviced, and any work performed to improve performance will invariably destroy the design characteristics of the component. Nozzle-cleaning kits are available for all sizes of nozzle, and they contain the items needed to clean a nozzle adequately. More detailed service techniques to recondition nozzle seats can be applied in a proper diesel injection shop, where special grinding equipment is available to form the seat laps and needle cones and correct the nozzle needle lift.

Failure to control needle lift will shorten overall nozzle life. It is, therefore, an essential part of procedure involving removal of metal from either needle or nozzle seat. The plant mechanics are not expected to perform this type of work. (Fig. 15.1).

15.2 INJECTION EQUIPMENT SERVICING

The area where service work on fuel injection equipment takes place must be used for that purpose alone. It must be kept clinically clean and well ventilated. The use of common diesel fuel as a test medium in fuel injection equipment poses a fire hazard and a health risk. All injectors should be set up and proved with regular nonflammable injector test fluid.

Reassembly of injection equipment should take place dripping wet. The parts should *never* be dried before assembly. The reassembled and tested equipment, be it injectors, pumps, or fuel pipes, should be protected by polyethylene caps placed in all orifices open to the atmosphere. If not enough polyethylene caps are available, there is no objection to completely wrapping the unit in polyethylene sheet or even using aluminum cooking foil. Anything that keeps dirt out is good.

15.3 USE OF INJECTOR TEST EQUIPMENT

The use of diesel fuel for testing injector nozzles is not recommended. Safer and better procedures involve the use of specially formulated test fluids.

It is not possible to have 100 percent of all tested nozzles without drip immediately before the opening pressure is attained. However, there should be no sign of leakage over 10 s at 10 kg/cm^2 (150 psi) below the normal nozzle opening pressure. Because of the wide variety in nozzle designs, the discharge of some nozzles may not have the expected sharp chattering sound that characterizes most nozzles. In such cases a check should be made against a new, unused injector. (For example, Pintle-type nozzles frequently make a low-pitched grunting sound.)

The spray pattern developed by any nozzle should be uniform and evenly spaced and without a visible stream of fuel from any one hole. Be aware that a few nozzle models are specially manufactured to inject in an offset pattern.

When an injector has been removed from the cylinder head of a heavily carbonized engine, there may be some difficulty in extracting the nozzle from the nut because of the carbon buildup. Before any attempt to strip the injector is made, and before the nozzle is removed from the nut, the nozzle exterior must be cleaned of as much carbon as possible. The nozzle must be handled carefully. To remove the nozzle from its nut, a bridge-type tool that will straddle the nozzle hole area must be used. Be sure to pull the needle out first. Never hit the nozzle directly with a hammer or any other tool, nor must the nozzle be cleaned with a steel wire brush.

15.4 INJECTOR CHANGEOUT

Spare injectors are normally stored in purpose-built racks in the vicinity of the injector test bench. Sometimes there are two racks: one for injectors awaiting attention and the other for the injectors that are ready for use.

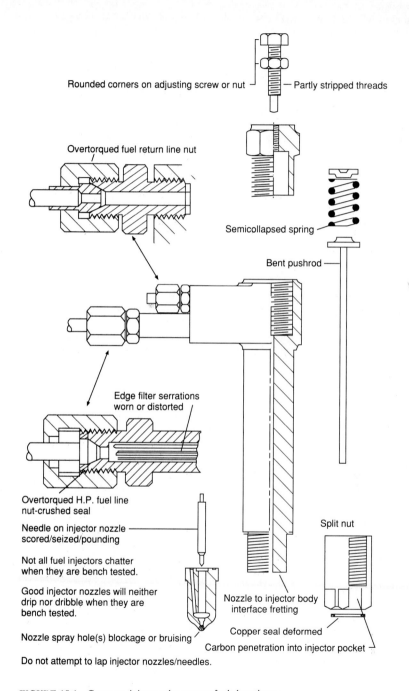

FIGURE 15.1 Common injector damage or fault locations.

In many diesel plants, the operator is expected to change out the odd injector if one injector fails during his shift. In such cases, the operator must select an injector from the ready-to-use rack and thoroughly inspect it before inserting it in the engine. He may wish to test the injector first. In that case, he must be aware of the correct test procedure and routine, and he must also be familiar with the nozzle chatter characteristics of the particular model of injector.

The injector must be inspected externally before insertion. The threads of the high-pressure pipe connection and the drain connection must be in good condition; the nozzle tip must not be visibly damaged in any way; and the whole body must be clean. In a well-organized diesel plant, the ready-to-use injectors are fitted with plastic caps over the fuel and drain connectors. The caps are removed immediately before the fuel pipes are connected. The used caps are then returned to the injector bench, where they can be reused after washing. The caps not only stop dirt from entering the injector body but also protect the threads.

After the defective injector is removed and before the new injector is installed, the copper seal ring must be removed from the bottom of the injector pocket. Be sure that not more than one ring is removed. One of the main causes of injector removal difficulties is that old seal rings remain in the injector pocket and deform to the point at which they trap the injector nozzle. Alternatively, they fail to seal properly and permit carbon formation around the nozzle nut. Ensure that the new copper seal rings used on the nozzles are annealed; hard-copper rings will not provide a good gas-tight joint. On certain engines the height of the injector nozzle spray holes varied by extra copper washers can radically affect the engine's performance.

Having removed the suspect injector from the engine and before installing the replacement, test on the injection bench the injector that was removed to see if, in fact, it was at fault. In most instances the diagnosis and remedy will be correct, but the little extra effort is worthwhile and may save some trouble later.

15.5 INJECTOR NOZZLE SIMILARITY

There is often a great similarity in the external dimensions of injector nozzles used in different engines. It may happen that the superficial appearances of the injector nozzles for two or more engines in the same plant or organization may be the same; therefore, caution must be used in the selection and insertion of replacement nozzles.

Although the shapes of two nozzles may appear to be the same, there may well be differences in the spray hole angle, number of holes, diameter of the holes, configuration of the needle, and other variables. The part numbers will differ for each variant. The only possible way to ensure that the correct injector nozzle is used is to visually check the part number that is stamped on each and every injector nozzle and compare it with the related engine's parts book. In many cases, alternative nozzles made by one manufacturer may be cross-referenced and interchangeable with another (e.g., Bosch 0434-250-027 is interchangeable with CAV 5643014).

The use of an incorrect nozzle will cause poor combustion, black smoke, exhaust temperature excursions, burned valves, burned pistons, fuel dilution of the lubricating oil, poor turbocharger performance, knocking, loss of power, rough running, and a multitude of other disabilities.

15.6 NOZZLE CHATTER

The ability of a nozzle to chatter is often open to controversy, and it is a subject on which it is difficult to generalize.

A nozzle which chatters on the test bench may, contrary to popular opinion, be unsatisfactory in terms of engine performance. The pressure setting may be incorrect, the operating temperatures of the engine, which are quite different from those of the test bench, may affect the injector performance, and the injector nozzle holes may be eroded. A correctly set up injector will have a chatter or stutter akin to that of all of the others in the same engine.

There are some injector nozzles that operate in a lower injection pressure range, or have a pintle configuration which will atomize the fuel with a much-reduced chatter effect.

In all cases where there is doubt about the nozzle chatter, check the nozzle number, the injection pressure, and the nature of the test fluid being employed.

Nozzle chatter depends on many factors, some of which are nozzle pressure flow characteristics, nozzle seat geometry, accuracy of differential angle, and the fit of the needle guide diameter in the nozzle body. Furthermore, good and bad nozzles can be made to exhibit various degrees of chatter by variations in test pump operation. It is advisable to consult a specialist or read specific instructions before rejecting a particular nozzle for reason of chatter performance.

Injector nozzle opening pressures are relatively simple to adjust, but they should not differ from the maker's specifications. Extreme opening pressures can result in spring spindle buckling or nozzle spring failure. Similar failures can be caused by using nozzles with excessive needle lift. On the other hand, if opening pressures are allowed to drop or are set too low, cylinder gas may blow back into the injector and build up heavy carbon and lacquer deposits through the HP pipes back to the delivery valve of the fuel pump.

15.7 INJECTOR TESTING FLUIDS

There is some danger in the use of common diesel fuel for bench-testing fuel injector assemblies. There is a risk of fire and explosion resulting from accidental ignition of the atomized fuel particles, and inhaling the oil-laden atmosphere can be detrimental to health. Personnel who are testing fuel injection equipment must be aware of the hazards of skin penetration by high-velocity injector spray. *Never place a hand in the way of the nozzle spray pattern when qualifying an injector.*

Recommended test fluids will be similar to Shell Calibration Fluid B or C, Imperial Mentor 29, or Wakefield Dick Injector Test Oil No. 3. For Caterpillar engine injection equipment, SAE J967 Calibration Fluid is recommended.

15.8 NOZZLE OVERHEATING

Overheated nozzles can suffer hardness relaxation and subsequent failure. Overheating will also lead to carbon deposits in the needle guide. In either case, needle seizure will occur. Nozzle overheating happens when injectors are incorrectly fitted or the cylinder is overloaded.

Injector nozzles observed to have a blue color indicative of overheating may have been fed with gasoline. This condition is quite rare. It is unlikely to be encountered on stationary units, but it may be found in diesel engines that are refueled irregularly or for which the fuel is delivered by casually employed tanker trucks from multiple sources.

15.9 COOLED INJECTOR NOZZLES

In engines burning heavy fuels it is necessary to provide cooling for the injector nozzles. If the injector nozzle is allowed to get too hot, carbon deposition on the nozzle tip will lead to blockage of the spray holes and disfigurement of the spray pattern. On the other hand, if the nozzle is overcooled, corrosion of the nozzle surfaces exposed to the combustion will take place. The corrosion will be brought about by the action of the exhaust gas and its sulfur content contacting the nozzle that has cooled to the exhaust gas dew point temperature. It is, therefore, most necessary to monitor and control the injector nozzle coolant temperature, particularly at startup and shutdown and during periods of fluctuating loads. If the nozzle corrosion is uncontrolled, there is a great risk the nozzle tip will weaken and will detach from the nozzle itself. The result will be costly damage to the turbocharger.

15.10 COLD CORROSION OF INJECTOR NOZZLES

Cold corrosion occurs when injector nozzles are overcooled as a result of incorrect engine operating temperatures. A combination of water condensation from the exhaust gases, carbon deposits, and sulfur in the fuel gives rise to sulfuric acid attack on the nozzle. Equipment will deteriorate beyond the point of reclamation unless the operating temperature defects are recognized and corrected.

In extreme cases of cold corrosion, the nozzle tip will spall to the point at which the wall between the combustion chamber and the nozzle cooling passage is penetrated. Combustion gases will then pass into the cooling spaces with resultant overpressurization. At that stage, it may be assumed that all of the nozzles presently in the engine will be suffering from the effects of overcooling. It will be necessary to examine all the nozzles before severe damage to the exhaust valves and turbocharger blading is incurred.

Overcooling has no benefit to the operation of the engine. On a 5.0-MW engine operating on heavy fuel the cost of overcooling and the resultant injector nozzle damage can well cost $2 per operating hour (1989 figures) for parts and labor even when there is no damage to the turbocharger and without taking the unplanned downtime into account.

15.11 WATER IN FUEL

Should any water pass through the fuel system as far as the injectors, it will cause momentary, intermittent, or gross loss of power. It has been said that the heat present in an engine's combustion chamber will cause water in the injection nozzle to flash into steam and blow the tip off the injector nozzle. That may be somewhat arguable, but there is no doubt that the presence of water in a fuel injector nozzle or fuel injection pump will have ruinous consequences.

15.12 FUEL INJECTION PIPES

The installation and care of fuel injection pipes is of great importance from both an engine performance and a plant safety point of view. All modern diesel engine fuel injection pipes are preformed, and their original profile must be maintained. Deformed pipes can offer resistance or obstruction to the flow of fuel and thereby affect engine tune (Fig. 15.2).

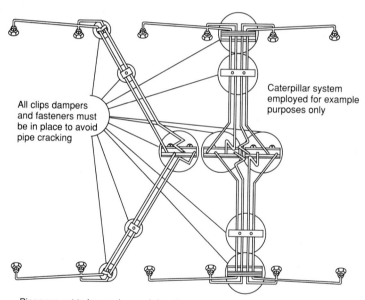

FIGURE 15.2   H.P. fuel pipes.

When removing either fuel pumps or injectors for whatever reason, the entire connecting fuel pipe assembly must be removed. Under no circumstance must the pipe be bent or twisted aside and left hanging. After removal, the pipe must be carefully examined for evidence of wear at either end where the pipe nuts or pipe sealing rings bear on the pipe. Also examine the areas where the pipe clamp or damper contacts the pipe. If there is any sign of metal loss in these areas, the pipe must be discarded. No high-pressure fuel injection pipe that is worn should be brazed, soldered, or welded. Injector pipes are not to be polished with emery paper or other abrasive material.

Fuel pipes may be deformed by mechanical damage or by overtightening, particularly with the swaged end type of connection. Special care is also required when remaking Ermeto-type high-pressure pipe connections. Heavy wrench loads are *not* required to effect an efficient joint. It is essential that the pipe and Ermeto fitting be correctly aligned with the union. The union nut should be tightened and secured firmly; and only a further one-third to two-thirds turn with a wrench is required to effect an efficient joint. Overtightening invariably leads to pipe deformation and possible fracture.

Replacement pipes must always conform strictly to the maker's recommendations regarding material, fittings, and dimensions. Any deviation from specification can result in poor combustion and reduced service life. Pipe clamps and dampers must not be discarded or left slack after servicing if premature pipe fracture and catastrophic fire hazards are to be avoided.

When a fuel pipe seal ring leaks, the pipe nuts can be retorqued, but the manufacturer's torque specifications must not be exceeded. When the nuts are tightened, the pipe seal ring will crush a certain amount. If it does not seal by the time the recommended torque is achieved, the pipe assembly must be replaced, because further tightening will only inflict damage on the fuel pump or injector coupling threads.

Rapid cavitation damage in the fuel injection system can be the result of excessive injection pressures brought about by deformed injection pipes or blocked nozzles. Injector leakoff pipes must be kept clear and undamaged. A blocked or deformed pipe can create excessive pressures in the fuel injection system by imposing a hydraulic lock above the needle, thus supplementing the nozzle spring force. Any abnormal pressures will promote leakage at joints or possibly fracture part of the system.

Over a period of time, considerable damage can be done to the high-pressure (HP) fuel pipes if they are habitually used as grab handles. It is not unusual for operating personnel to use the fuel pipes as handles when they want to climb up on the engine to inspect the engine top, adjust the governor, or do something of similar nature. If this sort of damage is to be avoided and routine inspections are to be facilitated, purpose-built platform or steps should be provided.

Certain manufacturers made HP fuel pipes that have identical external appearances but varying bores. All of the HP fuel pipes on an engine must have identical bores. Precautions must be taken to see that any replacement HP fuel pipes are the correct bore. Bore size can be gauged with suitably sized twist drills.

15.13 FUEL INJECTOR CONNECTION

When more than one pipe connecting the fuel injection pump to the injector is removed, care must be taken that the reconnections are made correctly. It is possible, particularly on V-engines, to inadvertently cross-connect an injector to the wrong fuel pump. Upon starting, one or more of the cylinders will not fire because the injection timing will not correspond to the valve timing. The engine will often apparently run smoothly, but upon application of load, seizure of the misconnected cylinder(s) will occur. The seizure will take place because the unburned fuel in the cylinder will wash away any lubricating oil on the cylinder liner.

15.14 ENGINE TIMING

When any work is done on the timing gear train, the timing chain, the camshaft, one or more fuel pumps, or the fuel pump drive and its couplings, the engine fuel timing must be checked before any attempt is made to restart the engine. Gross mistiming errors are probably more easily made than minor timing errors, but the results of a gross error can be catastrophic. The person who works on the fuel system or its timing is failing in his duty if he neglects to thoroughly check the fuel injection timing at the end of the work exercise.

15.15 PURPOSE OF FUEL INJECTION PUMP

The fuel injection pump shown in Fig. 15.3 is a typical constant-stroke plunger type, and it is operated by a motion transmitted by the camshaft. The fuel injection pump has a triple purpose:

1. To elevate the pressure of the fuel oil to a value that will allow the most efficient atomization of the fuel leaving the injector nozzle to develop
2. To supply the correct quantity of fuel to the injector(s) commensurate with the load as sensed by the governor
3. To initiate the injection of the fuel at a precisely designed point in the engine's cycle and within a specific sector of camshaft rotation

15.16 INJECTION TIMING

The accurate timing and period of fuel injection are of critical importance to the engine's fuel consumption, load-carrying ability, and overall efficiency. In an engine turning at 1800 rpm the fuel pump and injector are required to pass fuel to the cylinder 15 times every second, and the period of each injection is 0.0007 s long. The injection pressure may be anything up to 10,000 psi, but it is exactly specified for each model of engine.

The correct timing and period of the fuel injection may be affected by several factors. Timing of injection will be late if the injection pressure is set too high. The commencement of injection will be late; the cessation of injection will be early; and the engine will lack power because insufficient fuel will be injected when the rack is in the full-fuel position. On the other hand, if the fuel injection pressure is too low, the commencement of injection will be early and cessation of injection will be late. It is probable that the engine will make black smoke because it is exhausting amounts of unburned fuel.

15.17 FUEL INJECTION PUMP HEAT

Fuel injection pumps generate a certain amount of heat in their operation. Occasionally, a pump will become excessively hot, and this condition will usually be

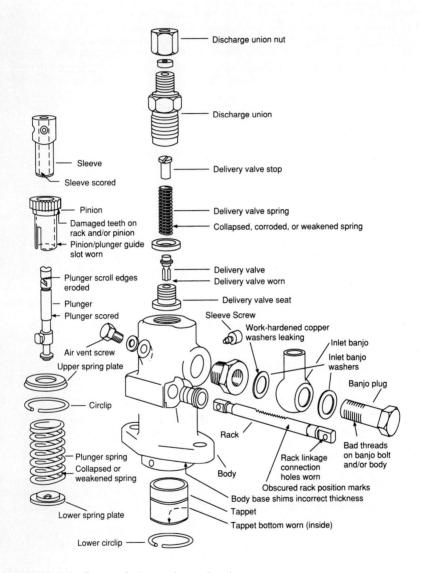

FIGURE 15.3 Common fuel pump damage locations.

caused by the failure of the delivery valve. The exhaust temperature of the related cylinder will be observed to be below normal. A broken delivery valve spring may be at the root of the trouble, but delivery valve spring corrosion, attributable to water in the fuel, is a possibility.

Water in the fuel will eventually corrode various parts of the fuel injection system. An early indication of water contamination is discoloration of the pump plunger and barrel bore; the surfaces assume a machined appearance. If the pump elements are seen to be in that condition, immediate remedial action must be taken to prevent further water entry. Inspection of components by qualified and trained journeymen specialists at scheduled intervals will greatly reduce injection equipment failure or damage and avoid future problems.

15.18 FUEL PUMP SEIZURE

The fuel pump will seize through overheating, lack of lubrication, dirt passing the filters, water in the fuel, or corrosion. In normal circumstances, the fuel passing through the fuel injector pump provides sufficient lubrication between the plunger and the barrel. If the fuel quality is changed, perhaps because of severely cold weather, there may be insufficient lubrication in the fuel. That will lead to some fuel pump difficulties. It can be said that fuels which retain their pourability at subzero temperatures forfeit some of their lubrication properties. It may be necessary to add approximately 1 percent clean lubricating oil to the fuel at the daily service tank when low-temperature fuel is being used (Fig. 15.3).

In warm weather, any water supplied with the bulk fuel will usually precipitate quite quickly in the bulk storage tank, but if water reaches the daily service tank, problems can be expected at the fuel pump. The fuel pump plungers and barrels will suffer some scoring, and there will be evidence of corrosion when particles of water stand in the fuel pump of an idle engine.

On some engines using lighter grades of residual fuel, starting difficulties together with fuel pump problems will be experienced if the changeover from the light grade starting and stopping fuel to the running heavy-grade fuel is mishandled. If an engine is still hot after a run, it may be possible to make an early restart on the heavy fuel. This is very poor operating practice, and it must not be allowed to take place except under the most dire emergency conditions. Pump seizure and scoring are almost certain.

Dirt will enter the fuel pump only by bypassing overloaded or distorted fuel filter elements or through the assembly of fuel injection equipment and connecting pipework that is less than 100 percent clinically clean. Poor or defective fuel filtration will lead to nozzle blockage, which in turn can induce heavy cavitation of the fuel pump plunger. Poor filtration will also be indicated by extensively abraded areas developing on the plunger surfaces.

15.19 STANDBY UNIT FUEL PRIMING

Infrequently used engines such as those employed in standby duties will not start readily if their fuel systems are not fully primed. Much battery power or starting air can be wasted in attempting to start an unprimed engine. Electric starter burnout is not uncommon in these circumstances. If an engine is not thoroughly primed before a start attempt, it will run roughly after starting. Perhaps the different noise character will disguise or camouflage other noises warning of trouble.

15.20 PRIMARY FUEL PUMPS

Primary fuel pumps that are directly engine driven are sometimes known as low-pressure or fuel transfer pumps.[1] Many primary fuel pumps are fitted with shear pins to protect the engine drive system. Primary fuel pumps are of the positive-displacement gear type, and the pumps are frequently located in such a position that an accumulation of solids or a slug of water can form either in the pump itself or in the pressure relief valve. In cold weather the water becomes ice and the ice or dirt accumulation will prevent the pressure relief valve from operating. In such conditions, upon starting the engine, the pump will be overloaded, the shear pins will separate, and no fuel will be delivered to the fuel injection pump. On pumps not fitted with shear pins, the pump shaft will probably break or there will be damage to the engine drive system.

Primary fuel pumps direct-driven by the engine timing gear train are often a cause of fuel leaking into the engine lubricating oil. The actual leak through the fuel pump drive shaft gland is quite difficult to detect with any great certainty, and changeout will be necessary if a leak in this area is suspected. Gland failure may often be attributable to misalignment or substandard bearings in the pump or drive system.

---

[1]The term *fuel transfer pump* should only be applied to the pump used to move fuel from the bulk storage tank to the day service tank.

CHAPTER 16
TURBOCHARGER

16.1 PURPOSE OF A TURBOCHARGER

The purpose of a turbocharger is to deliver air at above atmospheric pressure to an engine and thereby permit the combustion of a larger amount of fuel with a consequent increase in power output. In broader terms, it can be said that a diesel engine fitted with a turbocharger will be capable of producing about 50 percent more power than a normally aspirated engine of the same basic configuration.

The turbocharger accomplishes the increase in power through the utilization of energy extracted from the engine's exhaust gases (Fig. 16.1). The exhaust gas flow turns a turbine wheel which in turn drives a centrifugal air compressor. The pressurized air is then fed to the cylinders. Being pressurized, a greater weight of air enters the cylinder on each stroke, and that allows a larger weight of fuel to be burned. The power output of the engine is thereby increased.

Normal wear of a turbocharger is confined to the bearings. Turbocharger bearing life is quite short in terms of operating hours owing to the high rotational speeds and the temperatures imposed upon any exhaust gas turbine rotor. No turbocharger bearings should remain in service for more than 8000 h. Effective lubrication of turbocharger bearings is vital, and it is safe to say that most turbocharger failures have their origin in lubrication defects. It is, therefore, essential that turbocharger lubrication systems and lubricating oil levels be regularly checked by the responsible personnel. The operator must also be aware and take heed of the lubrication requirements of turbochargers on an engine that is being slowed down to a stop.

16.2 TURBOCHARGER RUN-DOWN

Many turbochargers that receive their lubrication from the main engine lubrication system suffer damage through the unloading and stopping of the engine too hastily. When the engine comes to rest, the lubricating oil flow ceases but a turbocharger that has been spinning at anything between 20,000 and 100,000 rpm takes a little time to slow down and stop. During its rundown, the bearings must still be lubricated, but they will get no more oil after the engine has stopped. The operator must therefore take the time to reduce first the load and then the engine speed sufficiently slowly to allow the turbocharger to come to a standstill before

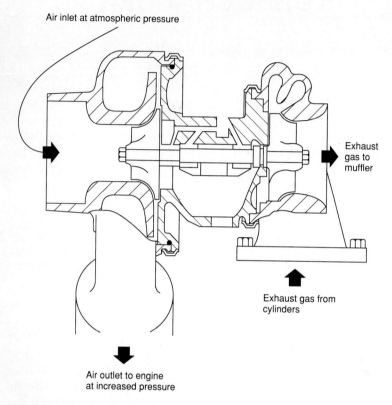

FIGURE 16.1 Typical turbocharger.

or at the same time as the engine. He must be aware that the stopping practices and habits that apply to a large engine are not necessarily the best for a smaller engine, and vice versa.

16.3 TURBOCHARGER LUBRICATION

Many turbocharger breakdowns are attributable to failures of the turbocharger lubrication system, and all apparent lubrication leaks should be reported as and when they are observed (Fig. 16.2). If the leak is sudden or severe, the engine should be stopped forthwith. It is possible for the oil seals on the turbocharger shaft to fail and allow quantities of oil to flow out of the bearings into either the compressor or turbine cases without being seen. Because of the drop in lubrication pressure, the bearings will be endangered. The only warning of this condition will be increased lubricating oil consumption. Loss of oil pressure at the turbocharger bearings will probably not show on the engine oil pressure gauge. Again,

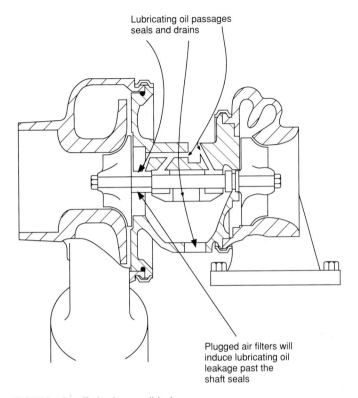

FIGURE 16.2 Turbocharger oil leakage.

it will be necessary for the operator to constantly compare lubricating oil consumption quantities from one shift to the next.

Some turbochargers have their own lubricating oil systems with individual sumps, pumps, dipsticks, and sight glasses. During each inspection tour, the operator is advised to look very closely at the sight glass and satisfy himself that the oil is in fact fluid. If oil completely covers the sight glass, there is a possibility that the turbocharger shaft seals are leaking and the lubricating oil is being severely aerated or is foaming.

16.4 TURBOCHARGER BALANCE

Any turbocharger must be properly maintained and cared for if it is to produce a sustained optimum performance. The internal parts are extremely sensitive; they are, of necessity, finely finished to very close tolerances. Because of the ultrahigh rotational speeds (40,000 to 100,000 rpm), the balance of the rotating assembly is critical. Imbalance of the rotor can be caused by wear, distortion, dirt buildup, blade cracking, or foreign object intrusion. If the situation is allowed to develop, destruction of bearings, seals, blading, and the rest of the rotating assembly may occur. The beginnings of imbalance can be detected by use of a vibration meter.

Plant personnel should not attempt to clean a turbocharger rotor assembly. Chipping, scraping, or some other manual method of removing caked-on carbon deposits will only give imperfect results and lead to an imbalance of the rotor. The only satisfactory method of cleaning the exhaust turbine or the compressor wheel is by blasting them with glass beads followed by dynamic balancing.

Deficient performance of turbocharger rotors may also occur when the engine is not properly tuned. In that event, uneven exhaust pulses may cause rotor flutter. Abnormally high exhaust temperatures coupled with rapid and intermittent overload may cause thermal shock damage or high-temperature cracking of the turbine blades. Incorrect tuning of the engine may also contribute to the buildup of excessive amounts of carbon on the blading, which will cause imbalance and interfere with the rotational speed of the rotor. The common practice of setting fuel pumps by observing the color of the exhaust smoke is wrong; it must not, under any circumstance, be carried out on a turbocharged engine. The fuel pump rack can be set only by following the instructions and procedures in the engine manufacturer's manual.

16.5 TURBOCHARGER DEFECTS

Turbochargers are not complicated mechanisms, but they do need careful upkeep. Particular attention must be given to their bearings and lubrication. They work under severe conditions with rotor speeds of up to 100,000 rpm and exhaust gas temperatures reaching 540°C (1000°F).

When there is a turbocharger failure on a V-form engine, it is imperative that the engine be stopped at once even though the unit appears to be carrying the load successfully. It is highly probable that the bank of the engine still being blown by the good turbocharger will be overloaded while the cylinders of the other bank suffer from fuel wash that is due to the incorrect fuel/air ratio. If the situation is allowed to continue, early piston seizure can be expected.

16.6 TURBOCHARGER OVERHEATING

All turbochargers become very hot when working. The casing on the turbine side will heat up to temperatures approximating those of the exhaust gases [500°C (930°F)]. On the compressor end, temperatures of up to 175°C (350°F) can be expected. In view of the temperatures likely to prevail, the operator or any other diesel plant personnel inspecting an operating turbocharger will do so cautiously. After an engine has stopped, the inspecting person must be aware that the turbocharger takes some time to cool down.

When overheating of a turbocharger is recognized, steps must be taken first to normalize the situation and then to eliminate the cause of the problem (Fig. 16.3). If the situation cannot be brought under control, the engine must be stopped before total failure of the turbocharger occurs. Continued running of an overheated

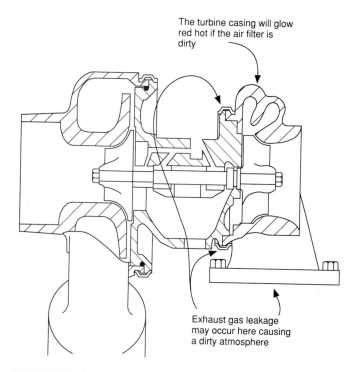

FIGURE 16.3 Turbocharger overheating.

turbocharger will result in damage that will be costly to repair. The air filter elements must be examined first. Do not attempt removal of the element while the engine is running or run the engine without filter elements properly in place (Chap. 19, "Filters".)

If the air filter is plugged, it is possible for lubricating oil to be either pulled or pushed past the turbocharger rotor shaft seals and result in smoke, carbon, or lubricating oil starvation. Permanent damage to the turbocharger shaft seals can be incurred through the strains imposed on the seals by a dirty air filter element.

The air intake filters should be visually checked daily for evidence of dirt buildup, and they should also be examined for any signs of warp or inward collapse, which is indicative of almost total element plugging. In such cases, turbocharger bearing failure will follow if the air filter elements are not changed out.

An overheated turbocharger will identify itself by radiating an unusual amount of heat, by possibly glowing a dull red color, and by instrument indication when so equipped. Overheating will lead to cracking of the turbine wheel or its blading; some of the blade tips may deform or be lost altogether. The housing also may crack. The lubrication system may become carbonized and fail.

Persistent and excessively high exhaust gas temperatures will eventually damage a turbocharger. The high temperatures have an effect on the turbocharger lubricating oil, which tends to coke. Carbon fouling of the turbine wheel can be expected. Elevated exhaust temperatures will contribute to erosion or pitting of the turbine housing. Abnormally high exhaust gas temperature can be caused by incorrect fuel pump rack settings, dirty air filter elements, dirty aftercooler cores, leaking exhaust valves or cracked exhaust valve seats, slow rotation of turbocharger, overload, or impeded exhaust gas flow (Fig. 16.3).

16.7 JACKETED TURBOCHARGER

Turbochargers generally rely on surface cooling in the smaller sizes, but liquid cooling is commonly used on turbochargers installed on engines of 1 MW or greater capacity. Liquid-cooled turbochargers must be protected from freezing in a manner similar to the engine jacket. Without adequate cooling and frost protection, there is a danger of cracking the coolant casing.

16.8 TURBOCHARGER NOISE

Sometimes turbochargers will emit a high-pitched scream or whistle that may cause concern. Before stopping the turbocharger to inspect the bearings or the rotor, check that all of the inlet and outlet connections are tight. A small leak can

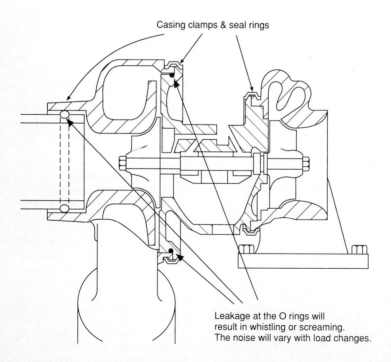

FIGURE 16.4 Turbocharger noise.

cause a great deal of noise. A gasket or rubber boot connection can leak because of loose bolts or insufficiently tightened gear clamps.

Careful examination of the expansion connections in the turbocharger system is necessary at intervals so that any perforations can be detected. These holes, splits, tears, or cracks can permit much dirt to bypass the air filters. Air passing through any of these anomalous apertures will add to dirt buildup in the turbocharger and cause excessive cylinder wear (Fig. 16.4).

16.9 CARBON DEPOSITS IN TURBOCHARGERS

All turbochargers eventually become coated with carbon on the turbine side. That is quite normal, and the deposits can be removed by solvent or glass bead methods when the turbocharger is overhauled. It may happen that inordinate amounts of carbon build up on the turbocharger rotor and casing and reduce turbocharger performance. The source of the carbon can be deduced from the nature of the carbon formation. Soft-carbon deposits are derived from incompletely burned fuel; hard-carbon deposits have their origin in lubricating oil.

Although the deposits will eventually reduce turbocharger performance, the origin of the deposits will not necessarily be in the turbocharger itself. They may be the result of a defect in the fuel system, the cylinder heads and valve gear, or the pistons.

16.10 TURBOCHARGER INLET AND OUTLET CASING POSITIONS

When replacement turbochargers are installed, it often happens that the positions of the compressor or turbine casing have to be adjusted relative to the existing mountings on the engine so they mate with the manifolds properly. It is important, particularly on smaller turbochargers, to apply the correct torque to the casing bolts after the positional alterations have been made. If the bolts are overtorqued, there is a possibility of the casing being distorted, followed by a turbocharger failure shortly after start-up. The correct torque is cited in the OEM service manual.

Before the turbocharger is placed on the engine, the bearings must be prelubricated by flooding the oil inlet and outlet orifices and then spinning the rotor by hand. That will ensure a thorough prelubrication of the turbocharger bearings.

16.11 TURBOCHARGER OUTPUT DEFICIENCY

A performance deficiency in a turbocharger is directly reflected in the performance of the engine. In general terms, the use of a turbocharger in conjunction with increased fuel delivery rates enables the output of an engine to be raised some 50 to 60 percent. It follows that if the turbocharger(s) on a pressure-charged engine fail, the engine may continue to operate at 40 percent of capacity. In extraordinary circumstances, that may be necessary.

If there is a turbocharger failure on an engine that is equipped with more than one turbocharger and it is imperative that the engine continue to be used, all of the turbochargers must be locked up. That will enable the engine to run in a balanced naturally aspirated manner, although at a much reduced capacity.

16.12 TURBOCHARGER INSULATION

Some engines are retrofitted by their owners with insulating blankets that cover the turbocharger more or less completely. This practice has some considerable disadvantages in that the exhaust side of the turbocharger must be ventilated so the turbine side casing can dissipate as much exhaust heat as possible. At the same time, the compressor side of the turbocharger must be as cool as possible when the engine is running on load.

The design of any turbocharger allows for a certain amount of expansion in both sides of the unit; insulation will interfere with the expansion calculations and characteristics. The compressor side casing of the turbocharger must be allowed to shed as much heat as possible so the air arriving at the aftercooler can be cooled more efficiently. When an engine is fitted with turbocharger insulation and is running at full rated load or 10 percent overload, there is a good possibility of the aftercooler being overloaded with a consequent loss of efficiency. In an extended case, there may be damage to the valves and piston crowns through overheating. When an engine is fitted with a turbocharger insulation blanket, it is recommended that the blanket be removed so the engine can operate as originally intended by the designers.

CHAPTER 17
AFTERCOOLER

17.1 PURPOSE OF AFTERCOOLER

The purpose of the aftercooler is to extract unwanted heat from the aspirating air after the air has been compressed by the turbocharger and before it enters the cylinder. Cooling the air also permits a greater weight of air to enter the cylinder. The cooling medium of the aftercooler on a stationary engine is almost always liquid. In some engines the aftercooler is in circuit with the engine jacket cooling water, but on engines of 1 MW or more it usually has its own cooling circuit.

The ability of an aftercooler to perform its function is governed by its state of cleanliness and the temperature and rate of flow of entering coolant. Any foreign material entering the cooler will reduce the cooler's heat transfer capability and thereby inhibit efficient engine operation (Fig. 17.1).

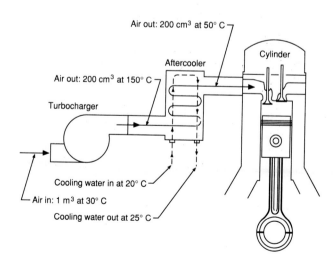

FIGURE 17.1 Typical aftercooler circuit.

17.2 EFFECT OF AMBIENT AIR TEMPERATURES ON AFTERCOOLERS

If the aftercooler becomes dirty on either the air or the coolant side, the aspirating air being passed to the cylinders will be inadequately cooled. The first indication that something is amiss will be when the exhaust temperatures become disproportionately high for the load being carried. The engine will not be able to carry full load without exceeding the maximum permissible exhaust temperatures if an aftercooler is fouled.

Figure 17.1 shows the approximate aspirating air temperatures that can be expected at full load. Under normal conditions, with the thermostatic regulators controlling the coolant flow through the aftercooler, changes in the ambient air temperatures should not affect the final aspirating air temperature at the inlet valve.

In the case of air-to-air aftercoolers (which are found only in automotive applications), changes in ambient air temperatures will have some effect on the aspirating air temperature. It is necessary to pay close attention to the cooling air side of the aftercooler. It must be kept free of debris and oil fumes, either of which will quickly impair the cooler's efficiency.

17.3 FOULING OF AFTERCOOLERS

Both the air side and the coolant side of an aftercooler can get dirty. The sources of air-side contamination are usually a leaking oil seal on the turbocharger rotor shaft, excessive amounts of educted oil from the crankcase, or defective air filters. The sources of coolant-side contamination are oil–ethylene glycol slime following a lubricating oil cooler breakdown, natural scale when using untreated water, and water-borne bacterial growth when using an open-circuit raw water source. Sand, mud, weed, snails, and fish also are found in aftercoolers when their entry is not prevented.

In the latter 1980s the proliferation of the Zebra clam in the Great Lakes of North America began to cause immense underwater pipe fouling problems. A solution to the phenomenon has not yet been found; any power plant drawing water from the Lake system can expect trouble from this infestation sooner or later.

Water-borne bacterial organisms generally do not originate on cooling surfaces when the flow velocities are greater than 2 ft/s. The initial growth starts in stagnant conditions and can be propagated in moving water. Cooling water drawn from a perforated culvert in the bed of a slowly moving river will provide good conditions for initiating bacterial growth.

Fouling of either side of an aftercooler will have a detrimental effect on the efficiency of the engine. In gross cases, it will lead to severe engine damage.

17.4 AFTERCOOLER CLEANING

In its normal condition clean surfaces on both the air and coolant sides of the aftercooler permits an adequate and continuous flow of tempered air to enter the cylinder for scavenging and combustion purposes. The aftercooler has no moving parts and is essentially the same as an ordinary radiator. The problems encountered on aftercoolers are very similar to those found in any common radiator, namely, plugging of either the coolant or air side with dirt or corrosion through the circulation of unconditioned coolant media.

Oil and dust are the principal causes of aftercooler fouling on the air side. Dust will be drawn in past the air filter element seals if the seals are damaged or if the elements are improperly fitted or perforated in any way. If the filter elements are plugged and are overdue for change, there is a good chance that oil from the turbocharger bearings will be sucked past the turbocharger shaft seals in considerable quantities and will devolve onto the aftercooler core. The outward signs of this condition will be dark exhaust smoke, inability to carry full load, an overheated turbocharger (possibly even glowing!), and high exhaust temperatures. The immediate remedy, of course, is to stop the engine and change the air filter element. If the symptoms then moderate, the engine should be monitored for increased lubricating oil consumption, because the turbocharger shaft seals may continue to pass oil. If oil loss through the turbocharger seals continues, the aspirating air temperatures will gradually rise as the aftercooler core becomes coated with more and more oil. The exhaust temperatures on one bank also will rise.

The aftercooler requires cleaning at regular intervals; the coolant side of an aftercooler in a closed circuit loaded with treated coolant should be cleaned at approximately 5000-h intervals. If the aftercooler is served with sea, river, or lake water in an open circuit, cleaning at 3000-h or shorter intervals will probably be needed. The air side of the aftercooler will get dirty only on an incidental basis. If a situation that could possibly lead to air-side fouling develops, inspection and preventive work must be carried out.

As in ordinary radiator cleaning, no screwdrivers, welding rods, putty knives, gasket scrapers, or such tools should be used to poke dirt out of the fins or tubes. The manufacturers of large-engine aftercoolers often provide the correct cleaning brushes and tools. As a general rule, however, wooden rods and slats, together with bristle brushes, a solvent bath, and low-pressure air (not more than 35 psi) are all that is needed to clean any aftercooler. At no time must any object be hammered into or through the tubes or fins to displace dirt. Whatever solvent is used must be verified for compatability with the aftercooler metals.

Cleaning an aftercooler is best effected by removing the cooler from the engine and stripping all casing sections from the core, which is then immersed in a solvent bath. Caution must be exercised in the selection of the cleaning solvent because some solvents contain caustic which will attack any aluminum components. Agitation of the solvent will speed the cleaning process.

After the cooler has been immersed in the solvent bath for some hours, it can be removed and blow-dried by using compressed air at not more than 35 psi (230 kPa) and then checked for cleanliness by sighting through the fins and tubes with a portable light. When the OEM provides one, a cleaning tool kit should be used and its components replaced as they wear out—and certainly immediately after the cleaning work is finished.

Aftercooler cores are constructed from quite thin metals, and they require careful handling during the cleaning process. They must not be lifted by wrapping ropes, chains, or wires around the core, nor must they be jolted or dropped. Aftercoolers seldom leak, but when they do, the engine must not be used until the aftercooler has been repaired or replaced. Leakage can develop through vibration, induced fatigue, or corrosion of the tubes, but corrosion is the most common cause. Having and following regular vibration-monitoring routines will go a long way toward preventing metal fatigue damage of the core, and the control of coolant ph values will prevent corrosion of the tubes taking place.

If some cooler tubes are found to be leaking and it is essential to return the engine to service, not more than 10 percent of the tubes may be temporarily plugged. The plugs are either tapered wood or rubber, and they are frequently available from the OEM. Use of such plugs is a purely temporary expedient pending the installation of a new or exchange cooler core.

CHAPTER 18
GOVERNOR

18.1 PURPOSE OF A GOVERNOR

The purpose of the governor is to regulate the speed of the engine in conformity with the engine's application. In a diesel engine, speed control and regulation can be obtained only by metering the supply of fuel; there is no practical way to control the amount of air inducted into the engine. There are several ways to meter the fuel passed to the individual injector nozzles, but the most common one is semirotation of the injection pump barrels which are controlled by movement of a rack. The rack is operated by the governor output arm.

The basic mechanism of a governor relies on the principle of the centrifugal force exercised by rotating flyweights. In mechanical governors the centrifugal force generated is balanced by the compression of a spring mounted on the shaft about which the flyweights rotate. Changes in engine speed alter the rotative position of the flyweights, which in turn alters the position of the output arm. The same mechanism is applied to hydraulic governors, but the flyweights' rotative positions regulate an oil pump which supplies hydraulic pressure to the output arm.

Probably one of the most frequent causes of real or artificial difficulty experienced by diesel generating plant operators is an ignorance of basic governor functions and governor controls. Before an operator can be put in charge of a generating plant, he must be fully familiar with the statements which follow. They should be read slowly and carefully while paying maximum attention to the terminology and explanations. Adherence to the principles stated will go a long way toward eliminating what may be regarded as habitual but needless difficulties experienced with prime-mover controls in so many generating plants. Successful governing is a major factor in the quality control of the end product, namely, electricity for the consumer.

18.2 GOVERNOR FUNCTIONS

The functions of governors can be placed in four basic classes:

1. *Load-limiting* governors are arranged to limit the load that may be applied to an engine.
2. *Variable-speed* governors will maintain any preset speed of an engine regardless of load.

3. *Speed-limiting* governors limit the minimum and maximum speeds of an engine. The intermediate speeds are controlled by an external, usually manual, means.
4. *Isochronous or constant-speed* governors maintain an engine at one constant speed regardless of load.

Note: The term *engine* includes diesel engine, gas turbine, steam turbine, water turbine, windmill, or any other form of prime mover.

Many governors are designed to have a basic and a subsidiary function. Thus, an isochronous governor may incorporate a load-limiting function. Also, many governors have a ramp feature whereby full operating speed cannot be obtained after start-up until the engine has been fully warmed through at reduced revolutions.

Load-limiting governors will be found installed on compressor engines; variable-speed governors will be found on agricultural tractors or equipment with hydrostatic transmissions; speed-limiting governors are used on trucks; and constant-speed governors are used on generator sets.

18.3 GOVERNOR TERMINOLOGY

Hunt. A rhythmic variation in speed which can be eliminated by blocking the governor manually but which will recur when the engine is returned to governor control.

Isochronous. A term meaning single speed. It is applied to governors fitted to engines that are required to operate without speed variation, that is generator sets.

Jiggle. A high-frequency vibration of the governor fuel rod end (or terminal shaft) and fuel linkage.

Response time. The length of time the governor requires to react to a change in the engine's rotational speed.

Speed droop. A decrease in rotational speed of an engine from a no-load to a full-load condition. Speed droop is usually expressed as a percent of the engine's speed.

Stability. Relates to the competence of the governor to maintain a constant engine speed.

Surge. A rhythmic variation in speed which can be eliminated by blocking governor action manually and which will not recur when the engine is returned to governor control unless speed adjustment is changed on the load changes.

18.4 GOVERNOR OPERATION

The load determines the power factor. In single-engine operation, the power factor will be unity or less. In parallel operation, the power factors of the loaded units should be the same. In a parallel operation using two engines, the lead engine should be the larger. It should be set to zero droop, and the smaller unit will be set with a percentage droop. For practical operating purposes, droop alteration can be accomplished only on governors that are provided an external droop-setting facility.

It is not possible to run engines in parallel satisfactorily when there are no external droop adjustments on the governors. It is possible to run engines in parallel when one engine has external droop adjustment and the other does not, but the engine with the external feature will of necessity be the lead engine and set to zero droop.

18.5 GOVERNOR CONTROLS

The *speed droop control* is provided so the load can be automatically divided and balanced when more than one engine is driving a shaft or when multiple engines are paralleled into a generating system. When operating ac generating units in parallel with other units, set the droop sufficiently (30 to 50 percent on the dial) to prevent interchange of load between the units. Any one unit in the system that has enough capacity has its governor at zero droop, and it will regulate the frequency of the entire system. This unit will then take all the load changes and control the frequency provided its capacity is not exceeded. Adjust the system's frequency with the synchronizer knob on the governor with zero droop. To distribute load between units adjust the synchronizer knobs of the governors with speed droop.

The *load limit control* limits the load that may be applied to the individual engine; in effect, it is a variable stop acting on the fuel pump rack(s). The load limit is proportional to the capacity of the engine, and it has nothing to do with the capacity of the system. Never attempt to move the governor output arm on even a static engine without turning the load limit control to maximum. Return the load limit control to less than 100 percent—preferably to 50 percent or less—before attempting a start. If the load limit control is at maximum at startup, there is a likelihood the engine will go into overspeed.

The *synchronizer control*, otherwise known as the *speed adjusting control*, serves to alter the engine speed when the engine is running alone or to change the load carried by one or more units running in parallel.

The *droop setting* of a generator set running alone and carrying the total load should have the speed droop control set at zero. In that mode the unit will be able to handle load variations without changing speed, and it will thereby maintain the correct voltage and frequency.

Note: System frequency is adjusted by operating the speed adjuster or synchronizer of the governor having zero droop. The distribution of load between units is accomplished by operating the synchronizers of the governors having speed droop.

18.6 GOVERNOR OIL

Subject to overriding OEM instructions, the oil used in a hydramechanical governor should be in the SAE 10 to 50 range, the grade being determined by the operating temperature. For almost all practical purposes, SAE 30 grade is adequate. Whatever oil is used must be clean, and the container used to feed the oil to the governor must be clinically clean. Do not overfill the governor, and log any makeup oil added. The oil level must be inspected at least every 8 h of operation as a standard operating procedure.

18.7 GOVERNOR CONTAMINATION

There are some governors (Woodward PSG, 1361, etc.) that utilize lubricating oil from the engine rather than have an independent reservoir (Woodward UG8, UG40, etc.). Governors utilizing engine oil may exhibit some sluggishness of response if the lubricating oil filters are bypassing or if there is an antifreeze leak into the crankcase.

If a crankcase leak becomes apparent, it will be prudent to remove the governor and have it serviced at the same time the leak rectification is carried out. Flushing of the governor will be insufficient, because the glycol from the coolant will get into the oil in all passages and a varnish-like film will form on pistons, cylinders, and valves within the governor.

18.8 GOVERNOR TROUBLES

The modern hydramechanical governor is an extremely reliable piece of equipment, and it will give many thousands of trouble-free hours of service if properly handled. The observance of a few basic rules will contribute greatly to a maximum reliability.

- The manual attention given to any governor must be limited to the external controls only.
- The governor should be flushed out by trained journeymen at regular intervals.
- The governor should be overhauled at an OEM-accredited shop at not less than 15,000-h intervals.
- The governor must be operated according to the manufacturer's recommendations.
- Never manually force governor controls or the output arm by using any form of wrench or lever.
- Whenever there is a difficulty involving a governor, stand back and think for a minute. Discuss governor problems with others often, and be sure to check that the conclusions reached during the conversation match the statements in the governor manual.

It is recommended that any work related to governor installation, linkage adjustment, or components only be carried out by skilled journeymen and that two persons be present at start-up and during running trials.

During the installation of a gear-driven governor, great care must be taken to ensure that the inherent side loading of a bevel gear drive is minimal. That will be achieved by careful and accurate shimming. If the shimming is done well, the gear system will have a correct amount of backlash and the working depth of the teeth will not be exceeded. Whenever the governor is taken off the engine or changed out, the checking procedure must be carried out because the thickness of the interfacing gasket may vary from that of the original.

It is important that diesel power plant personnel be able to identify governor problems correctly, but the repair of modern governors by plant personnel is not recommended. If the governors are to be effective, they must be tested and proved on proper test equipment. The plant that has governor-testing facilities, governor spare parts, the machine tools necessary for repair operations, and, above all, the skilled personnel to perform governor rehabilitation work is indeed rare. For those reasons alone, it is imperative that all internal governor work be carried out in an authorized dealer's workshop. The owner of a plant usually establishes such a policy. In doing so, he enhances his warranty position.

The greatest number of governor failures have their beginnings in the use of dirty vessels to add makeup oil. Any oil used in a governor must come from a closed container. If a jug or can is used to transfer the oil, it should be washed and dried before it is used. Remember that most governors do not have filters built into the oil circulation system, and any fabric threads, particles of insulation, wind-borne dust, metal cuttings, paint chips, splinters of wood, insect corpses, or other common power-plant-dwelling contaminant entering a governor by way of the makeup oil stays there until the governor develops digestion problems. In the troubleshooting process, the plant personnel should limit their physical activities to external adjustments of the governor only, and that work should be performed by a journeyman mechanic.

Any governor shipped for servicing or repair must be sent out in a close-fitting wooden box. Do not ship governors wrapped in paper or cardboard. The basic governor faults and remedies are listed in Table 18.1.

18.9 TYPICAL GOVERNOR ADJUSTMENTS

Normally, the only necessities when putting a new or overhauled governor into service are bleeding entrapped air and adjusting the compensation to obtain maximum stability. All other operating adjustments are made during the testing in accordance with the prime-mover manufacturer's specifications. They should not require further attention. Do not attempt internal adjustment of the governor.

Compensating Adjustment and Needle Valve

This valve must be correctly adjusted while the governor is controlling the engine even though the compensation may have been previously adjusted at the factory or on governor test equipment. Although the governor may appear to be operating satisfactorily because the unit runs at constant speed without load, the governor may not be correctly adjusted.

High overspeeds, low underspeeds, or slow return to speed after a load change or speed-setting change are some of the results of an incorrect setting of the compensating adjustment and needle valve.

TABLE 18.1 Basic Governor Faults and Remedies

Probable Cause	Correction
Engine hunts or surges	
Compensating needle valve opened too far.	Adjust needle valve while following OEM instructions.
Dirty or foaming oil.	Remove and drain governor; flush with fuel oil; check oil supply for air entrainment; fill with correct oil.
Improper relationship between governor output shaft and engine.	Governor shaft travel to power output should be approximately linear. Readjust or rework linkage to obtain a linear relationship between engine fuel pump and output shaft position.
Excessive backlash or binding in linkage.	Repair linkage.
Excessive backlash or binding in fuel control.	Repair fuel control.
Insufficient utilization of governor output shaft travel.	Readjust or rework linkage to use more output shaft travel.
Negative droop set into governor.	Set correct percentage of positive droop.
Governor internal parts worn or incorrectly adjusted.	Repair governor at dealer's shop.
Jiggle at governor output shaft	
Rough engine drive.	Check drive gear alignment; inspect gear teeth; check gear train for eccentricity or backlash; check timing chain tension; check vibration damper; determine reason for cyclic load variation.
Governor not sitting square on mounting.	Realign and torque to correct value.
Governor response slow during starting	
Low governor oil pressure.	Check governor oil supply system and level: look for foam; verify oil type.
Engine is slow to return to speed following a change in load or is slow to respond to a speed-setting change	
Compensation needle closed is too far, overloaded.	Adjust according to manual instructions; reduce load.
Internal parts of governor sticking.	Remove governor; flush out; reinstall; test and replace if still unsatisfactory.
Low governor oil pressure.	Check governor oil supply system; look for foam; verify oil type.
Fuel supply restricted.	Refer to notes on fuel systems.

TABLE 18.1 Basic Governor Faults and Remedies (*Continued*)

Probable Cause	Correction
No output from governor	
Dirty oil or sticking parts.	Remove governor; flush out; reinstall.
No governor oil pressure.	Check oil supply system and oil level; look for foam; verify oil type; reset engine cranking speed if too low.
Governor drive failed.	Effect repairs within engine.
Governor damaged internally.	Remove governor and ship to dealer for repair.
Linkage binding or maladjusted.	Repair or readjust.
Engine will not carry full rated load	
Fuel rack does not fully open.	Check and adjust linkage; inspect for binding or obstruction.
Low governor oil pressure.	Check the governor oil supply level and system; look for foam; verify oil type.
Voltage regulator at fault.	Readjust; repair as necessary.
Improper load division between paralleled units; one unit on zero droop, all others on droop	
Incorrect speed droop setting on one or more of the droop units.	Increase droop on affected units until load remains steady on each droop unit. System load variation is absorbed by the lead unit with zero droop. Droop units assist in correcting speed variations on large load changes, but return to their original loads after the load change is absorbed by the lead unit.
Different speed settings among the droop units.	Readjust speed setting on improperly set unit.
Improper load between paralleled units, all units on droop	
Incorrect speed droop setting one one or more units.	Adjust droop on each unit until desired division of load is obtained. Increasing droop results in the unit taking a smaller share of load charges, decreasing droop will result in a larger share.
Different speed settings between units.	Readjust so that all speed settings are the same.

18.10 GOVERNOR COMPENSATING ADJUSTMENTS (WOODWARD UG—TYPICAL)

When the engine is started for the first time or after the governor has been drained and cleaned, the governor must be filled with oil. Although the governor may appear to operate satisfactorily because the prime mover runs at constant speed (without load), it may need adjustment. High overspeeds and underspeeds after load changes and slow return to normal speed are the results of incorrect compensating adjustments.

Following start-up and when the temperature of the engine and that of the oil in the governor have reached their normal operating values, compensating adjustments can be made without load on the prime mover. That will ensure that the governor gives optimum control.

1. Loosen the nut holding the compensation-adjusting pointer and set the pointer at its extreme upward position (maximum compensation).
2. Remove the compensation needle valve plug and open (unscrew) the compensating needle valve three to five turns with a screwdriver. Be sure that the screwdriver fits into the shallow slot of the compensating needle valve and not into the deep slot located at right angles to the shallow screwdriver slot. Allow the prime mover to hunt or surge for about ½ min to bleed trapped air from the governor oil passages.
3. Gradually close the needle valve travel until the hunting just stops. Do not go beyond that position. Now check the amount of valve opening by closing its valve completely. Note the number of turns required to close the needle valve. Open the valve to the previously determined opening at which hunting stopped. Test the compensating adjustment by manually disturbing the prime-mover speed. If the engine speed settles out properly and the needle valve is more than one-eighth to one-quarter turn open in a governor that has but one compensating spring or more than one-quarter to one-half turn open in a governor with two compensating springs, the adjustments are satisfactory.
4. If hunting did not stop with the needle valve at minimum openings, raise the compensating pointer two divisions on the scale. Open the needle valve again and allow the prime mover to hunt.
5. Repeat step 3.
6. Repeat steps 3, 4, and 5 until adjustment is satisfactory. Desirable opening of the needle valve is from one-quarter to one-half turn on a governor with one compensating spring and from one-half to three-quarters turn on a governor with two compensating springs.

It is desirable to have as little compensation as possible. Closing the needle valve further than necessary will make the governor slow to return to normal speed after a change. Excessive dashpot plunger travel causes the compensation-adjusting pointer to move toward its maximum position and results in excessive speed upon load change.

18.11 FRICTION CLUTCH ADJUSTMENT

Adjust the friction on the clutch coupling to avoid a change that is due to vibration. Also, tighten the clutch enough to enable the synchronizer motor, if used, to turn the speed adjustment gear. Do not tighten it too much, because that will damage it and restrict the operation of the synchroniser motor. The motor may burn out if it is overloaded. Remove the top cover and, using the right size Truarc pliers, remove the retaining ring, the spring-loaded cover, and the spring from the clutch. Be careful the spring does not jump out and drop down inside the governor. To increase the friction, turn the nut clockwise and vice versa. Check the adjustment with a torque wrench. It is preset at the factory to between 4.5 and 5.5 lb-in on governors with speed-setting motors or 1.5 to 2.5 lb-in on governors with manual speed settings.

18.12 CAUTIONARY NOTE

The adjustments described can be carried out only by a skilled mechanic who is entirely familiar with this sort of work. In any event, the work done on any governor can be satisfactorily performed only with total reference to the OEM manual no matter what make or model of governor is worked on. The preceding notes on compensation adjustment are intended for study purposes only.

CHAPTER 19
FILTERS

19.1 FILTER DESCRIPTION

Pressures imposed upon the engine manufacturers by both the competitors and the purchasers continuously influence the production of engines with higher speeds, increased outputs, lower weights, and reduced fuel consumption. Thus, today's engine is very efficient, highly stressed, and far less forgiving of poor operation practices than the engine of yesteryear.

Any solid material that is permitted to come into contact with a moving part will cause a certain amount of wear. If the amount of solids entering or circulating within the engine can be kept under control, the wear rates and consequent overhaul routines will be forecastable with some accuracy. The life and performance of the unit will be well within predictable expectations.

Fuel filters, lubricating oil filters, air filters, and coolant filters have one common purpose: to collect and retain solid particles from the passing media. Every filter has a certain efficiency that is a measure of the solids removed from the passing medium in relation to the total amount of impurities in transit. Thus a filter with a 95 percent efficiency retains 95 percent (by weight) of the impurities in the medium. The filter life is measured in hours from the time the filter is newly installed to the time when it cannot retain any more solids without the filter bypass opening.

The majority of all filters used in today's engine installations work on the principle of mechanical straining. The quality of the filtration is determined by the uniformity of pore size in the filter fabric through which the medium being filtered must pass. Filter fabric selection is in accord with the medium that is being passed through, the efficiency of filtration required, and the extraordinary conditions applying to the individual use.

Other filter types are in use, and they are often specified for particular reasons. These include centrifugal-cyclonic, oil bath, adhesive membrane, catalytic, and magnetic. The use of centrifugal filters in engine lubricating oil systems is becoming quite widespread (Fig. 19.1). Cyclonic filters are frequently applied to air induction systems working in extremely dust conditions.

Vane-type primary filters rely on baffling and low exit pressure to remove moisture and solids from an induction air stream. They must have their vanes precisely positioned, and the air velocity must be maintained at specified rates if they are to be effective.

Pleated paper filters have the advantage of low cost. The pleating gives the porous paper a greatly enlarged filtering and catchment area relative to the filter

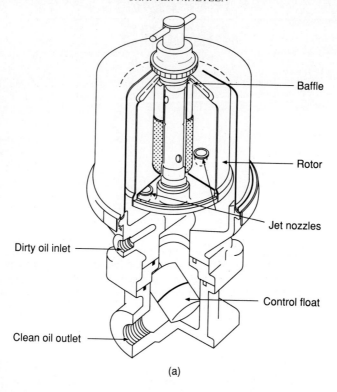

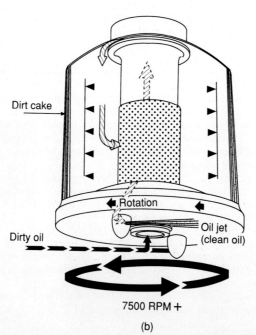

FIGURE 19.1 (*a*) Centrifugal oil filter; (*b*) centrifugal filter rotor.

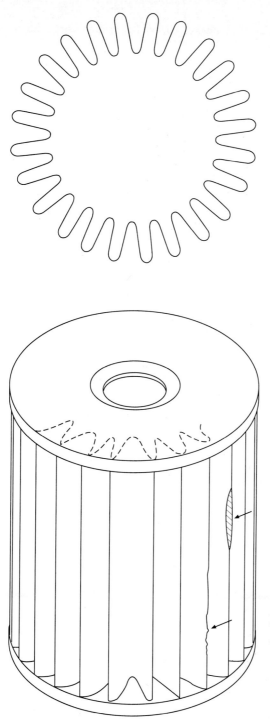

FIGURE 19.2 Pleated paper filter. The part of the diagram at right illustrates the increase in filter surface area obtained by pleating. The nominal diameter of the element is 15 in, but pleating yields 45 in. In actual practice the pleats are much closer together.

Look for evidence of crushing or splitting on new elements.

length and channels (Fig. 19.2). Pleated paper filter elements are generally used in full-flow filters, and they will collect debris down to about 10-μm size. Unfortunately, any rupture of the element while in service will have very damaging effects because unfiltered fuel or lubricating oil will then pass through to wearing surfaces.

The various engine filter elements will need to be replaced at a certain frequency of engine hours. As long as conditions remain normal, the frequency of change should not alter. When it is found that a filter is exhausted much sooner than expected, the conditions must be reviewed to see if the deterioration trend will continue, what can be done to normalize the filter use, and what the consequences are of doing nothing. In general terms, it can be said that poor filter element performance will contribute to poor engine performance.

19.2 LUBRICATING OIL FILTERS

The purpose of lubricating oil filters is to remove dirt of a specific size from the circulating lubricating oil. If 100 percent of the dirt generated in or inducted into an engine could be caught and retained by the lubricating oil filters, the wear rate of the engine's moving parts would be reduced almost to zero.

Notwithstanding the latest chemistry and the makeup of today's lubricants, the basic characteristics and ingredients of lubricating oil have never changed greatly. Solids are still borne in the oil flow, and because of the small clearances between lubricated parts designed into today's engines, a higher proportion of the dirt particles will be approximately the same size as the lubricating oil film is thick. Whenever a particle contacts a lubricated metal surface, wear will occur.

It is, therefore, natural that the lubricating oil filters on the modern engine must be able to trap and retain a range of particles down to the smaller sizes. To do that, filter surfaces are larger and have a finer permeability. Because full-flow filters will plug up quite quickly, bypass filters are installed. Although they pass only about 10 percent of the circulating volume of oil, they have a large capacity for retaining solids. The full-flow filters are fitted with a bypass valve so oil will continue to circulate through the system after the filter element is fully loaded but the wear rate will sharply increase. If and when the bypass filter becomes plugged, lubricating oil will still pass by or through the full-flow filter. The provision of lubricating oil filters, be they full-flow, in-line, bypass, or secondary, has nothing whatever to do with the service life of the lubricating oil.

Lubricating oil filter elements are constructed in a manner similar to that of fuel filters but with higher pressure and temperature ratings. The material of the element must be proof against sulfuric acid attack, and the element must be capable of retaining much larger quantities of solids in comparison with the fuel filters on the same engine. In a given engine the lubricating oil filters during their life will pass approximately 40 times the quantity of fluid that goes through the fuel filters.

The filtration methods used on late model engines must be able to collect much smaller particles; to do that, it is necessary to provide greatly increased filtering surfaces. The centrifugal bypass filter is rapidly gaining favor for its ability to remove 0.5-μm particles. These units are clean and easy to service, and many of them are now being retrofitted to older engines.

19.3 LUBRICANT CONTAMINATION AND ORIGINS

Dirt may originate from both inside and outside the engine. Most dirt originates in the combustion chamber either as products of combustion or as the heavy end residue of unburned fuel; a large proportion of the dirt is carbon which comes from the combustion chamber in the form of smoke. This carbon is a very soft and pure form; and although the particles are very small, they rapidly combine to form the basis of sludges.

Raw fuel and unburned heavy ends carry sulfur compounds which in the presence of water are converted to sulfuric acid. Water is continuously produced in any engine because of the rapid thermal fluctuations and subsequent condensation. An acid condition in the lubricating oil will induce wear in various engine components, and such conditions are more prevalent in engines operating on heavy fuel. High-speed, light-fuel engines will be less likely to suffer from acid attack. New lubricating oil contains acid inhibitors which gradually become spent during their service life.

Crankcase dirt may have almost any origin. Some may be abrasive; some may be organic. The abrasives will be any form of mineral dust, and the organic matter will be from plants, animal or insect refuse, pollen seeds, and so on. The mineral dirt particles will have hardnesses ranging from spongy to diamond-like, and they are most damaging. Diamond is defined as a colorless mineral composed entirely of carbon, and it is the hardest substance known. The close relationship between these two materials is indicative of the undesirability of carbon in the lubricating system.

Abrasive material usually retains its own identity; being inorganic, it does not react with other materials. Some proportion of the hard particles will fall to the bottom of the oil pan; some will be entrained in the oil drawn up by the lubricating oil pump. On passage through the pump gears, there is a minor milling action which will reduce the average size of the hard particles. The lubricating oil filters should retain all this solid debris.

The effectiveness of the lubricating oil filter is somewhat linked to the condition of the engine. If there is a lot of blowby (high crankcase pressure), the amount of solids originating in the combustion process will be higher than in a good engine. The gases passing down the cylinders will tend to carbonize any lubricating oil film that may be on the lower cylinder walls. No matter whether the engine is old or new, the lubricating oil filters must collect all the solid matter that may be circulated by the lubricating oil. If the filters are overloaded, broken, or bypassed, the wear on the moving parts of the engine will be greatly accelerated.

Whatever their origins, most of the lubricating oil contaminants will have a detrimental effect on the engine's mechanism if they are permitted to circulate through the lubrication system. Therefore, an essential feature of engine design is the provision of the correct lubricating oil filter elements.

19.4 SPIN-ON ELEMENTS

The spin-on type of filter element continues to gain popularity on stationary engines for both fuel and lubricating service. It is easy to change without spillage, and it is also durable in both storage and transit. Its principal disadvantage is that the pleated element within the canister cannot be inspected without removing the top of the canister. In some cases an ordinary domestic can opener is suitable for the job, but a purpose-designed tool which is now available does a better job.

It is just as important to inspect a canister type of lubricating oil filter as any other type, and particularly so if there has been an incident such that there is any possibility of any metal fragments being detached from the parent metals of the engine. Fragments of bearing metal are collected more easily on the filter than anywhere else in the engine, i.e., the pan.

19.5 CENTRIFUGAL FILTERS

Because of its efficiency, a centrifugal filter can separate the smallest particles from the fluid being circulated. In the case of lubricating oil, this results in a greatly reduced wear rate in the machine. Such filters are not new, but the reduced clearances of and higher loadings on modern engines call for their use more and more often.

The additives in the lubricating oil tend to survive better if they are not repeatedly exposed to the same contaminants as when the usual filter element is used. Centrifugal filters have a limited throughput capability compared to the ordinary full-flow filter; accordingly, it is fitted in a bypass mode. These filters spin out solids as small as $1\mu m$, which results in the formation of a dense mass or cake in the filter housing. It is readily removed, and when it is examined in good lighting, any metallic debris is easily seen because it is not obscured by fluid oil films. If the cake is soft and spongy, it is possible that there is a water leak in the engine.

A reasonable amount of care in handling the components is required when centrifugal filters are being cleaned. Plastic or wooden tools are recommended for cutting out the accumulated cake. These units revolve at very high speeds, and rough treatment will cause the moving elements to bind or run in an unbalanced state. To operate effectively, they require an oil pressure of not less than 30 psi (Fig. 19.1).

19.6 LUBRICATING OIL FILTER INSPECTION

Before any new filter element is inserted into the filter housing, it should be carefully examined for any sign of damage. If any is seen, the element must be discarded. The lubricating oil filter element installed in any engine must be of OEM supply or be specifically approved by the OEM. The use of jobber filter elements may appear to be economically attractive or otherwise expedient, but the engine owner or operator must be aware that the use of nonstandard elements can jeopardize warranty positions.

Each OEM filter element is carefully matched to the engine's lubricating oil flow rate and to the type of lubricating oil specified for the particular engine. A non-OEM filter may not have the correct design characteristics. Oil filter change frequency must follow the engine manufacturer's recommendations, but in the event of a bearing failure, coolant leak to the lubricating oil, or fuel oil dilution of the lubricating oil, the filter elements must be changed. If any metal is observed in the filter, a bearing inspection will be necessary before the engine is returned to service.

Lubricating oil filters are designed to pick up and retain all contaminant particles of a size that will damage the engine components. As the engine ages and wears, it is quite possible that the lubricating oil filter elements will accumulate a full load of solids before the scheduled oil change is due. In such cases, it will be necessary to increase the frequency of lubricating oil filter element change if an acceleration of engine wear rate is to be avoided. The alert operator will identify the changing conditions by careful and continuous regard for the condition of the lubricating oil filter elements at changeouts and by taking notice of the oil pressure gauge and the filter differential pressure gauge readings.

If lubricating oil filter elements are not changed out in a timely manner, there is a grave risk the lubricating oil filter will plug up completely. That will force the filter bypass valve to open and allow unfiltered oil to pass through the engine's lubrication system. If that happens, rapid and greatly damaging engine wear will ensue (Fig. 19.3).

19.7 LUBRICATING OIL FILTER ELEMENT DISPOSAL

Any lubricating oil filter element that is taken out of service must be destroyed and disposed of in an acceptable manner so there is no possibility of its deliberate or inadvertent reuse. It is not unknown for a spin-on type of filter element in particular to be reused, and it is suspected that certain pleated paper elements have been sent around for a second time on occasions. *Never attempt to wash out and reuse a disposable lubricating oil filter element of any description.*

19.8 LUBRICATING OIL FILTER—PLUGGING

Lubricating oil filters do not plug as long as the oil remains suitable for use in the engine. If a filter plugs, it is doing the job it is designed to do: It is protecting the engine by removing particulate contamination from the lubricating oil. Filter plugging can be considered a symptom of a problem within the engine, and corrective action should be taken before serious damage to the engine occurs. There are several reasons for filter plugging:

1. *Excessive oil contamination*, indicated by a heavy, thick, loosely held sludge in the filter element. This type of contamination consists of fuel soot, oxidation products, and combustion products.
    a. Excessive fuel soot is generally caused by defective injectors, overfueling, a restricted air intake, a generally poor mechanical condition of the engine, or failure to change the lubrication oil at the appropriate intervals.
    b. Oxidation by-products are usually brought about by operating at excessively high engine temperatures, running with inadequate cooling, gener-

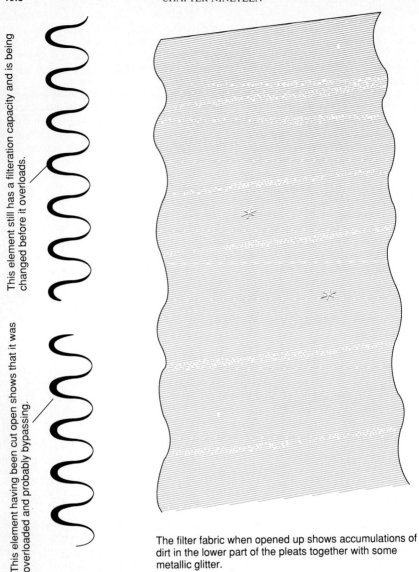

FIGURE 19.3 Fuel and oil filter element inspection.

ally poor mechanical condition of the engine, or overextending the oil change interval.

    c. Combustion by-products are caused by combustion gas blowby, high-sulfur fuel, poor combustion, or overextending the oil change interval.

2. *Coolant leakage and moisture condensation* are indicated by wavy pleats in the filter element and deterioration of the paper outer wrap. The additive

package will separate from the oil and become ineffective when a small amount of coolant contamination is present in the oil.
   a. Some of the reasons for coolant leakage are:
      (1) Leaking valve cage assembly,
      (2) Leaking precombustion chamber
      (3) Cracked cylinder head
      (4) Cracked cylinder block
      (5) Leaks in the oil cooler
      (6) Leaks past the cylinder liner sealing rings
   b. Some of the causes of condensation include:
      (1) Excessive engine idle running
      (2) Low engine operating temperature
      (3) Inadequate crankcase ventilation
3. Gel or emulsion formation in the pleats of the filter may be observed, but the filter may appear to be clean. When the filter is tested, it will show a restriction if moisture is still present. The problem often occurs when fuel oil in a bulk storage tank has been contaminated with small amounts of water (less than 0.5 percent). Fuel filter plugging can take place in as little as 5 min after the contaminated oil has been passed to the engine. This condition is prevalent when main storage tanks are replenished at subzero temperatures.
4. *Fuel dilution* is indicated by a brown or red color of the filter. Fuel dilution may be a result of overfueling, incorrectly calibrated and/or incorrectly adjusted injectors, restricted air intake or exhaust system back pressure, excessive engine idling, cracked injector cup, mutilated or missing injector O-ring, seized injector nozzle, or sticking valves.

19.9 FUEL FILTER SELECTION

Although the user is urged to use OEM filter elements, there will always be instances when jobber filter elements will be used. When non-OEM elements are purchased, they must satisfy several conditions:

1. The filter element selected for fuel oil service must have a flow rating and a pressure rating above those of the primary fuel pump.
2. The surface areas must be large enough to pass clean filtered fuel continuously from one lubricating oil change to the next.
3. The filter must be of a grade that will retain all particles of 3 μm or larger.
4. The material of the filter element must be able to withstand the regular operating fuel temperatures.
5. None of the material used in the filter element may be affected by the fuel being passed.

   Fuel filter elements members are made from cellulose fiber paper impregnated with an epoxy resin. The membrane is folded in multiple pleats and is mounted on an inner support frame to which the end cups are fastened. Examine all elements for evidence of crush, particularly at the cup. The use of poorly selected fuel filter elements or the purchase of "economy" elements can have a severe impact on fuel injection equipment repair costs. "Cheap" elements are simply not worth the risk.

19.10 UNFILTERED FUEL DAMAGE

If unfiltered fuel passes the fuel filter elements, there will eventually be problems with the HP fuel pump. Scoring of the barrel and plunger will take place; possibly the plunger will seize in the barrel. Erosion on the plunger scroll will affect the accuracy and quantity of fuel delivered by the pump. There may be injector nozzle needle seizure.

Water in the fuel can cause similar difficulties. If it is known that water has been passed through the fuel system, any paper filter elements must be changed out. The elements can hold considerable amounts of water, which can migrate to nowhere but the HP fuel injection system.

19.11 EDGE FILTERS

Edge filters are frequently installed in the fuel inlet stem of the injector, and these all-metal inserts must be removed for cleaning whenever the injector is serviced. Because the carefully machined filter surface, to the uninitiated eye, looks like a coarse turning job, edge filters are often handled roughly. The result is damage to the filtering edge. Edge filters should be washed only in diesel fuel. If they show evidence of corrosion, they must be replaced. Although they may be brushed with a brass-wire brush, they must not be cleaned with a scraper or steel-wire brush.

19.12 AIR FILTERS

The purpose of the air filter is to prevent airborne dirt from entering the engine. During the filter's operation, very large volumes of air are consumed by any engine. For example, a 500-kW generator set working at full load will use an air volume of approximately 45,000 to 55,000 ft^3/h. That volume can, depending upon the engine's geographic loation, carry quite measurable amounts of dirt into the engine if no filters are provided. Any dirt entering the engine, be it sand, pollen, insects, soot, fog, or other contaminant will cause wear. The importance of the air filter system is obvious.

19.13 AIR FILTER TYPES

Several types of air cleaners are in general use, and each has a different filter element. The most common filter is the type in which a disposable pleated paper dry element is used. The elements are easily and quickly changed by unskilled labor and are relatively cheap. The paper used in these filters will permit the passage of air but will not pass solid particles. The paper eventually becomes so loaded with solids that the passage of air becomes impaired, at which time filter element replacement is necessary. Tapping or banging the filter to remove the accumulated dust is grossly incorrect and must not be done.

In very dusty conditions, a cyclonic self-cleaning air filter is often used. Such a filter is so designed that the dirt extracted from the airstream falls into a container which is readily removed and emptied at frequent intervals. In the oil bath air filter, the aspirating air is cleaned by turning the airstream abruptly in the vicinity of an oil bath. Any particles in the airstream impact the oil surface and are retained.

Eventually the oil bath must be cleaned out and charged with fresh oil. A certain amount of care is needed on the part of the servicing attendant when replacing the oil in this type of filter. If the oil level is too high, there is some danger of excess oil being inducted into the engine; if the level is too low, the dirt-trapping effectiveness of the filter is impaired. Some oil bath air filters have a second-stage steel gauge element which will require washing at intervals.

On larger engines operating in very dusty conditions, continuously cleaning curtain filters are used. The filter elements are connected to each other in an endless chain and rotate continuously through an oil bath. The solids washed off the elements collect in the bottom of the bath and must be manually removed at intervals. Again, the interval between servicing depends on the environment in which the engine operates.

All the various air filter configurations are acceptable, and all work very well provided that the filter elements, the housings, the seals between the elements and the housing, and all the ducts between the filter elements and the engine remain in good order. Any damage to any of the components in the air system will result in the leakage of unfiltered air into the engine.

19.14 AIR FILTER ELEMENT SELECTION

All air filter elements used must be OEM-approved. If the use of other filter elements is proposed, the potential supplier must furnish warranty statements regarding the recourse available to the buyer in the event of poor filter performance. Whichever air filter element is selected, its ability to filter the local contaminants from the air must be considered. There may be seasonal variations in the amount of solids carried in the atmosphere, and the element chosen should have a year-round capability.

19.15 CHANGING AIR FILTER ELEMENTS

Many engines are fitted with vacuum indicators to warn when the air filter elements are becoming restricted. They are useful, but visual inspections must nonetheless be made at frequent intervals. The operator must not wait until the indicator shows the filter is totally plugged before changing it out. On the other hand, the operator must be aware that a filter with a dirty surface appearance is not necessarily spent. He must refer to the vacuum or restriction gauge.

No attempt must be made to change filters while the engine is running, because there is a great danger of foreign objects being dropped, sucked, or otherwise ingested into the turbocharger. The term *foreign object* includes fingers, which are just as damaging to turbochargers as nuts, wrenches, or wiping rags.

All replacement air filter elements must be carefully inspected before being installed in the air filter housing. Discard any element that appears to be split in the pleats, to be deformed in the seals, to be perforated, or to have been soaked in either oil or water. Be sure that the installed element is properly seated on its seals so that all air entering the engine must pass through the element.

Should uncleaned air get past the air filter, extreme cylinder liner wear will commence; if no air filter is in place, the wear rates will be greatly accelerated. If the air filter is totally loaded with trapped dirt, the engine will not get sufficient air to carry its full load. It will make black smoke, and its exhaust temperature will climb. The turbocharger will overheat, will probably glow, and will eventually destroy its bearings if left unattended.

19.16 AIR FILTER ELEMENT CLEANING

Cleaning and reusing dry-type air filter elements should not be permitted under normal circumstances, but it is acceptable in particular instances. In such cases, the manufacturer's manual must be consulted for guidance on the washing procedure. Some filters may only be washed once; others twice; still others many times over. In each and every case in which washing is permitted, the cleaning medium flow may be only in reverse to normal airflow. The clean filter must be very carefully examined against a strong light for perforations, cracks, or other defects, and it must be discarded if there is any doubt as to its efficiency. All discarded filters must be destroyed forthwith. Any particles of dirt penetrating either a new or a used filter will cause some engine wear.

Some engines utilize air filter elements composed of a metallic wire mesh. The mesh is coated with an oil film and when it becomes dirty, is readily washed. Caution must be taken to ensure that the mesh does not corrode while in storage. In those isolated cases in which air filter element washing is permitted, great care must be taken to use only approved cleaning materials. Gasoline must not be used under any circumstance, and any other solvent contemplated must be OEM-approved. Air filter elements are *not* to be cleaned by either beating them or banging them against a solid object.

19.17 AIR FILTER APPEARANCE

An air filter element may look dirty, or it may look clean. Its appearance will very much depend on the nature of the matter being extracted from the aspirating airstream. If, for example, there is soot in the atmosphere, it is probable that the filter will quickly become black and greasy; on the other hand, if a rock crusher is operating nearby (as often happens in a mine power plant), it is likely that the filter element will assume a light-colored coating.

It will be seen that the color of the air filter is not necessarily a good guide to the state of the element. In all cases, it is better to refer to a restriction or differential pressure gauge. This instrument indicates the difference between the atmospheric pressure and the plenum pressure as drawn by turbocharger. As the filter plugs up with airborne debris, the difference between the two zones of air pressure will increase and the indicator will register the vacuum.

19.18　EFFECT OF DIRTY AIR FILTER ELEMENT ON AFTERCOOLER

A plugged air filter on a turbocharged engine can be a major contributor to fouling of the aftercooler. The air filter element, once it is fully laden with dirt taken from the passing air, will have a reduced ability to pass the needed quantities of air to the engine. The turbocharger will continue to be driven and attempt to draw air into the engine. (The problem is compounded on engines in which the exhaust gases from bank A drive the turbocharger serving bank B, and vice versa.) The low-pressure situation created by the dirty air filter will lead to an anomalous pressure imbalance between the air side and the lubricating oil side of the seal at the compressor end of the turbocharger shaft, and large quantities of lubricating oil will enter the compressor and be passed on to the aftercooler fins.

There is always a possibility, therefore, of a high aspirating air temperature remaining in effect because of the inability of the now-fouled aftercooler to extract unwanted heat from the compressed airstream even after a new clean air filter element has been fitted.

19.19　EFFECT OF DIRTY AIR FILTERS

Dirty air filters will contribute to high exhaust temperatures, lack of power, high cooling water temperatures, overheated turbochargers, sluggish response to load change, heavy fuel consumption, and dirty exhaust smoke. Attention to air filter conditions is very much a part of an engine operator's duties.

Hard-carbon deposits can be developed in any turbocharged engine through the continuing use of dirty air filter elements. When a turbocharger cannot draw its design quantities of air through the air filter elements, a partial vacuum in the air intake plenum occurs. Because of the abnormally low pressure on the air side of the compressor end bearing seals, much of the lubricating oil fed to the bearing escapes into the aspirating airstream. It then becomes carbonized on the cylinder head air passages and the air inlet valves.

19.20　SOURCES OF DIRT IN AIR FILTERS

The modern air filter is very efficient, and it will readily become plugged with airborne dirt or oil if such is present in the engine's environs. The rate of dirt buildup on the element will not necessarily be constant. Provided the filter medium and the element seals are undamaged, the element will continuously remove dirt from the passing air effectively to the point at which the engine efficiency is impaired. However, a change in the atmospheric content adjacent to the engine can dramatically alter the changeout requirement and frequency. That factor must be recognized and acted upon as soon as the cause for alteration is identified. Air filters will trap oil or fumes coming off leaking fuel lines, lubricating oil lines, crankshaft seals, warped crankcase doors, oil spills on the floor, exhaust pipes, open access doors, unswept floors, paint sprayers, and similar sources. Thus, it will be seen that it is important to keep all engine leakage, whatever the source, under control.

19.21 USE OF DAMAGED AIR FILTER ELEMENTS

Air filter elements with perforations, cracks, distortions of the seal faces, or evidence of dirt on the suction side of the element are not acceptable for use. Damaged air filter elements will not filter air properly. Any unfiltered air entering the engine will cause extraordinary amounts of wear directly in the cylinders, pistons, valves, and valve guides. Secondary wear effects will materialize in bearing surfaces, and premature degradation of the lubricating oil can be expected. No damaged or used air filter elements should ever be retained in the plant.

19.22 AIR FILTER ELEMENT STORAGE

The care of air filters is of great importance to the life of an engine. This care must commence upon delivery of a new supply of filter elements. New air filter elements must remain in their original unopened packaging until they are needed for installation. They must be stored in a dry, draftless area, and they must not be stacked in a manner that would lead to crushing or other damage.

19.23 VENTILATION AIR FILTERS

Air filters used in building ventilation are commonly made of woven wire mesh. Multiple layers of the wire mesh coated with a thin film of oil will prevent quantities of atmospheric or wind-blown dust from entering the building. Other filters are made from fiberglass enclosed in a large mesh wire and sprayed with an adhesive oil. These filters are not recoverable. Wire mesh filters require a routine of washing, reoiling, draining, and reinstallation. All too often, these filter panels are neglected. After they are plugged with dust, further incoming air is largely unfiltered, being sucked through whatever aperture in the building shell it can find. Entry doors become heavy, are difficult to open, and then slam shut; seams and joints in the building shell whistle; and the interior walls become streaked with dirt where the outside air leaks in. Turbocharger performance may be affected by bad ventilation filter conditions.

Ventilating air filter panels must to be washed regularly in a noncorrosive solvent to release the trapped solids and fume condensates. Having been allowed to thoroughly drain, the filter is then immersed in a mineral oil having good anticreep qualities. The saturated filter is again drained of surplus oil and returned to service.

Most ventilating air filter panels in industrial use employ a light-metal frame which will not stand a great deal of rough handling. With care and proper attention, however, ventilating air filters will work effectively for many years.

CHAPTER 20
ASPIRATING AIR

20.1 ASPIRATING AIR MANIFOLD

The aspirating air manifold serves to pass air from the air filters to the engine air intake ports. The air manifold is sometimes known as the *scavenge trunk* or *air box*. On pressure-charged engines, the air manifold connects the turbocharger or blower to the air intake ports of the engine.

20.2 ASPIRATING AIR TEMPERATURES

Any engine is so designed that each component subject to heating is able to withstand the thermal stresses imposed on it during normal operation. Certain components, such as pistons, cylinder liners, and cylinder heads, must be able to transfer or dissipate large quantities of heat. If the heat transfer or dissipating abilities of individual components are exceeded, the components become thermally overloaded and their failure can be expected. Thus, the effects of introducing overheated air into the cylinder can be quite damaging.

It is most important that the temperature of the aspirating air entering the cylinders be carefully regulated, in the 50 to 60°C range (120–140°F).

The most common and noticeable result of allowing overheated aspirating air into the engine will be high exhaust temperatures relative to a given power output. Where aspirating air temperature gauges are fitted, care must be taken that the temperature of the air entering the cylinder does not exceed the manufacturer's stated limits. These limits will vary with the aftercooling system and the load of individual engines (see Fig. 17.1). Refer to the OEM manual for governing temperature figures.

Another and very dangerous effect of introducing overheated air to the cylinder is overheated pistons. On engines fitted with aluminum pistons, the consequences can be especially severe. The aluminum absorbs so much heat that it reaches a semiplastic state that is followed by separation of the piston crown from the skirt or a breakage in the area of the wrist pin bores. Subsequent damage to the engine caused by the flailing connecting rod will be catastrophic.

There will be little warning of this sort of event by either sound or sight other than a high reading on the air manifold temperature, but that should be sufficient warning to alert the operator.

20.3 ASPIRATING AIR TEMPERATURES; HEAVY FUEL

In engines burning heavy grades of fuel, the aspirating air temperature has a marked effect on the efficient combustion of the fuel. Air should enter the cylinder at a temperature between 50 and 60°C (120 and 140°F) unless decreed otherwise by the OEM.

Attention must similarly be paid to the temperature of the coolant circulating in the aftercooler. If the temperature is sufficiently low (seawater, river water, or lake water), there is always a possibility the charge air will be overcooled and poor combustion will result. One sign of this condition will be a dark exhaust smoke being seen even at low loads.

20.4 AIR BOX DRAINS

Careful attention must be given to the air box drains of two-cycle engines. Condensate derived from the atmosphere, oil mist from the engine, and leakage from the blower lubrication system cause an accumulation in the air box of a carbonized oil and water mix which must be drained off continuously. In some cases there are manually operated cocks; in others there are pressure-operated valves. When the engine is in operation, the drains must be open and the operator must check them periodically to see that they are clean. He must also take note of any change in the drainage pattern. If the oil-water flow increases, it may indicate the beginning of a problem such as failed blower bearing seals, excessive crankcase pressure, failed piston rings, or poor injectors.

It is recommended that the air box covers be removed at intervals of, say, 3000 h and the air box interior be visually inspected. If necessary, the air box should be wiped clean with rags. At the same time, the air box drain tubes should be blown out with 30-psi compressed air. Failure to maintain a clean air box can result in a localized fire which will be exceedingly detrimental to the engine.

20.5 AIR BOX CLEANING (SCAVENGE TRUNK)

In two-cycle engines, it is not uncommon to develop amounts of an oil-carbon mixture in the air box or scavenge trunk. Drains are usually provided to allow excess fluid oil to run off. The quantity of oil derived from the drains should be observed, and any change from normal should be noted. Maintenance schedules call for the regular manual cleaning of the air box. If cleaning is not done, there is a possibility of an air box fire taking place.

An engine with an air box fire will lose power and may trip its load. Without load, it may continue to run on the minimal amount of air available to it. The unit must be stopped immediately to prevent the spread of damage, but it is probable that substantial damage will have occurred to liners, pistons, and turbocharger. A detailed inspection will be necessary.

In four-cycle engines, overheating of the air manifold may be caused by:

1. Leaking air inlet valves
2. Dirty aftercooler on air or water side
3. Incorrect engine timing

Following an air box fire, no attempt to open the engine may be made until the engine has completely cooled down. (Sec. 7.3).

CHAPTER 21
EXHAUST SYSTEM

21.1 EXHAUST TEMPERATURES

Indications of overly high exhaust temperatures are given on exhaust pyrometers (when they are fitted), by the exhaust manifold showing a redness, and sometimes by sparks or flame being ejected from the exhaust stack. The signs may also presage turbocharger problems. When unusual conditions develop and exhaust temperatures show an abnormality, the load must be transferred to another generator and the defective unit must be stopped for repair or adjustment without delay. Exhaust temperature excursions are indicative of problems in cylinder heads, fuel injection systems, and aspirating systems.

Abnormal exhaust temperatures may be observed on one or more cylinders; all the temperatures may be rising together; all the temperatures may be falling together; or individual cylinder exhaust temperatures may be rising or falling independently of each other. Generally speaking, engines that are fitted with individual fuel pumps operating independently of each other have each cylinder exhaust outlet fitted with an exhaust thermometer. Engines with multielement single fuel pumps will normally have only one exhaust thermometer in the exhaust manifold. Small engines (300 kW or less) commonly have no exhaust temperature thermometers fitted at all. The typical exhaust temperatures quoted in Sec. 21.2 are those of engines at full load under normal and general conditions and are cited only as a rough guide.

When it is possible to read individual cylinder exhaust temperatures, the variation in temperature between one cylinder exhaust and another should not be more than ±30°C (50°F). If variations are exceeded, it will be necessary for a mechanic to inspect the engine and do whatever is necessary. Frequently it will happen that the exhaust temperature, when it rises on one cylinder, will fall on an adjacent cylinder, particularly if the high-temperature cylinder is being overfueled.

If all the cylinders are doing the same amount of work, the individual exhaust temperatures and firing pressures should be quite closely matched. However, if the exhaust temperature of one particular cylinder rises above the norm, it is possible that cylinder has developed an exhaust valve problem. On the other hand, if the exhaust temperature of an adjacent cylinder simultaneously falls, it is probable that the rising exhaust temperature is caused by overload and the cylinder with the falling temperature is suffering from a fuel injection difficulty.

21.2 EXHAUST TEMPERATURE EXAMPLES

The following are examples of typical exhaust temperatures that can be expected at full load on selected engines:

	°C	°F
Mirrlees K8	425	800
Caterpillar D399	470	880
Caterpillar D398	440	830
Caterpillar D353	470	880
Ruston RK16	450	850
Waukesha	380	720
Cummins KT2300	500	930
Cummins KT1150	425	800
DD71	550	1020

The engine operator should verify the figures for the engines in his care by reference to the OEM operator's manual.

21.3 OBSERVATION OF EXHAUST TEMPERATURES

On engines that do not have exhaust temperature gauges the individual cylinder exhaust temperatures may be compared with each other by observation of the heated exhaust manifold color. This is done using a piece of pipe of convenient length and diameter. For most practical purposes the pipe will be about 3 ft (1 m) long and 1 in (2.5 cm) diameter. By placing one end of the pipe so that it is almost touching the exhaust outlet from the cylinder and then sighting through the pipe, the exhaust outlet will be seen to have a red color. The color at each exhaust outlet should be the same if the load is steady and all the cylinders are firing correctly. The color will be brighter as the load increases.

An unbalanced brightness will indicate that something is amiss in one or more cylinders. A cracked, burned, or leaking exhaust valve will result in a high exhaust temperature. A damaged exhaust valve seat will produce a similar result. A stuck injector nozzle may give a depressed exhaust temperature, as will a seized fuel pump.

This method of monitoring cylinder conditions is quite inexact, but it is very useful in the absence of regular instrumentation. It is recommended that safety glasses be worn when conducting this test because particles of scale and dirt usually trickle down the pipe. Do not use plastic (PVC or ABS) pipe; it will not tolerate exhaust temperatures. Practice and familiarity with the individual engine are essential ingredients for successful use of this technique. It can be said that any exhaust system or part thereof, including the turbocharger, that is visibly red under normal lighting conditions is probably in trouble and in need of instant action.

21.4 HEAT-SENSITIVE CRAYONS

Heat-sensitive crayons permit the visual gauging of elevated temperatures in the absence of conventional thermometers. These crayons are supplied in sets covering temperature ranges from 200 to 670°C (400 to 1250°F). Selectively applied to heated surfaces, the crayoned mark will go through a very distinct color change which is then compared with an accompanying chart. Beyond the temperature range of the crayons, the metal will naturally begin to show color anyway, and below the temperature range of the crayons the surface temperatures can usually be judged by a practiced hand contact (using the back of ones hand).

21.5 EXHAUST TEMPERATURE VARIATIONS

The causes of rising exhaust temperature are listed below.

1. Individual cylinders
    a. Cracked, burned, bent or sticking exhaust valve
    b. Cracked, burned or loose valve seat insert
    c. Cracked cylinder head
    d. Broken valve spring
    e. Overfueling
    f. Bent pushrod(s) on valve (or injector)
2. In all cylinders
    a. Aspirating air too hot
    b. Air filters plugged
    c. Turbocharger pressure too low
    d. Aftercooler dirty
    e. Fuel timing incorrect
    f. Engine overloaded (consider a seizure load)
    g. Engine overheated
3. On an individual cylinder
    a. Seized injector nozzle
    b. Seized fuel pump plungers
    c. Fuel rack or tappet backed off
    d. Fuel cam follower bearing failure
    e. Fuel cam failure
    f. Bent pushrod(s)
    g. Dirty injector edge filter
    h. Leaking injector pipe
    i. Fuel starvation
    j. Air in fuel
    k. Water in fuel

Excessive exhaust temperatures will shorten the life of exhaust valves. Overheating for a short period will probably cause the valve seat to lose its sealing ability; gross overheating will lead to distortion of the valve head. Furthermore, exhaust valve seat inserts become loose or displaced and frequently crack when subjected to temperatures beyond their design limits.

21.6 EXHAUST TEMPERATURE ADJUSTMENT

At no time should exhaust temperatures be balanced by means of fuel pump rack adjustment. To do so will essentially mistime each adjusted pump and also create a second deficiency. If the misadjustment of one or more of the fuel pump racks is extreme, it is possible for the engine to go into an overspeed condition upon startup or shutdown. When an abnormal exhaust temperature is noted on an individual cylinder, the peak pressures of all the cylinders should be checked.

The pressure check is best done when the load is steady and is either full or nearly full. The peak pressure test will show if the cylinder with the errant exhaust temperature is firing correctly. As a general rule, the peak pressures should be within 5 percent of each other, but the average pressure will vary with the make, model, and rating of the individual engine. In many cases the correct cylinder peak pressure is stated in the original engine test data sheets.

21.7 EXHAUST EMISSIONS

The operator arriving at the power plant to take up his shift should pause before entering the building and study the exhaust emission. He should review the exhaust fumes for color and give attention to the uniformity of the fume density. He should notice if there is any sparking or, in extreme cases, evidence of flame, if there is any slobber or sparking, and when inspecting in the hours of darkness, if there is any glowing of the exhaust system that would indicate a valve problem. Any unusual condition noticed must be recorded in the log sheet.

Intermittent color changes in the exhaust are signs of trouble in injector needles or fuel pump plungers and barrels, dirty fuel filters, water in the fuel, or sticking inlet or exhaust valves. The color change may be repeated quite regularly or may occur in an erratic pattern. An inspection of the exhaust temperatures prevailing may help in identifying which cylinder(s) is at fault. The temperatures on only one cylinder will be errant if there is a valve or fuel injection problem, but if the fuel filters are dirty or there is water in the fuel, there will be quite a ragged pattern of exhaust temperatures.

21.8 EXHAUST SPARKING

Some of the causes of sparking may emanate from a failing turbocharger, overload, dirty waste heat boiler, incorrect fuel timing, piston crown damage, leaking or burned exhaust valves, and cracked valve seats. If sparks are seen emerging from the engine exhaust, it is advisable to shut the unit down immediately, because in most cases the sparks have their origin in distressed metal.

21.9 BLACK EXHAUST SMOKE

The darkness imparted to the exhaust gases is caused by the presence of incompletely burned fuel. Unburned fuel appearing continuously in the exhaust of an engine represents an inefficient operation. In the combustion process, particles of fuel are cracked into molecules of hydrogen and carbon by the high temperatures in the cylinder. The hydrogen molecules totally combine with oxygen, as do most of the carbon molecules. The uncombined molecules of carbon leave the cylinder in the exhaust gas and give the exhaust plume the appearance of smoke.

A certain amount of incompletely burned lubricating oil may also contribute to the exhaust smoke. Unburned fuel will give the appearance of black smoke, and lubricating oil will impart a bluish color. Such exhaust colors are literally the color of "money going up in smoke" besides which it is environmentally inconsiderate.

In a well-tuned diesel engine, the fuel air ratio will be in accordance with the manufacturer's designs and specifications throughout the engine's load range and the emitted exhaust will be clear. If the engine falls out of tune and the proportion of fuel relative to the air consumption increases, the fuel will not be completely burned and the result will be dark-colored exhaust gas emission.

Black smoke will also be developed under overload conditions. Most engines have a capability of overfueling their own cylinders, but the aspirating system, whether natural, turbocharged, or blown, will not supply more air than is needed to produce slightly more than full load under ordinary conditions because of the limitations of the air manifold capacity.

21.10 WHITE EXHAUST SMOKE

White smoke will sometimes be generated when a diesel engine has just started and the combustion chambers have not attained full operating temperature. Until the combustion chambers are fully heated, the fuel will partly vaporize or incompletely burn. As the engine and its combustion chambers warm up, the fuel will be totally consumed and the white smoke will fade away.

If the white smoke persists after the engine has reached full operating temperature, the timing may require a check. If it is retarded, the reason for it being in such a position must be found. White smoke can also be caused by unbalanced injectors, burned valves, worn rings or pistons, and cracked or badly worn liners.

Inspect the aspirating air system for leakage. This is particularly important on two-cycle units. The cetane number of the fuel is often cited as the cause of white smoke, but it is rarely the culprit. Nevertheless, it is worth looking at when all other avenues have been explored. The cetane number of the fuel can be verified by the local fuel test laboratory if necessary. For direct injection engines the cetane number should be 40, but unfortunately the owner of the engine is usually stuck with the fuel on hand. In any case, the cetane number is rather unlikely to vary much either way, and most engines will tolerate some variations in the cetane rating.

21.11 SLOBBER

Engines that run without load or with very light loads for long periods will develop signs of exhaust slobber at the manifold and/or at the exhaust outlet. The condition applies to both two- and four-cycle engines. Slobbering is not necessarily indicative of a defective engine, but the engine is almost certainly being improperly utilized. It is a housekeeping problem as well. At low loads there will be some overfueling and incomplete combustion that will introduce unburned fuel vapors into the exhaust system. There the prevailing low temperatures permit early condensation of the vapors. The condensate mixes with the previously deposited soots and carbons to produce a dense black penetrating fluid which readily leaks past the exhaust manifold gaskets. The developed mixture of fuel and soot may not be harmful to the engine, but the cosmetic effect is not tolerable.

A slobbering engine will become extremely dirty both inside and outside. It is probable that the generator will eventually become coated with a soft black slime, the air filters will have a short life, there will eventually be evidence of fuel dilution in the lubricating oil sample, and the lubricating oil filters will plug up far too quickly.

Engines that are permitted to run continuously at light loads will suffer from glazed liners, sticking piston rings, loss of compression, heavily carboned exhaust passages, poor turbocharger performance, and an inability to carry full load when required to do so. Exhaust fires, particularly on two-cycle engines, are also a good possibility under such conditions. When light loading is unavoidable, artificial loading should be applied at hourly intervals for a few minutes. Transferring the load temporarily from another engine for short periods of time is one method.

21.12 CARBON DEPOSITS IN EXHAUST PASSAGES, TURBOCHARGERS, AND CYLINDER HEADS

For the purpose of engine troubleshooting, it may generally be assumed that soft-carbon deposits are indicative of incorrect fuel/air ratios. Too much fuel is being supplied, and the exhaust color will almost certainly be dark. Hard-carbon deposits are generally caused by excessive lubricating oil being burned. Bluish exhaust smoke may be seen; lubricating oil consumption may be rising.

Similarly, liquid-cooled mufflers or waste-heat recovery boilers tend to get loaded with soft carbon, particularly if the engine is run for extended periods at light load factors. Rising back pressures will cause the turbochargers to slow down and the exhaust temperatures to rise. The fouling of waste-heat boiler tubes is very dependent upon the type of fuel being used in the engine; heavier fuels often create more soot.

Spark arresters may also become excessively coated with semi-burned fuel and carbon and thereby affect the engine's performance. In the event of a fire developing in the exhaust line or the muffler, the engine must be stopped immediately and personnel must stand by with portable fire extinguishers. If a CO^2 fire extinguisher is available, it may be possible to apply its nozzle to the muffler drain or air manifold drain connection with some useful effect.

The cleaning of waste-heat boiler tubes should be part of the regular unit maintenance routine, but this work should not be done by the shift-keeping operator. It should be done by regular dayworkers or contractors. The removal of carbon deposits from the waste-heat recovery boiler should be done whenever possible within a plan and timetable so that the cleaning does not have to be carried out under emergency conditions. Very few waste-heat recovery boilers are fitted with exhaust bypassing valves, so each engine arranged for waste-heat recovery must be shut down while the boiler cleaning operation is in progress. It is recommended that waste-heat boiler cleaning operations be planned to coincide with oil change downtime and planned overhauls.

21.13 EXHAUST GAS PRESSURES

The exhaust gas pressure at the exhaust valve will be about 75 psi, and that pressure will drop to approximately 15 psi in the exhaust pipe and muffler. It will be seen that the exhaust gas pressure is quite adequate to blow out any improperly fitted gasket. Impediment to the free flow of exhaust gas may be caused by excessive carbon in the exhaust system, a slow or plugged turbocharger, a flooded or iced exhaust pipe, or a fouled waste-heat boiler.

21.14 EXHAUST MANIFOLD GASKETS

When an exhaust joint leaks, a steel gasket, if fitted, will still give a partial seal, but a copper asbestos gasket will disintegrate and blow right out. A common cause of exhaust gasket blowout is through overfueling of the engine at startup. Ideally, the engine should start with the governor load limit set at 50 percent or less. Most often, engines are started with the load limit set at or near 100 percent. When the engine starts up with a high load limit setting, a considerable exhaust back pressure is built up in the exhaust manifold. It is not relieved until the turbocharger builds up speed. During this period, the exhaust manifold joints will seem to leak. With normal steel gaskets the leakage will be seen at each start-up, even at low load limit settings, but as the manifold warms through, the joints will gradually close up. When copper gaskets are fitted, the wire-drawing effect of the exhaust gas, coupled with the gas heat, tears the gaskets apart. The subsequent retorquing of the manifold has no effect on closing the gap.

Tightening the flanges immediately following the detection of leakage will not be beneficial more than once. Overtorquing the bolts will seize the threads, break the bolts, or warp the flanges. On any engine, one should not attempt to retorque the exhaust manifold flanges until the engine is thoroughly warmed up. All of today's engines have torque specifications for the tightening of exhaust manifold fasteners. Do not exceed the values quoted: Overtightening will only result in damage or distortion.

When properly installed and cared for, exhaust gaskets will give long trouble free service. If any make or model of engine appears to have exhaust line problems, check to see that the correct gasket is being used, and that the flanges are flat and correctly torqued. There should be no impediment to the passage of exhaust gas.

It can be said that the expenditures incurred in the maintenance of exhaust systems, combined with good operating practice, will lead to large savings in plant cleaning costs, improved working conditions, better lighting, and a safer, more pleasant working environment. An unrepaired leaking exhaust gasket is the fastest known way to obtain an all-black decor in the power plant. The cost ratio between gasket repair and plant cleanup has to be something like 1:1000. No exhaust leak has been ever known to cure itself, no matter how long it has been left unattended.

21.15 EXHAUST EXPANSION PIECES

For any practical and permanent purpose, it is not feasible to repair-weld expansion pieces that have split on the corrugated section. The welding techniques required and the content of exotic alloys used in their manufacture render any field repair of expansion pieces impractical.

When an engine warms up, it goes through a thermal growth. The exhaust pipes are particularly affected by this expansion, so they are firmly fastened to the individual cylinder heads in sections. To accommodate the expansion and contraction movements, intermediate bellows pieces are fitted. The bellows are made of thin-walled, high-temperature-resistant steel pressed in a radial corrugated form. Because of the elevated temperatures to which the bellows pieces are subjected and to protect the inner portion of the corrugations from the effects of exhaust gas scouring, a tubular shield is fitted and fastened to the entry flange only.

Care should be exercised in the installation of cleaned or replacement bellows pieces. On a cold engine there will be a "rattling" fit between the exhaust pipe sections, and a slight tension will be imposed upon them as they are bolted up. When the engine heats up, the expansion pieces will be relieved of their slight tension.

It must be clearly understood that exhaust expansion pieces are not to be regarded as flexible joints. Utilization of expansion pieces as flex joints for alignment purposes will only result in large and futile expenditures. Early cracking can be expected, and large quantities of carbon will permeate the engine house.

21.16 CARBON LOADING OF EXPANSION PIECES (EXHAUST BELLOWS)

Over a period of service, quantities of carbon, sometimes mixed with sulfur, will accumulate in the corrugations of the bellows and behind the shield. The free movement of the whole expansion piece will be restricted, and other parts of the exhaust system will become unduly stressed. It is, therefore, necessary to remove the expansion pieces during the engine overhaul and clean out the deposits. This is usually done by a solvent-soaking process. Replacement of the expansion pieces by new ones should not be necessary (Fig. 21.1).

21.17 EXHAUST SYSTEM FITTED WITH CATALYTIC CONVERTER

It will be necessary for the operator to check the pressure gauges on the exhaust system on each side of the catalytic converter if the engine is so equipped. Cleaning—washing the catalyst in an oil-breaking solvent—will be required about ev-

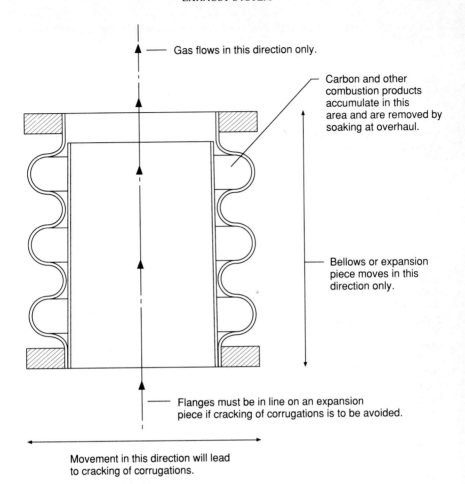

FIGURE 21.1 Expansion piece.

ery 2000 h. Piston ring wear, broken rings, leaking turbocharger seals, and defective injectors will reduce the effectiveness of emission control and also affect the engine's efficiency by imposing an exhaust back pressure.

21.18 INSPECTION OF MUFFLERS AND STACKS

An external examination of mufflers and stacks exposed to the weather should be carried out at least annually. The exhaust system drains must be seen to be clear; the muffler (silencer) casing should be closely examined for any sign of corrosion perforation; and all the supporting brackets, frames, and fasteners must be secure.

External corrosion causes the paint to deteriorate quite rapidly, and to some people the appearance is aesthetically displeasing. However, the real value of a painted exhaust system lies in the high visibility of any developing perforations or leakage. Dark vertical streaking of the muffler is generally indicative of holes in the muffler shell. The corrosion may attack the exhaust system metal from either the inside or the outside. Inside corrosion may be caused by allowing the engine to operate with exhaust gas emissions at or below dew point temperatures. The dew point temperature is typically in the 100 to 150°C (200 to 300°F) range. Exterior corrosion may result from stack fallout of sulfuric acid or from atmospheric salt (when the plant is located near the sea) or from locally generated pollutants. The rate of interior corrosion will also depend in part upon the sulfur content of the fuel used.

When a plant is supplied by sea or barge, a lead time of between 6 to 12 months for a muffler replacement will be required.

21.19 WASTE-HEAT RECOVERY

If an engine is habitually run on light loads, it is probable that quantities of soft carbon will accumulate in the exhaust system and waste-heat recovery boiler. This carbon will affect the efficiency of the boiler because the transfer of waste heat to the surrounding water will be considerably inhibited. Not only will the heat transfer be reduced but exhaust gas back pressures may impair the turbocharger performance.

The water side of any exhaust gas boiler is prone to fouling by the solids borne in the feed water. Depending on the size of the installation, the treatment of the feed water may be anything from simple to elaborate. The pH of the water must be kept in the vicinity of 7.0 if acid attack of the boiler shell and tubes is to be avoided. If there is too great an alkalinity over a prolonged period, there may be a metallic alteration effect on the boiler shell or tubes. Any plant operating fired or waste heat boilers will have at least a basic water testing kit and water treatment chemicals so the pH of the boiler can be adjusted as necessary.

CHAPTER 22
FUEL OIL

22.1 FUEL USAGE

The fuel used in a diesel power plant is the greatest single item of expense; it amounts to at least half of the total plant operational costs. The control of the plant fuel inventory is of obvious importance, and the need for accurate fuel metering, both into and out of the fuel storage tanks, cannot be overstressed.

Fuel arriving at the plant must be measured. In the case of water-borne fuel, the sheer quantities involved in any single delivery may render metering impractical. In that case, tank dips and fuel-temperature readings will have to be taken before and after delivery of the new supply. The tank drain must be checked for the presence of water before and after delivery. Following these exercises, the actual fuel delivered can be calculated. Whenever fuel supplies are carried by water transport, some water in the fuel can be expected. It is a rare vessel that is 100 percent watertight.

In the case of road-delivered fuel, there is no reason why the fuel should not be metered off the tank truck with checks for water taking place before and after delivery. The owner of the power plant is interested only in purchasing fuel and does not expect to pay for anything else. If water arrives with the fuel, its quantity must be determined and subtracted from the total delivery and the fuel supplier's representative must be advised immediately. The superintendent and the operator in charge at the time of fuel delivery must safeguard the owner's interest in all matters of fuel receipts.

The total fuel quantities delivered must equal the amount of fuel consumed over a given period of time, making allowance for the fuel quantities presently and previously in storage. Any discrepancy must be accounted for so plant efficiency and security can be maintained.

The total power generation over a period of time must be within a certain efficiency ratio to the amount of fuel consumed. In the very best plant with ideal conditions of load, altitude, and temperature, together with modern equipment in first-class order, a generation-to-consumption ratio of 17 kW/imp gal (14.1 kW/US gal, 3.74 kW/L) can be expected. Whatever the conditions or type of equipment, however, each plant should have an established level of fuel efficiency below which deviations should not be allowed. A performance of less than 15 kW/imp gal (12.5 kW/US gal, 3.3 kW/L) must be regarded as unsatisfactory. Poor fuel economy is commonly the result of bad engine utilization, but often the load simply cannot be matched to the capacities of the individual engines available.

For example, a diesel power plant with an output averaging 1000 kW/h and operating at 16 kW/imp gal (13.3 kW/US Gal, 3.52 kW/L) will consume in one month:

$$\frac{744 \times 1000}{16} = 46{,}500 \text{ imp gal}$$

while the same plant operating at 15 kW/imp gal (12.5 kW/US gal; 3.20 kW/L) will consume in the same period:

$$\frac{744 \times 1000}{15} = 49{,}600 \text{ imp gal}$$

which is a difference of 3100 imp gal (3840 US gal, 14,092 L).

22.2 FUEL METERING

For the comparison of one engine's fuel efficiency performance with other units, manufacturers generally state the fuel consumption that can be expected. This is expressed in terms of a weight of fuel, its heat value, and the rate of consumption per brake horsepower hour at an altitude not exceeding 500 m (1500 ft).

In actual practice, the fuel consumption and efficiency for almost all diesel power plants is calculated on the basis of information gathered from the engine's fuel meter and the unit's kilowatt-hour meter. Using the two readings which are taken at fixed intervals (i.e., at the end of each shift), the ratio of power produced to fuel consumed is easily calculated and compared with the expected norm.

Again, the cost of fuel is the largest expense of all in the operation of a power plant. It is most important that the fuel consumption be accurately metered and the installed units be so dispatched that the best possible efficiency can be reflected in the consumer's power bill.

22.3 EFFECT OF UTILIZATION FACTOR ON FUEL CONSUMPTION

Some considerable savings in fuel expense can be had by paying continuous attention to the equipment utilization factor:

$$\text{Utilization} = \frac{\text{total generation}}{\text{total capacity operated}}$$

Example 1

Suppose the output of the plant is 10,000 mW per year, the engine gives 16 kW/imp gal of fuel, and the utilization factor is improved from 65 to 75 percent:

$$\frac{10{,}000{,}000}{16 \times 0.65} - \frac{10{,}000{,}000}{16 \times 0.75} = 96{,}153 - 83{,}333 = 12{,}820 \text{ imp gal}$$

A 1 percent increase in the utilization factor will yield an approximate 0.3 percent improvement in the engine's fuel efficiency.

If the same power plant operating individual equipment with utilization factors

FUEL OIL 22.3

in the region of 65 percent of rated capacity is able to improve the factor to 75 percent, there should be reduction of about 3 percent in the amount of fuel needed to obtain the same generated output. The net amount of fuel saved will be the fuel saved by improved utilization less the fuel saved by improved efficiency. The net saving will be 12,820 − 2,399 = 10,321 imp gal, or 10.7 percent.

Example 2

Consider the difference in fuel consumption when a diesel generating plant formerly operating at 14.75 kW/imp gal (12.65 kW/US gal, 3.25 kW/L) is improved to 15.25 kW/imp gal (13.10 kW/US gal, 3.35 kW/L).

$$\frac{15.25 - 14.74}{15.26} = 3.2\% \text{ improvement}$$

If the plant has an average output of 1000 kW/h the savings in fuel oil over a 1-year period are valuable.

$$\frac{100 \text{ kW} \times 8760}{15.25 \text{ kW/imp gal}} = 574,426 \text{ imp gal}$$

$$\frac{1000 \text{ kW} \times 8760}{14.25 \text{ kW/imp gal}} = 614,736 \text{ imp gal}$$

a difference of 40,310 imp gal, 48,372 US gal, or 103.007 liters.

The 3.2 percent improvement in fuel efficiency can in most cases be initially obtained simply by matching the generation equipment more appropriately to the prevailing load. No equipment operating at an average utilization of better than 75 percent would be expected to show a fuel efficiency of less than 15.00 kW/imp gal if it is in good repair.

22.4 DIESEL FUEL QUALITY

The quality of the fuel used is a major factor affecting the performance, life, reliability, and exhaust emissions of any diesel engine. The power plant staff can play some part in the quality control of the fuel by regular attention to the strainers, water separators, and water drains. In plants where more than one grade of fuel is used, the operator with good knowledge of the fuel systems, their interconnecting points, and the dangers of supplying incorrect fuel to an engine will be more successful than the operator who is ignorant of the system. Two or more grades of fuel will be used in plants operating oil-fired boilers or large engines that consume residual fuels but normally use a distillate fuel for start-up purposes.

The quality and performance of any distillate fuel will be impaired if it becomes contaminated with heavier fuels, other miscible fluids, or water. New bulk fuel may give trouble shortly after delivery because its insertion into the storage tank will have stirred up quantities of previously accumulated and settled sludge, emulsions, or water.

Diesel fuel oil may be divided into two main classifications:

1. *Distillate fuels* are obtained from crude oil which has been vaporized and condensed in a distillation process and will have a fixed boiling point. Distillate fuels do not contain any asphaltic components.
2. *Residual fuels* contain the residues of a distillation process and may contain any proportion of the grade of fuel submitted to a distillation process.

Clean fuel assures maximum engine service life and performance. Dirty fuel and fuels that do not meet minimum fuel specifications will adversely affect combustion, filter life, injection system performance and service life, startability, and perhaps service life of valves, pistons, rings, liners, and bearings. Fuel costs can represent 50 percent or more of total engine operating costs; it is good economics to carefully consider proper fuel selection.

22.5 FUEL OIL TERMINOLOGY

Ash

The ash content of fuel oil causes wear through abrasion. The amount of ash expected in a distillate fuel will be 0.1 percent by weight or less. Included in the ash content are the metallic elements vanadium, sodium, and nickel, but they appear only in residual fuels. They have a corrosive effect that reduces turbocharger, upper cylinder, and exhaust valve life. As with the sulfur content, distillate fuels contain only minimal ash quantities, whereas residual fuels may contain much higher percentages.

Carbon Residue

The carbon residue is the amount of carbon remaining after evaporating and burning off all the volatile matter from a laboratory sample. The carbon residue figure indicates the facility of the fuel to deposit carbon on engine surfaces subject to combustion gases. A maximum carbon residue will contribute to deposits of carbon and consequent sticking of piston rings and exhaust valves.

Cetane Number

Between fuel injection commencement and ignition there is a time lapse, and that delay is expressed as a cetane number—the rate at which the fuel ignites. The cetane number is a measure of the ignition quality of a diesel fuel as determined in a laboratory. The higher the cetane number the better the ignition quality and the lower the tendency of the engine to knock. Higher cetane numbers indicate a shorter ignition lag; they are associated with better all-around performance in most diesel engines, especially high-speed engines. Generally speaking, the cetane number for a fuel used in a high-speed diesel engine with direct injection should be not less than 40; for engines with indirect injection, a minimum of 35 is acceptable. European engines are generally designed for a fuel with a higher cetane number than engines originating in North America.

Difficulties in starting an engine may be met with if the cetane number of the fuel is too low. Engine knocking and puffing of white smoke may be observed during the engine warm-up, especially in cold weather. Prolonged operation of an engine on fuel with a cetane number that is too low will provide conditions for excessive carbon deposit formation in the combustion spaces.

Burner fuels (i.e., light boiler fuels and domestic heating fuels) require an open flame for satisfactory combustion. The cetane number of such fuels is poor in comparison with that of regular diesel fuels, and thus the use of such alternate fuels must be avoided whenever possible.

Vendors may offer fuels with a variety of titles or names. In all cases the new supply should be carefully checked for diesel engine use and suitability, along with the proposed fuel specifications.

Cloud Point

The cloud point is the temperature at which fuel oil becomes clouded by the formation of wax crystals. When the cloud point temperature of the fuel is not higher than the ambient temperature in which the engine is expected to start and run, fuel filter wax retention will not pose a problem. The selection of low-cloud-point fuel is an important consideration in polar regions.

Distillation End Point

The distillation end point is the temperature at which a sample of the fuel is completely vaporized. The distillation or boiling range of a diesel fuel must be low enough to permit complete vaporization at combustion chamber temperatures. Complete combustion of the fuel is largely dependent upon total vaporization prior to the commencement of ignition. Poor vaporization is likely to occur in very cold weather, during light-load operation, or in periods of idle running.

Flash Point

The flash point is the temperature at which the fuel vapors will ignite when exposed to an open flame. It has little bearing upon the fuel's performance in an engine, but it may be a factor in the legalities of handling and storage. The pertinent local fire codes should be consulted as a guide to the upper temperature limits of heated fuel before any fuel is ordered.

Specific Gravity

The specific gravity of a fuel is the ratio of the weight of a unit volume of a fuel compared with the weight of an equal volume of pure water. The specific gravity of water is 1.000, but for petroleum products the American Petroleum Institute (API) method is normally used. In that case, the specific gravity of water is 10.00 and

$$\text{Degrees API gravity at } 60°F = \frac{141.5}{\text{specific gravity at } 60°F} - 131.5$$

As the specific gravity decreases, the API gravity increases. The gravity may be determined at any reasonable temperature in conjunction with temperature correction charts. Temperature correction charts are usually furnished by the major fuel suppliers.

Because the specific gravity varies with temperature, all gravities are expressed in conjunction with a given temperature. The heavier fuels, those of high specific gravity, contain more heat units per volume than the lighter fuels. The light-distillate fuels are used on high-speed engines, whereas heavy or residual fuels are used on large, low-speed engines (e.g., 600 rpm and slower).

For example, diesel fuel with an API gravity of API 44° and a specific gravity of 0.8063 will give 18,600 Btu/lb or 155,000 Btu/US gal. At the other end of the scale, fuel with an API gravity of 10° and a specific gravity of 1.000 will give 17,500 Btu/lb or 146,200 Btu/US gal.

Pour Point

The pour point is the lowest temperature at which an oil will flow. A pour point of 6°C (10°F) lower than the ambient temperatures in which the engine is expected to start will permit easy starting, and fuel filter wax retention will be avoided. A high pour point will interfere with the movement of fuel from bulk storage to the day service tanks in cold weather and will also detract from the engine's ability to start. If the bulk fuel has been served with a pour point depressant, the cloud point temperature of the fuel becomes the significant factor.

Sulfur

The sulfur content of the fuel should be always as low as possible in order to minimize engine wear, inhibit the formation of deposits, and limit the sulfur dioxide exhausted into the atmosphere. Limited amount of sulfur can be tolerated, but the quantity of sulfur in the fuel will have a direct influence on the life of the engine components. The detrimental effects of burning high-sulfur fuel will also be reflected in shorter lubricating oil change intervals.

In most instances, the percent of sulfur in diesel fuel is usually not more than 0.5 percent but sometimes the percent may be very much higher. Economic pressures, source of crude oil, and geographic locations of users and refineries have a considerable impact on the quality of fuel supplied. Because of the effects on the engine that variations in fuel sulfur may have, it is advisable to send samples of fuel for analysis at about 2000-h intervals. The intervals may well be adjusted according to the frequency of bulk storage replenishment. Obviously, if the bulk storage tank is filled only annually, as is sometimes the case in isolated communities, only one sample per year need be analyzed. Distillate fuels with very low pour points (qv) contain only minimal quantities of sulfur, but the heavier residual fuels will contain much higher percentages.

Viscosity

Viscosity is a measure of a fluid's resistance to flow; it is expressed as the time required for a given quantity of oil to flow through an orifice of known size at a standard temperature. There are several standards in use worldwide. Those more commonly employed are Redwood, Engler, and Sayboldt, but usually fuel viscosity is reported today in centistokes (cSt). Heating the fuel lowers its viscosity, and residual or semiresidual fuels require heating before introduction into the fuel injector system; in fact, the heavier fuels cannot be pumped at all without heating.

Light fuels suitable for use in high-speed engines will usually have a viscosity rating of about 1.4 cSt. If the viscosity of a proposed fuel approaches 1.0 cSt, it may be prudent to add 0.5 percent clean lubricating oil to the daily fuel supply for the benefit of the fuel injection system. (Used lubricating oil must not be used.)

Water and Sediment

The amount of water and sediment in the fuel is mostly dependent upon the manner in which the bulk fuel is transported, handled, and stored. All new deliveries should be checked upon arrival for evidence of water and sediment, and the bulk storage facility should be test-drained frequently, particularly just after replenishment. Day tanks must be checked daily for signs of water.

High-viscosity fuels always retain some water, but it is almost entirely precipitated by preheating in the storage tank and the settling or day tank. Over a period of time, considerable quantities of a fuel-water emulsion may accumulate in the bottoms of the various fuel tanks used to contain heavy fuel. The use of an emulsion-breaking agent will separate the water from the fuel, enabling the one to be used and the other drained off. Sludge deposits which are apparent at tank cleaning time should not be removed and disposed of until a full attempt has been made to chemically treat and recover the fuel content of the sludge.

22.6 TYPICAL DISTILLATE FUEL

Table 22.1 illustrates typical blended residual fuel analysis. See Table 22.1 for the specifications of the fuel used in North American engines.

TABLE 22.1 Typical Blended Residual Fuel Analysis

Property or component	Specification
Viscosity	180 cSt (1500 Redwood #1)
Cetane number	45 minimum
Flash point	38°C (100°F)
Vandium	100 ppm
Sodium	100 ppm
Aluminum	1 ppm to engine
Silicon	1 ppm to engine
CCR%, 10% bottom, by weight*	15.0%
Water and sediment, % by weight	5.0%
Sulfated ash, % by weight	1.0%
Ashphaltenes, % by weight	10.0%
Sulfur, % by weight	2.5%

*Conrad carbon residue

TABLE 22.2 Typical Distillate Fuel Specifications for North American Engines

Specification	ASTM designation	Caterpillar	Cummins	Detroit	Waukesha	EMD	MLW
Speed, rmp		1200–1800	1200–1800	1200–1800	900–1200	900	900
Ash	ASTM D482				0.02	0.02	0.02
Carbon residue	ASTM D524 ASTM D189	0.15–0.35	0.25	0.35	0.25	0.15	0.35
Cetane number	ASTM D613	40D1/35 PC	40	40	40	40	40
Cloud point	ASTM D97	As per pour point					
90% distillation	ASTM D86	288°C (500°F)	357°C (675°F)	338°C (640°F)	357°C (675°F)	340°C (650°F)	357°C (675°F)
Flash point	ASTM D93	52°C (125°F) or legal					
Specific gravity	ASTM D287	35	32–40		30		
Pour Point	ASTM D97	6°C (10°F) below ambient					
Sulfur	ASTM D1552 ASTM D129	0.5	1.0	0.5	0.7	0.75	1.0
Viscosity	ASTM D445 ASTM D88	1.4	1.5–5.8		30–50	1.8–5.8	1.8–5.8
Water and sediment	ASTM D1796 ASTM D96	0.1	0.1		0.1	0.05	0.1

22.7 FUEL ADDITIVES AND MODIFICATIONS

Additives

Numerous products on the market are often referred to as fuel additives, fuel supplements, fuel treatments, fuel extenders, and fuel improvers. The introduction of any one of these products into the purchaser's fuel can be expected to give a certain predictable result. There is no doubt that certain objectives can be attained when certain specific conditions exist. However, it will be found that the continuous use of a given fuel additive will provide no real economic advantage in the long term. The engine owner will find that it is much cheaper and more beneficial to the engine to eliminate the cause of the condition rather than try to counter the effects of a particular fuel-related difficulty by rather illogical methods.

Few if any engine manufacturers will endorse any of the many fuel treatment products, and none is produced or used by any of the fuel refiners. Claims by the fuel additive producers concerning extended mileage or reduced fuel consumption are generally quite spurious and are usually founded on single-case incidents with little validity. However, certain fuel conditions *can* be usefully treated on an individual basis. Heavy fuels containing water, bunker tank sludges, or heavily emulsified fuels can be induced to precipitate their water content through the use of an emulsion-breaker additive, but that additive will not provide any other direct improvement. Other fuel additives that will assist in the removal of gums and varnishes from fuel injection equipment may be used. Some additives will reduce carbon adhesion to combustion surfaces, but again they do not alter the condition leading to excessive carbon and soot generation.

If a plant considers it expedient to use a fuel additive, a reasonable amount of care must be taken to see that the dose ration is consistently correct and that the addition is fully blended with the main fuel body. When use commences, a log entry must be devised to show what improvements are being achieved, thereby permitting the owner to assess the economic benefits of additive use.

Modifications

In winter operations, difficulties with fuel flow and filtration are sometimes encountered. Under no circumstances should methyl hydrate, gasoline, gasohol, or alcohol be added to the fuel. Severe fire and explosion hazards will result; damage to the fuel injection system can be expected; and future engine starts will be more difficult. Any of the above additions will reduce the fuel efficiency and increase costs. Fuel movement difficulties will be overcome by ensuring that the fuel-heating provisions are in order.

22.8 USE OF METHANOL (METHYL ALCOHOL) IN DIESEL ENGINES

Methyl alcohol (CH_2OH), or methanol, is a colorless, toxic, flammable liquid, miscible with water, ether, and alcohol. It is used in the preparation of formaldehyde, antifreeze, and solvents. It will *not* dissolve in diesel fuels.

Methyl alcohol may not be added to a diesel engine fuel, and to do so will probably void any OEM warranties. Various damages may occur: fuel system seals will decay, and the injection pump will commence leaking. The transparent plastic sediment bowl of the water separators will discolor and gradually become opaque. It is not unknown for the plastic bowl to disintegrate under prolonged exposure to methyl alcohol.

22.9 SLUDGE-BREAKING ADDITIVES

Numerous additives have been formulated for the express purpose of demulsifying sludged fuels. These additives work quite well, particularly if it is possible to agitate or circulate the body of the fuel. The quantities of additive required to treat a particular tonnage of fuel will be stated by the vendor. In many cases demulsification is assisted and hastened by warming the fuel. Many of the additives offered today are intended to fulfill more than one purpose. The most common secondary purpose is to act as a combustion improver, but this benefit is difficult to assess.

22.10 DISPOSAL OF USED LUBRICATING OIL THROUGH THE FUEL SYSTEM

Legislative and ecological pressures lead to the consideration of burning used lubricating oil through engine fuel systems. There is little if any economic advantage in disposing of waste lubricating oil in this manner, but it can be done only if certain conditions are rigidly observed. Only *diesel* engine crankcase oil can be mixed with diesel engine fuel oil. Waste oil from gasoline engines, transmission oil, hydraulic oil, and diesel crankcase oil contaminated by glycol, water, cleaning solvent, and so on, are specifically disallowed.

The ratio of fuel to used lubricating oil must be 250:1 or greater, and the fuel and oil must be thoroughly mixed in a 50-50 blend before the mixture is added to the fuel tank. Improper mixing will result in filter plugging. The lubricating oil should be processed through a filter of not greater than 5-μm capacity before it is blended. The regular filter of 2-μm capacity must be retained.

Because of control difficulties, the disposal of lubricating oil in the preceding manner is not recommended unless the plant is manned by journeyman class personnel.

22.11 FUEL SERVICE TANK FUEL TEMPERATURE

More fuel is supplied to the fuel injection system than is actually used, and the excess is returned to the fuel service tank (day tank). On older engines the proportion of fuel returned will be quite small, perhaps not more than 5 or 10 percent. On the more modern engines which employ very high injection pressures, much heat accumulates in the injection equipment and must be dissipated. For that reason excessive quantities of fuel are supplied to the injectors for cooling purposes. The excess delivery may be more than twice the actual needs. The temperature of the return fuel becomes elevated as heat is collected, and the temperature of the main body of fuel in the service tank also increases.

If the temperature of the fuel in the service tank rises, the temperature of the fuel delivered to the engine also will rise and a loss in power at the engine relative to the amount of fuel consumed can be expected. For a rise of 10°C (19°F) in the fuel inlet temperature above 32°C (90°F), the power loss will be approximately 2 percent. Alternatively, the increase in fuel consumption for a given load will be about 2 percent for every 10°C (19°F) increase in the temperature of the fuel.

Translated into kW/gal or kW/L, Table 22.3 illustrates the importance of fuel tank temperature control.

TABLE 23.3 Fuel Tank Temperature Control Data

°C	°F	% Power Loss	kW/Gal	kW/L
32.2	90	0	16.00	3.52
37.8	100	2	15.68	3.44
43.3	110	4	15.36	3.37
48.9	120	6	15.00	3.30
54.4	130	8	14.70	3.23

The operator must take notice of the temperature of the fuel entering the engine and make all possible efforts to stay in the 32°C (90°F) range. Continuous makeup of the service tank is usually the best way to moderate the service tank temperature. Attention must be given to the control of the fuel cooler when one is installed.

22.12 ENGINE DAMAGE CAUSED BY FUEL SULFUR

In the past, fuel sulfur wear in engines has been minimized through the use of correctly selected lubricants, adherence to established oil change intervals, and the efforts of the fuel refiners to provide a consistently good product. Of late, there have been indications that the wear rate caused by fuel sulfur is increasing.

Fuel sulfur wear is caused by the corrosive effect of sulfuric acid; that acid is developed from the sulfur content of fuel during combustion. Severe wear occurs in piston rings, liners, and exhaust valve guides most noticeably. Unlike abrasion-induced wear, fuel sulfur wear does not produce crankshaft wear, although some bearing damage could be caused by lubricating oil bypassing soot-laden lubricating filters. Fuel sulfur wear cannot be eliminated, but it can be kept down to reasonable proportions by two means:

1. The correct lubricating oil must be used, and the correct oils are listed in the appropriate pages of the manufacturer's manual.
2. Increased frequency of oil change will lower the wear rates. If the sulfur content of the fuel rises to above 0.5 percent, the oil change interval should be halved, and so on.

There are several reasons for the sulfur-induced wear increase during the past few years. Diesel fuel standards generally have always permitted 1 percent sulfur content, but the fuel suppliers have provided, in the past, a product well under that limit. It is now doubtful if the low level of sulfur can be maintained as crude oil prices increase more and more. Poorer-quality fuel is being brought onto the market, and often the consumer has no choice but to accept a lower quality.

In an engine running on sulfurous fuel and having a heavy lubricating oil consumption, it is possible that the wear rate will present no problem. Following an overhaul which reduces the lubricating oil consumption, however, wear rates may accelerate. This situation may seem paradoxical, but it is explained by the concentration of sulfur, which builds up because of the much reduced quantities of makeup oil.

There is a great difference in the effects of using fuel containing 0.5 percent and 1 percent sulfur. Sulfuric acid is derived from the condensation of sulfur trioxide, which is a product of combustion. The condensation (or dew) point varies with the sulfur percentage. At 0.4 to 0.5 percent sulfur range, most of the sulfur trioxide exits in the engine as a gas; as the sulfur percentage increases, the dew point becomes as high or higher than the engine operating temperature and more acid condenses. Therefore, it is important that the engine cooling system temperature should be kept as high as possible and on any engine not less than 75°C (165°F). The wear caused by sulfur does not follow a straight-line variation. The wear caused by 1 percent sulfur will be at least 4 times that caused by 0.5 percent sulfur.

A program of fuel sampling to monitor fuel grade and content serves also to provide advance warning of preventive steps necessary and allow revision to overhaul schedules. At this time, Canadian fuels have been fortunate in having a comparatively low sulfur content but the fuel quality can be expected to deteriorate in the future. For that reason, at least one fuel sample per plant per year should be taken, and that sample should be drawn immediately after annual resupply. Only by this means will the possibility of accelerated wear patterns and trends be determined. When such trends are identified, it will be necessary to adjust overhaul plans and budgets to allow for such changes.

22.13 DISTILLATE FUEL WAXING IN FILTERS

When the temperature of the fuel being supplied to the engine or being brought in from the bulk storage approximates the cloud or pour point, filter waxing may be experienced. In that event, an effort must be made to heat the fuel to a temperature several degrees above the cloud or pour point.

For practical purposes the fuel should have a cloud or pour point 6°C (10°F) below the expected ambient. The fuel ordered for summer use may thus be a little different from that needed for winter use. If the bulk supply is frequently replenished, it is possible that much of what is seen as wax may actually be ice crystals. Should any of this ice be pumped to the daily service tanks, accumulations of water will eventually result.

22.14 WATER IN BULK FUEL

Water in the fuel oil is not an uncommon occurrence, particularly at resupply time. It is quite fair to say that any difficulties caused by water reaching the engine fuel injection system indicate a lack of operator vigilance. Since most of the water contained in fuel oil is delivered with the fuel that arrives by either barge or by truck, very little of the water content in the stored fuel can be blamed upon condensation. Whatever the source, it is the duty of the plant operator to prevent water from invading the engines. Almost every diesel fuel storage and delivery system has water drains at the main storage tank and at the daily service tank. In many plants there is also a fuel-water separating facility which also has drains. These drain points should be tested at the beginning of each shift; if that is done, there is no reason why any water should progress as far as the engine fuel filters and on to the injection equipment.

Should water be passed to the fuel injection pumps, damage will certainly ensue. At best, it will be limited to plunger and barrel scoring; quite probably some plungers will be seized by the time the problem is identified and the engine stopped. The injector nozzles will be similarly affected. If an engine stands idle and water is allowed to gather in the engine-driven fuel pump(s), a corrosion problem will develop and the fuel injection equipment will quickly be ruined. In cold weather, there is the added danger of the water freezing, with inevitable damage when an attempt is made to start the engines.

All possible water must be drained from the fuel before the fuel is passed to the engine. Generally this can be accomplished in more than one stage. It may happen that substantial quantities of fresh water or seawater will be present in the bulk storage tanks. The water drain cock must be opened and the effluent must be checked for water every time before fuel is drawn from the bulk storage for use until there is no sign of water whatever. If bulk fuel is again delivered, the drain cock routine must be repeated as before.

Water will precipitate more readily from warm fuel, and thus it may be found that further water will appear in the daily service tanks or in the settling tanks. These tanks also are fitted with water drain cocks, and testing of tanks for water should be done every shift as part of the operator's routine.

The fuel drawn from the daily service tank is then frequently passed through a static centrifugal separator, where a whirlpool effect causes water to be precipitated from the fuel (Fig 22.1). The fuel next goes through a filter, which retains any solids, and then moves on to the primary fuel pump. The collected water is visible in the glass bowl of the separator, from which it is drained as it becomes evident. The time taken for water to show in the bowl may vary from minutes to weeks, and so continuous operator vigilance is essential.

In larger plants and where heavier grades of fuel are used, dynamic centrifugal separators are used to extract water from the fuel. It is common with such equipment for the fuel system to incorporate one or more settling tanks. The fuel is brought from the bulk storage tank to a settling tank, from which it passes to the separator and then goes to the daily service tank. It is quite normal for the separator to remove large quantities of solids as well as water from the fuel, and frequent cleaning of the separator elements is required.

22.15 HEAVY AND RESIDUAL FUEL

On very large engines of, say, 5 MW and above, one expects the use of heavy residual fuels to be commonplace. These fuels have viscosities up to 2500 cSt (10,000 Redwood) at 38°C (100°F). Residual fuels frequently carry solids, salts, and water. Because of the damage and excessive wear induced by the contami-

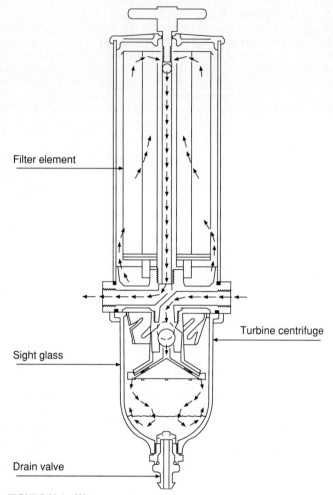

FIGURE 22.1 Water separator.

nants, it is necessary to condition the fuel before use. This is done by centrifuging the solids and water out after the fuel has been heated. The oil must be heated as hot as the grade of fuel will allow without cracking or gassing. The maximum temperature to which the fuel may be heated is very dependent upon the specification of the fuel, and that specification must be reviewed whenever there is a resupply of fuel. Preheating of heavy fuels is usually accomplished through the use of waste heat derived from the engine. Quite often an auxiliary boiler that may utilize oil firing and engine exhaust gas provides steam for fuel heating.

Residual fuels are very susceptible to sludging. The residual fuels supplied are usually blended with lighter fuel to conform with a given user's specification. Problems are introduced because the ashphaltenes and other high molecular weight compounds of the basic residual stock are precipitated by the cutting action and dilutants of the lighter blending fuel. This creates a sludge which settles in the main storage tanks and possibly in the preseparation tanks. If the sludge travels as far as the fuel separators, the separators will quickly become overloaded and ineffective. There may well be a substantial loss of usable fuel in the form of dumped refuse at the separator waste discharge. A further cause of sludge development may be the mixing of fuels with different origins.

The problems posed by fuel sludges can be mitigated somewhat by good operating practice and care, but there is no doubt that the use of residual fuels in a diesel power plant needs far more attention on the part of power plant staff than is necessary when using distillate fuels. Because of the differences in the specific gravity of distillate fuels and that of water, fuel and water separate from each other fairly readily. The water can be drained from the bottom of the daily service tank without any difficulty. However, the residual fuels have a higher specific gravity that may, in some cases, be equal to that of water, and separation by gravitational means then becomes exceedingly difficult.

Water is usually mechanically removed from residual fuel oils by passing the fuel through a centrifugal separator after it has been heated (Fig. 22.1). Separation also has the effect of removing quantities of solid matter from the fuel.

22.16 HEATING OF RESIDUAL FUEL

The circulating pumps in a residual fuel heating system may produce an intermittent chattering sound. This noise will occur when the fuel oil is being overheated enough to cause it to crack. The phenomenon is most likely to take place when the fuel oil is being recirculated through the heater during periods of light load or idle running. In these circumstances, the operator may find it necessary to adjust the heat supply and review the efficiency of the thermostatic heat controller. If the heating system for either residual or crude fuel utilizes engine jacket water as a heat source, the fuel pressure must be left at a higher pressure than the jacket water pressure.

22.17 CRUDE OIL AS A FUEL

Crude oils contain light ends (i.e., gasoline, kerosene, and multiple gases), which ordinarily are distilled off in the refining process and become distillate fuels. The remainder is known as a *residual fuel*. Light crude oils in their natural state can be used as an engine fuel oil. After any water that it may contain has been drained off, light crude oil can be satisfactorily used in an engine with but little modification. Heavy crude oil must be preheated, centrifugally cleaned, and trimheated before being passed to the injector. The injectors on engines using heavy crude are usually water-cooled.

The burning characteristics of crude fuel tend to be rather different from those of distillate fuel. Combustion tends to start sooner and end sooner if the timing is unaltered. Smoke may develop if combustion of the heavy ends of the crude fuel is incomplete. Before any attempt is made to use crude fuel, the supply must be held in a settling tank so that any inherent gases can evaporate and be led away to a wind-swept dispersal area. The fuel must be heated so that any entrained water is precipitated. The crude is drawn off to the conditioner, where it is filtered and heated and the final remains of water are dropped out. Depending upon its nature, the crude fuel must be heated to a certain degree: first, so that the filters do not get plugged with wax and, second, so that proper atomization occurs when the fuel is injected into the combustion space.

22.18 FUEL DAY TANK

The fuel day tank serves as a reservoir for a limited supply of ready-for-use fuel for one or more engines (Figs. 22.2 and 22.3). In most cases, the fuel day tank contains sufficient fuel for only one day's operation; fresh fuel is pumped into it from the main storage tanks. The pumping may take place under an intermittent automatic control, with the upper fuel level being maintained on an almost continuous but intermittent flow. Manual controls that are provided in the filling system will override the auto controls. If and when manual control of the filling pumps is necessary, the operator must guard against any distraction of his attention while fuel is being moved.

The fuel day tank is also commonly known as the fuel service tank or the fuel settling tank. In large plants where residual fuels are utilized, it is normal practice to have two fuel day tanks. The recently pumped fuel in one is heated and settled thereby permitting the final draining of water, while the other tank supplies the engines with warm water-free fuel. It is not desirable to have the main fuel storage facilities directly connected to the engines for reasons of safety, fuel quality control, fuel accounting, and fuel-switching arrangements.

The fuel day tank is usually arranged to give a static head pressure to obviate any cavitation in either the engine's low pressure, primary or fuel transfer pump, or the fuel injection pumps; its elevation also helps preclude the possibility of air locking in any of the fuel pipes between the fuel day tank and the engine's low-pressure fuel pump.

Fuel day tanks are frequently fitted with thermostatically controlled heating elements so that the fuel can be delivered to the engine at an optimum temperature. Heating also assists greatly in the precipitation of any water or solids that may be present in the fuel, particularly in periods of cold weather, immediately after replenishment of the main fuel supply, or when the main body of stored fuel has been otherwise disturbed.

Many fuel day tanks are utilized as receivers of return fuel from the engine's fuel injection system. This returned fuel is usually quite warm, and it adds heat to any newly arrived fuel. At no time should any valve or cock in the return line be closed when the engine is running. To close a valve or cock may result in considerable damage to the engine.

Every fuel day tank is fitted with a drain valve at its lowest point to permit the running off of any water that may accumulate in the bottom of the tank. The diesel plant operator must make a visit to the water drain valve as part of his inspection tour; he should open the drain valve and check for the presence of water.

Water will most often be seen following resupply of the main storage tanks, but substantial amounts of water may also be noticed after the temperature of the main fuel body rises from below freezing temperatures. Sometimes the water will emerge in a light brown cream consistency. This is emulsified fuel and water, and it must be cleared from the tank. When the station is equipped with more than

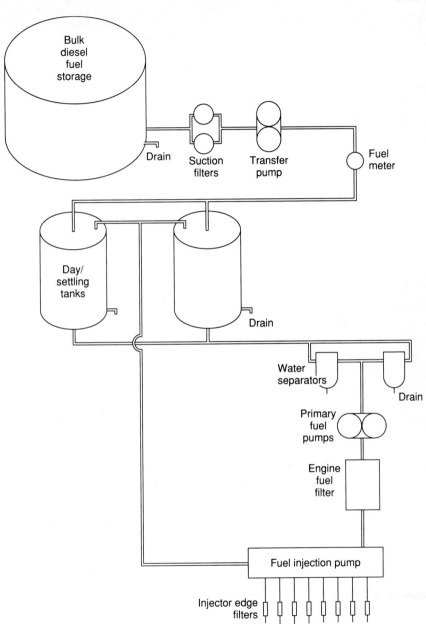

FIGURE 22.2 Distillate fuel system.

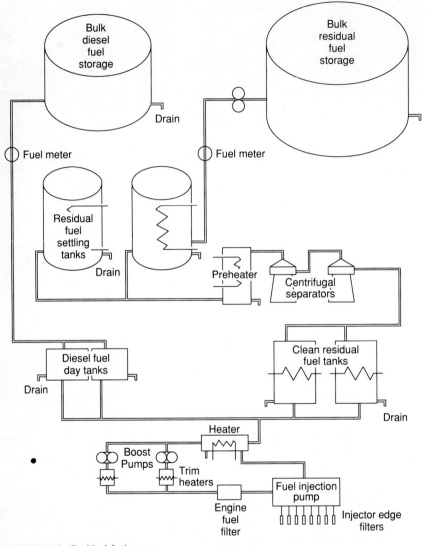

FIGURE 22.3 Residual fuel system.

one tank, the daily service tank should not be completely filled, because it will overflow when the new fuel warms up and expands.

22.19 FUEL TANK VENTS

The vents on any fuel tank, be it bulk storage, transfer, or day tank, must be clear and workable at all times. If the vents are not open, a vacuum will form in the tank when fuel is withdrawn and a partial to total collapse will result. It is recommended that the day tank vents be viewed every shift and the vents of the bulk storage tanks be viewed immediately before and after each replenishment and particularly when there has been a heavy snowfall or icing conditions are present.

22.20 FLEXIBLE FUEL LINES AND CONNECTIONS

The use of plain neoprene textile hoses in the engine's primary fuel system is not usually in accord with the fire protection authorities' regulations. Flexible connectors in pipe systems carrying fuel at low pressures are permitted if the hoses are a neoprene material with an outer woven wire armor.

It is required that all hoses be inspected at frequent intervals for any signs of crush, tension, compression, or other distortion. Corrosion, hardening, bulging, softening, or cracking may occur under adverse conditions of heat, cold, fumes, or stress. Damaged flexible connections represent a fire hazard.

Excessive vibration and externally imposed permanent or intermittent loads will damage the connections. At no time may any flexible connection be used as a grab handle or footstep, nor should any strap, clip, or tie be applied to a flexible connection to support another component. It is recommended that all flexible fuel line connections be renewed at not less than 15,000 service-hour intervals.

22.21 FITTINGS FOR FUEL PIPE SYSTEMS

If it is found necessary to repair or modify the pipe system that carries fuel oil, the new parts must be selected with care. The strength or schedule specification of the new or replacement items must be at least equal to those of the items remaining in the system. Under no circumstances should galvanized (zinc-plated) fittings be used. Because of the zinc content of brass, the use of brass fittings in a fuel system cannot be permitted. The sulfur in fuel oil reacts with the zinc to form a sludge which is very harmful to the fuel injection equipment.

22.22 FUEL SPILLS

Many spills are caused by the rupture of the pipe system connecting either the storage system to the supply point or the storage system to the usage point. These ruptures are brought about by overpressurizing the connective pipelines during pumping operations or by contact damage by vehicles or earth-moving equipment.

There is always a possibility of a fuel spill occurring, and these notoriously take place when an extraordinary activity is being undertaken: at annual resupply, when there is difficulty in getting fuel out of the tank, or when fuel is being transferred from one tank to another. It is imperative that when a fuel movement takes place, the person assigned to the activity be undisturbed and not be distracted. That person must know exactly what to do if a spill incident develops.

Overpressure damage can be forestalled by correct and continuous attention by the person handling the system during fuel movement operations. Contact damage to the pipelines can be prevented by the provision of high-visibility markers along the path of the pipelines, by the clear identification of pipeline crossing points, and by the secure fencing of the owner's fuel storage and power-generating facilities. Damage can also be avoided if the power plant personnel exercise a continuous awareness of events taking place in the vicinity of the power plant complex where the use of digging equipment, blasting, and earth moving is involved.

All primary fuel line valves should be kept closed when fuel is not in movement. That will limit the amount of fuel spilled if the pipeline is accidentally torn apart. When a pipeline breakage occurs on an open line, the valve at the bulk supply end must have the first priority of closure. All nonreturn valves in the fuel supply lines must be in good working order.

The support structures for the connecting pipelines must be correctly and securely positioned. At least annually these supports should be inspected, repositioned, adjusted, and otherwise maintained so that none of the bends, flexible connections, or other fittings are overstressed. Any leaks must be attended without delay, and any roadway crossings over or under the pipelines must be inspected for evidence of misalignment, crushing, or thinness of road surface dressing.

The prepared plant will have certain materials and equipment on hand in readiness to deal with an oil spill should one occur. The personnel will also have received instruction in the business of oil spill prevention, control, and cleanup. It is suggested that permanent written instructions for such a contingency be posted for the benefit of all.

Fuel leaks in the storage system may often be found at:

- Welded seams of bulk tanks
- Bolted flanges of sectional bulk tanks
- Ruptured or strained flexible connections
- Glands of shutoff valves
- Crushed sections of pipeline (often buried and out of sight)
- Fuel transfer pump glands
- Quick couplings on off-loading hoses.

All buried fuel pipelines must have surface markers to advise any heavy vehicle or digging machine operator of their location.

22.23 STANDING ORDERS FOR HANDLING FUEL SPILLAGE

In a well-organized plant there will be a series of instructions that will apply in the event of a bulk fuel spillage. These instructions should be approved by the owner, the relevant environmental control agency, and the fire prevention authority. Posted in a prominent position along with the other standing orders, they should be read and clearly understood by all power plant personnel. Discussion of oil spill prevention and handling should take place as part of the ongoing safety training program. It can be said that familiarity with the methods of oil spill prevention and cleanup on the part of the power plant staff is as important as knowledge of first aid and industrial safety.

A typical standing order for the handling of an oil spill could read:

Immediately following the detection of a fuel spill the operator in charge will:

1. Take the actions necessary to stop the flow of spilling oil (i.e., close both ends of the ruptured pipeline, stop the pump, etc.).
2. Take actions to limit the spread of the spill once the source of the spill has been shut off. Priority of action must be given to preventing the entry of spilled fuel into drinking water supplies, fire-fighting water sources, and drainage systems.
3. Advise the plant superintendent forthwith so that he may communicate with the various agencies concerned with oil spill matters. These agencies may include the fire-fighting, water resource control, environment pollution regulation, fish and wildlife conservation, highway, forestry management, and other authorities."

Fuel oil spills generally have a high political sensitivity and receive much unfavorable media attention. Regardless of the cause of the spill, the power plant owner always suffers some embarrassment together with considerable financial loss. It is incumbent upon every person concerned with the supply, handling, or consumption of fuel to exercise the same care and attention to the prevention of fuel spills as would be given to other work practices and property security matters.

22.24   PEAT MOSS

Many plants stock a supply of peat moss for the absorption of spilled fuel oil. This is an excellent absorbent material, but when saturated with oil, it is highly flammable. Furthermore, when soaked through, it is extremely messy if walked upon or driven over. Under no circumstances should peat moss be used inside a power plant for normal small-scale cleanup use. Any plant that attempts to use peat moss for housekeeping purposes will pay dearly for the error at some point in the future.

22.25   FUEL SYSTEM SECURITY

It is recommended that all valves outside the power plant (such as shutoff, filling, transfer, cross-connecting between one tank and another, and water drain valves) on the fuel storage facilities that are accessible to the general public be kept padlocked. The keys for all such padlocks should be retained on a silhouette board in the general office (and in the case of a multiply manned plant, issued only against a signature).

22.26   FUEL STORAGE TANK DIPS

Certain precautionary measures must be taken when dipping or sounding fuel tanks. Any tools used must be of nonsparking materials. Smoking on or in the vicinity of fuel tanks cannot be permitted. The cap on top of the tank must be returned to the closed position after the measurement is completed.

To determine the quantity of fuel remaining in a fuel storage tank, it will be necessary to take a depth measurement of the fuel and refer to the tank volume table in conjunction with the temperature conversion table. If there is any water accumulation in the tank, a depth measurement must be taken both before and after draining to determine how much water was run off. Otherwise, the fuel consumption figures will be wrong. In polar areas, it may be necessary also to check for ice, which will be in the bottom of the tank. This is done by comparing the winter full-depth reading with the summer absolute tank depth. The difference represents the thickness of ice in the bottom of the tank. This information can be entered in the fixed-data portion of the workbook.

22.27   FUEL STORAGE TANK ENTRY

Under no circumstances whatsoever should any person attempt to enter an empty fuel storage tank. If it is necessary to repair or clean a fuel storage tank internally, specially trained and equipped personnel who are familiar with the hazards of empty fuel tanks can be contracted for to do whatever is necessary. *Unauthorized entry into an empty fuel tank is usually fatal.*

22.28   FUEL STORAGE TANK FILLING

Fuel oil tanks should be filled only from the bottom. Splash or top filling should never be used, because there is a danger of static electricity being generated by the passage of air bubbles up through the loaded fuel.

Because of the air that may be entrained in the delivered fuel and the time needed for its natural release—during which time static electricity may be developed—no tank should be dipped during the filling process or for at least 5 min after filling has stopped. At no time should diesel fuel be loaded into a tank that previously contained gasoline without the approval of a knowledgeable and responsible person (e.g., the fire marshal). All fuel tank filling must be carried out in accordance with the fire and safety codes of the governing jurisdiction.

22.29   INSPECTION OF FUEL STORAGE FACILITIES

Once per shift or daily, in the case of semiattended plants, the operator should inspect the main fuel storage tanks. The water drain valve must be opened to check for the presence of water, and the main delivery and discharge valves should be confirmed as properly closed. The circumference of the tank must be examined for signs of leakage (Fig. 22.3 and Table 22.2).

22.30 FUEL TANK BERMS

1. The berm drain valve shall be kept closed and locked at all times except when it is being exercised.
2. Weed growth within the berm shall be treated with a defoliant on an as-needed basis to reduce the possibility of fire.
3. It is recommended that the condition of the berm be inspected by a designated person at approximately monthly intervals.
4. Undergrowth, erosion, or settlement are common earthen berm failings.
5. Splitting, spreading, and corrosion can be expected in the earth-filled metal cofferdam type of berm.
6. Weld separation and corrosion will be encountered in the all-sheet-metal type of berm.
7. Runoff and rain wash may damage earthen berms, particularly if they are in way of a natural water course.
8. Any condition that may cause spilled fuel to escape from the berm will require immediate attention.

CHAPTER 23
COOLING SYSTEM

23.1 PURPOSE OF THE COOLING SYSTEM

The purpose of the engine cooling system is to extract surplus heat from the engine and disperse it elsewhere. The heat surplus is due to the inherent inability of the engine to convert all of the thermal energy released through the combustion of fuel into useful work. The cooling system performs the basic function of circulating a fluid around the jacketed cylinder liners and the cylinder head in order to remove surplus heat so that it can be conventionally dissipated or utilized elsewhere.

In very general terms, it can be said that one-third of the heat content of the fuel is lost to the cooling system, one-third is lost to the exhaust gases, and the remaining one-third is converted into work or useful energy. The cooling system of an engine is responsible for a very high proportion of the difficulties experienced in an engine's operation, and for that reason it is important that every operator have very good understanding of the system and the problems he may face.

23.2 ENGINE OPERATING TEMPERATURES

Every engine is carefully designed to operate within a particular temperature range. The thermostat or regulator serves to permit surplus heat to leave the engine in a controlled manner. An engine is composed of many components made from differing materials and alloys. These components assume intended dimensions when the engine attains its designed operating temperature. Each individual part will be receiving heat from the combustion process, and that heat must be dissipated to the cooling medium in a precise manner if the engine components are to be maintained at their correct temperatures and thermal dimension and expansion relationships.

If the components do not attain their thermal dimensions, unusual effects will result. Valves will run with excessive clearances that will lead to accelerated seat wear, and some clatter may be audible. Piston rings will not totally conform to the liner bore, and increased blowby will develop and will shorten the life of the lubricating oil and the lubricating oil filter elements and contribute to accelerated liner wear.

Low-temperature running may cause excessive crankshaft and camshaft float.

If the crankshaft is designed to have a zero deflection at operating temperatures, it is probable that low-temperature running will induce some deflection that will stress the crankshaft and reduce its fatigue life. Alignment problems can be expected, particularly on generator sets that employ alternators with two bearings. The hogging effect often encountered on overcooled engines will provide conditions conducive to cylinder gasket leakage, notably on engines provided with one head to serve several cylinders.

23.3 ENGINE OVERCOOLING

Overcooling exists when the engine coolant circulates at abnormally low temperatures. Engines operated at temperatures below 74°C (160°F) will suffer accelerated cylinder, valve, and valve guide wear. On engines using fuel with a higher sulfur content, the wear rates will be even more pronounced. On engines that habitually run at low temperatures, piston ring gumming can be expected, and it will result in increased combustion gas leakage. Overcooling usually takes place when the automatic cooling system controls are not functioning properly or the radiator fans are run continuously on manual control.

Today's coolant regulators are extremely reliable, and cooling system problems, which are often attributed to faulty regulators, frequently have their origins elsewhere. Therefore, it is important to thoroughly check out any overheating or overcooling problem before discarding the original regulators. Any suspect regulator can readily be tested.

It may be found that the thermostats have been removed from the engine for some reason, the fans of remote radiators are switching in too soon or are running continuously, or the louver linkage of in-plant radiators is incorrectly set up. Changing the regulators to another temperature range will neither solve problems nor serve any real purpose.

23.4 HIGH JACKET COOLANT TEMPERATURE SHUTDOWN

The sensors for high coolant temperature protection devices are set to signal a shutdown at a particular temperature. That temperature is set when the engine is installed, and it should remain unaltered thereafter. When a high jacket coolant temperature condition initiates an engine shutdown, the cause of the situation must be identified and rectified before the engine may be restarted.

Before attempting to restart an engine that stopped in an overheated condition, the engine should be barred over to see if there is any sign of seizure. The lubricating oil filter elements should be checked for metal debris; it is possible that there may be some (aluminum) piston metal or bearing metal visible even if the engine is free to turn.

If following startup there is an increase of crankcase pressure from piston blowby, broken piston rings may be deduced to be a consequence of the overheating. Similarly, if when the unit is placed on full load, coolant is blown out of the header tank, cracked cylinders or cylinder heads can be expected.

If the engine will only partly bar around, there is a probability that one or more cylinder heads have cracked and the cylinder is flooded. *Do not attempt to start the engine in such a condition.* Review the circumstances under which the shutdown took place. By questioning the circumstances of the event, many of the features that can contribute to such a situation can be examined and either discussed or followed up further. By proper diagnosis of the problem, the correct course of action can be taken. Half an hour's consideration of a difficulty is worth a great deal more than five minutes of impetuous guessing in any problem-solving activity.

23.5 HEADER TANK

Coolant expands by approximately 5 percent of its original volume during the rise from ambient to full operating temperatures, and it is to accommodate this expansion that the header tank, sometimes known as the *expansion tank*, is provided. For the purposes of this discussion, however, the term *header tank* will be used exclusively.

The level of the coolant in the header tank should remain constant after the engine has reached its normal operating temperature. Changes in the load carried by the generating unit will alter the amount of heat that must be rejected, but the regulators (thermostats) and the radiator will maintain the heat content of the coolant at a designed quantity. There will be little expansion or contraction of the coolant during normal operations.

Air or leaking combustion gases must be eliminated from the cooling system, and thus the header tank is fitted with nonreturn vents. Usually, the vents are built into the filter cap, which permits excess coolant gas or air to escape. Frequently the header tank is fitted with a sight glass so the level of coolant can be seen from the operating floor and any contamination by lubricating oil, and so on, can be observed.

If the cooling system has been drained for any reason, the system must not be completely refilled. The replenishment must be stopped when the coolant just begins to show in the bottom of the header tank gauge glass. That will permit the coolant to expand after start-up of the engine without any overflow, and consequent wastage of coolant from the header tank, as full operating temperatures are attained.

The ejection of coolant from the header tank will be caused by one of three conditions: overheating, overfilling, or gassing.

Overheating

If there is a defect in the engine's cooling system or in the heat rejection system, the coolant within the engine can reach boiling temperatures. In that case, the high jacket coolant temperature switch should shut the engine down. After the engine has come to rest, the power plant personnel can examine it to determine the cause of overheating and make the necessary rectifications.

If the high jacket water temperature switch is malfunctioning, it may happen that coolant will boil. That will cause the header tank to overflow either through expansion of the coolant or through severe turbulence of the coolant.

Overfilling

The engine cooling system and its attached heat rejection system have a capacity to hold a certain volume of coolant. When the engine is at rest and is at a normal standby temperature, the whole system with the exception of the header tank will be filled with coolant. The header tank is provided to accommodate any expansion of the coolant as the coolant rises to standard operating temperature when the engine is run. If the header tank is filled with coolant when the engine is at rest and cold, then overflow at the operating temperature is inevitable.

Gassing

Gassing occurs when combustion gases leak into the coolant. Leakage takes place through cracks which develop in the cylinder head, cylinder liners, or precombustion chambers and also when the cyliner head is not fully seated on the gasket between the cylinder head and the cylinder liner. Improper seating may be caused by uneven torquing or breakage of the cylinder head fasteners.

The extremely fast expansion of the combustion gases which enter the coolant will give the coolant the appearance of boiling. Much coolant will be displaced, which will result in substantial quantities being ejected from the header tank. When the engine is gassing, it is highly unlikely that the high jacket water temperature switch will sense any temperature rise and will therefore not shut the engine down, nor will the jacket cooling temperature gauge show any increase in temperature.

Nothing will be achieved by draining coolant from the system in an attempt to stop overflow. If air appears to be present in the coolant at the engine, in the header tank, or in the radiator, it must be believed that it is in fact combustion gases. It is highly improbable that any symptom of boiling could be caused by air entrained in the cooling system through the coolant pump glands. In the event of gassing, the operator will have no alternative but to take the engine off load, shut it down, and initiate diagnostic procedures.

Gassing can be proved by attaching a plastic hose to a test cock mounted on the coolant outlet manifold and terminating in a bucket or container of about 20-L (5-gal) capacity. With the engine running on full load and the test cock open, bubbles will be seen coming out of the hose if there are combustion gas leaks. On smaller engines, the hose may be fitted to the radiator overflow pipe. On engines with individual fuel pumps the faulty cylinder can be identified by momentarily pushing the fuel rack into the full-fuel position, one cylinder at a time, while observing the coolant bubble flow.

To continue running an engine with evidence of gassing may lead to the multiple cracking of cylinder heads. It is possible for the leaking combustion gases of one cylinder to pressurize the engine cooling spaces to a degree that will prevent the entry of further coolant into one or more cylinder heads and thereby provide the sort of condition in which the cylinder heads will experience a thermal shock.

23.6 FALLING COOLANT LEVEL IN HEADER TANK

A falling coolant level indicates that coolant is being lost from the system somewhere. It is highly improbable that the loss is caused by evaporation; almost invariably it will be caused by leakage. That leakage may be in one or more of several places, the most likely of which are:

- Cylinder liner O-rings
- Cracked exhaust valve cages
- Leaking injector tubes
- Leaking radiator tubes
- Poor coolant pump glands
- Improperly closed drain valves
- Leaks in pipe system to remote radiators

It is not enough just to bring the level of the header tank back to its normal operating level. The reason for the loss must be determined and remedied. Whenever there is a coolant loss that is clearly not outside the engine, the lubricating oil level in the pan must be very carefully watched. If the lubricating oil level rises or fails to go down at its normal rate, it is probable that coolant is entering the lubricating oil system. In that case, an immediate shutdown is called for. The thinking operator will take heed of the amount of makeup oil that the engine normally requires, compare it with the present demand, and then relate it to the amount of makeup coolant that is added to the header tank.

23.7 RISING COOLANT LEVEL IN HEADER TANK

Under certain circumstances the coolant level in the header tank will rise, and another set of conditions is indicated. The coolant volume in the system will be expanded by the introduction of combustion gases or pressurized aspirating air. Commonly, a cracked cylinder head, a leaking precombustion chamber, broken or loose cylinder head studs, or a blown cylinder head gasket will permit gas to leak into the coolant spaces. As the gas expands, it raises the level of the coolant visible in the header tank before venting to the atmosphere.

In most instances of such gas leakage one can expect large quantities of coolant to be ejected from the header tank, which will leave no alternative but to stop the engine and repair the fault. If the engine is allowed to run under such conditions, cylinder head cracking will be inevitable. The leakage of combustion gas into the coolant frequently gives the coolant the appearance of boiling, and the operator must not be misled by this phenomenon.

23.8 COOLING FLUIDS

Because ordinary fresh water is corrosive at engine operating temperatures, the chemical condition of the coolant in an engine must be monitored. Fresh antifreeze media carry the necessary inhibitors, but the inhibitors decay after a period of time. Those in the coolant mix can thereafter be rehabilitated by the introduction of additives. *Caution must be exercised to see that the additives used are compatible with the coolant they are intended to treat.* If there is any doubt, ask the supplier and the OEM.

The term *fresh water* refers to water in a pH range of 6.5 to 8.0 and containing no more than 100 ppm chlorides. A nontechnical description would be that the water is drinkable (potable). Seawater refers to salt water, river water, lake water, and all waters that do not meet the fresh water description.

Some raw waters may present problems for the engine operator. Algae, bacteria, silt, and other materials may be in the raw water supply. It happens on occasion that portions of the system will become fouled and cleaning will be necessary. When the cleaning is done by chemical means, extreme care must be taken to preclude any chance of environmental harm by the effluent from the cleaning process. Some of the cooling system cleaning agents that can be applied without careful consideration are highly poisonous in rivers and streams.

Deposits will obstruct the flow of water, interfere with the flow of water, or interfere with the transfer of heat. Dirt may settle out and plug the water passages in an engine or its radiator, and large particles can completely block the flow of water through radiator tubes. Dissolved minerals will form deposits in the system if the coolant boils. These deposits are usually very hard, and they adhere tightly to the internal surfaces of jackets and cylinder heads. Lime, the most common deposit, is a very poor conductor of heat. A layer of lime $\frac{1}{32}$ in thick has the insulating equivalent of 2 in of cast iron. Many an operator upon seeing just an eggshell thickness of scale will conclude that the engine is nice and clean!

Minerals alone are not responsible for all water problems. Hydrogen sulfide (H_2S) and carbon dioxide (CO_2) are often present when water has been in contact with decaying organic or vegetable matter, and these substances will act very rapidly to corrode copper surfaces. Another common corrosive is the oxygen that is present in water from streams, lakes, and shallow wells. The logical preventive measure is to keep oxygen out of the cooling system. Air will enter through hose connections, pipe fittings, or water pump seals that are permitted to leak slightly. The connections to the inlet side of the water pump are particularly important because the inlet side is the suction side of the pump. Air leakage may also take place at the radiator cap, particularly if the engine is persistently overcooled.

Galvanic activity in salt water cooling circuits produces a corrosive interaction with metal that results in the deterioration of system components. Cathodic protection can be employed by installing sacrificial zinc rods in the seawater flow passages at numerous locations in the cooling system. In order to maintain this protection, the zinc rods must be inspected regularly and replaced when deteriorated.

Kits for testing alkalinity are available, and so are the additives necessary to restore the antiscaling and anticorrosive properties of the cooling media. Use of these aids greatly reduces the overall cost of ethylene glycol coolant because a well-maintained coolant will have a much longer service life.

23.9 FROST-PROTECTION COOLANTS

Ethylene Glycol

The liquid coolant of an engine in high-latitude service must be proofed against freezing at temperatures of $-46°C$ ($-50°F$). Protection is achieved by mixing 55 percent glycol with 45 percent water. The normal commercial type of glycol-based antifreeze product also contains various additives the most important of which are the corrosion inhibitors. Because of the secondary protections offered by antifreeze mixtures, it is better to operate year-round without altering the cooling mixture.

Raw water alone should never be used in an engine cooling system because of its harmful effects. Nor is the use of 100 percent glycol beneficial, because the glycol will start to gel as it cools to $-23°C$ ($-10°F$) and will greatly impede its own movement. The frost resistance of all coolants must be checked regularly, and in all engines the pH value of the coolant must be checked. It should be kept slightly to the alkaline side of neutral; hence, the optimum reading should be between 8.5 and 9. If a pH reading of less than 7 is present, there is a likelihood of corrosion occurring on the wetted side of liners.

Maintenance checks will occasionally indicate a necessity for small amounts of coolant makeup. The operator must check the strength of the main antifreeze mix and add either water or ethylene glycol according to the readings obtained. At normal operating temperatures there is always a certain amount of loss of water from the coolant mix through evaporation, and that tends to increase the ratio of antifreeze to water remaining. Over a period of time, the correct ratio can become quite disarranged if attention is not given to the correct makeup adjustments.

Ethylene glycol is a colorless dihydroxy alcohol used as a basic antifreeze agent. The color of the product varies from supplier to supplier, so it is not a reliable guide for future makeup purposes. At the same time, care must be taken to ensure that any makeup antifreeze used is compatible with the original and, similarly, that any conditioner material added is compatible.

Ethylene glycol has a deleterious effect on rubber seals; after approximately 30,000 h of immersion, the breakdown of liner O-rings can be expected. It is, therefore, important that the O-rings be changed out in good time so that there is no leakage of coolant into the crankcase.

Similarly, ethylene glycol has a weakening effect on rubber and fabric hose sections, hoses, or connections. All hoses and connectors should be examined for bulging, interior lining detachment, softening, or degradation at regular intervals and renewed as necessary, usually as part of a scheduled overhaul routine.

Ethylene glycol is expensive, and care must be taken in its storage, handling, and dispensing. However, there is no economy in providing a weak mixture, because the system will just freeze up that much sooner. Beware of the overstrength mixture, which will gel and refuse to circulate in the radiator.

When glycol gels, no damage is done in the first instance to either the engine or its cooling system provided the unit remains static. However, if the engine is started, although the coolant in the engine jacket will liquefy, the gelled coolant in the radiator will not circulate because it will not receive sufficient transferred heat for it to alter its state. The engine will quickly overheat and either will shut itself down as the overheat protection device is triggered or will seize.

Propylene Glycol

In a waste-heat recovery scheme by which heat from the engine's cooling system is used to temper potable water supplies, a nontoxic antifreeze medium must be utilized. Propylene glycol is usually substituted for the normally used ethylene glycol whenever there is any possibility of contamination through leakage. The same attention given to the standard coolant must also be given to propylene glycol–loaded system so that the correct level of alkalinity is maintained. The heat transfer properties of propylene glycol are inferior to those of the less costly ethylene glycol. It is possible, therefore, that some engine installations may experience cooling problems at full load if a conversion to propylene glycol use is made without some modification to the heat rejection capability of the system. The life of propylene glycol is quite short, and samples must be sent to the supplier for inspection regularly.

The operator must be sure that the cooling system(s) in his care uses either ethylene glycol (usually greenish-yellow in color) or propylene glycol (usually blue in color). At no time should one type of glycol be used as a makeup or a substitute for the other. Both ethylene glycol and propylene glycol supplies must be kept away from open flames, because both are quite flammable.

23.10 COOLANT CONDITIONERS

Scaling of cylinder liners and cylinder heads should not generally occur in engines whose systems are loaded with the correct glycol-water mix. If, however, scaling does occur, heat transfer will be inhibited and a cleaning operation must be implemented. Scale that is more than 0.050 in (1.2 mm) in thickness is unacceptable and must be removed.

Periodically, maintenance quantities of a corrosion inhibitor must be added to the cooling system. For testing purposes, a valve is sometimes installed in the engine block for the gathering of coolant samples. Do not take samples from the radiator or heat exchanger. Coolant conditioner elements provide an alternative means of corrosion protection. These elements are replaceable spin-on units similar in appearance to a lubricating oil or fuel filter. They are coming to be used on the smaller 1200- to 1800-rpm engines.

Do not use liquid or powder coolant water conditioner if the engine already has coolant conditioner elements fitted. Soluble oil is not to be used for any form of cooling system protection. Inhibitors containing chromate compounds should not be used in cooling systems. The optimum concentration of chromate solutions is difficult to control, and the solutions are extremely toxic. Before dilution, they can damage human skin. Furthermore, environmental regulations severely limit discharge of chromate solutions into inland and coastal waters.

Electrical systems are generally so designed that no continuous electrical potential is imposed on any cooling system components. The presence of any electrical potential can be the cause of cooling system material decay by electrolytic processes. All water-conditioning chemicals must be treated with respect for both economic and health reasons. The operator must familiarize himself with their correct usage proportions and the water-testing procedures. He must also acquaint himself with the proper methods for handling and storing conditioning materials. Some coolant conditioners are not compatible with glycol cooling solutions, and caution in their selection must therefore be exercised.

23.11 DECAY OF COOLANT INHIBITORS

Coolant inhibitors are often formulated to change or lose their color as they become spent. If they are used in conjunction with a frost-protecting agent and visual inspection is the method chosen to monitor the coolant effectiveness, a colorless ethylene glycol must be used. Undyed ethylene glycol can be obtained, but the suppliers are naturally anxious to supply a product that is tinted with a brand-identifying color.

23.12 CHROMATE INHIBITORS

The introduction of chromate inhibitors into an ethylene glycol–loaded cooling system will result in the deposition of chromium hydroxide, otherwise known as *green slime*, on the cooling surfaces. This deposit will result in poor heat transfer, and the engine will readily overheat. In the event of green slime forming, the engine must be thoroughly cleaned out by using an approved descaling agent.

Potassium bichromate, also known as potassium chromate ($K_2Cr_2O_7$), is a poisonous yellow-red crystalline substance that is soluble in water. It is used as an oxydizing agent and also as an analytical agent. At one time, it was in common use in engine cooling water systems as a corrosion inhibitor. Its use is still fairly common in engines operated in warm climates and without frost-protected coolants.

23.13 HIGH CONCENTRATIONS OF SILICATES IN COOLANT

In recent years there has been a large increase in the number of aluminum engine components, particularly in automotive and semiportable engines. Many manufacturers of ethylene glycol coolants have modified their formulas to accommodate the changing market demands, and that has involved an increase in the concentration of silicates in the glycol solution. This provides a better protection for the aluminum parts of the engine. But if the glycol is mixed with hard water and coolant supplements are then added, certain difficulties can arise.

- The total solids in the resultant solution may increase.
- Insoluble masses will deposit on critical surfaces and cause blockage of coolant passages or plugging of radiators and coolers.
- A silica gel or polymerized silicate will form on cooling heat transfer surfaces and precipitate to the radiator bottom tank or lower part of the cylinder jacket.
- The solids developed may cause excessive wear in the circulating pump and cause the seals to fail.

For diesel engines certain precautions can be taken. The intended coolant must be reviewed before its introduction to the engine. Any frost-protected coolant must have a low silicate content. Clean, soft water must be used for the mix. The likelihood of silicate precipitation increases with the water hardness. Products sold specifically for automotive use often state that they are designed for use in engines with aluminum components.

Silica gel, when wet, feels slippery or slick. It may have a putty-like texture. When dry, a white cake or powder remains. The deposits can be removed only by soaking in a caustic solution. That must be done with great caution because caustic will damage O-rings, gaskets, solder, and aluminum.

23.14 COOLANT BRAND CHANGE

Whatever antifreeze medium is used, and particularly if it is intended to be changed to another brand, the mix ratio and the lower protection temperature should be verified. Low-cost antifreeze often requires no water, and other antifreeze agents have varying heat transfer capabilities. Severe frost damage to the engine may be incurred if low-grade, low-priced antifreeze solutions are used.

23.15 COOLANT PRESSURES

An engine cooling system is always operated under pressure. That permits the engine to operate at more efficient temperatures without boiling the coolant, and with a pressurized system the chances of circulating pump cavitation is minimized. Because the system is pressurized, the header tank will be closed and entrainment of oxygen in the coolant will be much reduced. The use of a frost-protected coolant medium will also raise the boiling point of the coolant. The cooling system pressure varies from one installation to another and may be affected by the pump capability, the elevation of the header tank, the altitude of the installation, and the pressure setting of the filler cap. Consult the OEM manuals and the original test data sheets for specific details.

23.16 COOLANT TEMPERATURES

Most generating set engines are designed for a maximum coolant temperature differential of 5 to 9°C (10 to 15°F). When the temperatures have normalized after start-up, the coolant entering the engine should not be at less than 70°C (165°F). The coolant should leave the engine at 82°C (180°F), but that may be a little lower on engines with pressurized cooling systems. It is most important that the temperature differential be maintained within the recommended limits, because the thermal shock attendant on low entry and high exit temperatures will not be beneficial to the engine. Leaking liner seals and weeping cylinder head gaskets will be the most common of expected trouble, and cracked cylinder liners and cylinder heads, warped blocks, and so on, will not be unexpected in extreme cases.

23.17 COOLANT VELOCITY

Coolant flow in the system must be maintained in a velocity range that will achieve optimum heat transfer without erosion damage to system components; if the velocity is too low, overheating of the coolant will occur. When reviewing a cooling problem, the rate of flow can be gauged by the flow sight glass in the discharge line; a fall-off in flow rates may be caused by a loose or worn pump impeller, gelled coolant, or air locking.

23.18 WATER PUMPS

The water pump will fail to pressurize the cooling system for one of several reasons:

1. The pump impeller may be loose. Impeller bore scrolls or shaft damage may have taken place if this defect has existed for some time.
2. In the case of a dual-walled impeller, foreign bodies may be lodged therein with resultant plugging. If that is the case, the plugging material's origin must be identified and the material migration and its source nullified.
3. The pump impeller clearance may be incorrect. This may be caused by either wear or looseness of the shaft within the impeller, by excessive float of the impeller-shaft assembly, by the fitting of a gasket of incorrect thickness, or by failed bearings.
4. The pump draws air through the seal. It is probable that the seal that permits air to be drawn past it will leak substantial quantities of coolant when the engine is at rest.
5. Dezincification; eroded pump impeller or casing (Fig. 23.1).

23.19 IMPELLER CORROSION

Many engines employ a double-circuit cooling system embodying an intercooler (Fig. 23.2). In these systems, the engine is cooled by fresh water (i.e., not seawater or frost-protected water) in a closed circuit. The water passes through the cylinder heads and jackets and absorbs waste heat. It then passes to an intercooler where the waste heat is transferred to the second circuit, which is loaded with a frost-inhibited coolant. From there it goes to the radiator, where the waste heat is dissipated.

The objective of using this method is to protect the engine from the damaging effects of ethylene glycol leaks into the engine crankcase. Problems arise when the condition of the fresh water in the engine side of the system deteriorates. The pH must be checked regularly, and a pH value of 8.5 to 9 should be maintained by the use of suitable additives. If the pH falls below 7, corrosion of the pump impeller and cylinder liners can be expected.

The rate of pump impeller corrosion in fresh, unconditioned water is much faster than in an ethylene glycol–loaded system. That is because all of the antifreeze fluids in common use contain protection supplements. Nevertheless, the protections fade after a period of service, and the operator has to pay attention to the condition of both circuits if trouble is to be avoided.

23.20 WATER PUMP CAVITATION

During operation, it is possible for minor combustion gas leakage to accumulate in the cooling system if the radiator has an inadequate air venting and deaeration capacity when the thermostats are fully open. Any entrained air present in the cooling system also is drawn into the water pump and causes cavitation. Cavitation can also be brought about by suction pipe restrictions creating a partial vacuum at the water pump inlet and inducing the pump gland to pass air into the cooling system. A cavitating pump reduces the amount of water being circulated and thereby brings about numerous engine difficulties.

The effects of cavitation on the coolant pump itself can lead to breakdown. On

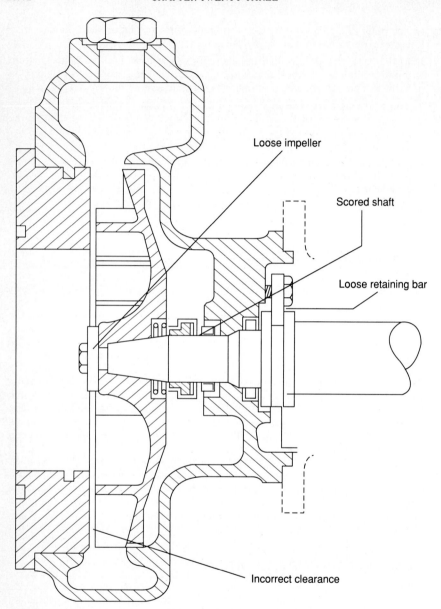

FIGURE 23.1 Cross section of a water pump showing possible faults.

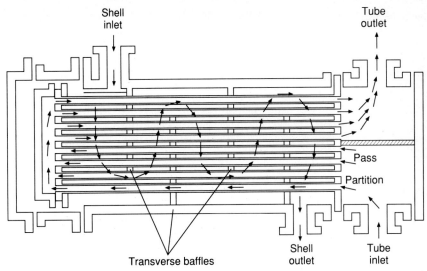

FIGURE 23.2 Flow diagram of a typical intercooler.

the wetted surfaces of the pump interior, the continuous formation and collapse of myriad, microscopically small air bubbles has a chemical action on the metal. Over a period of time, quantities of metal are eroded, and in extreme examples cavitation erosion can result in perforation of the pump casing. Cavitation may sometimes be suggested by an abnormally noisy pump or low output pressures.

23.21 COOLING SYSTEM OBSTRUCTIONS

Cooling system obstructions take the form of large quantities of solids or semi-solids (e.g., leak sealants, emulsified oil, broken-down soft scale, gasket cements, and disintegrated rubber hose debris). Generally, this will gather at the lowest point in the system or where the coolant passages are smallest, as in the radiator, lubricating oil cooler, or aftercooler.

Flow obstructions will occur where flexible connections or hoses are distorted. Collapse of hose linings may also lead to flow impediment. On new installations there is a possibility that an overlooked construction strainer is still in place and is loaded with debris.

23.22 MARINE GROWTH

Marine growth in sea, lake, or river (otherwise, raw water) cooling systems occurs in many areas of the world. The main inducement to such growth is increased temperature of the water as it passes through the system. Marine growth refers to minute marine plant or animal life which enters the raw water cooling system, attaches itself, and grows. Seawater strainers have minimal effectiveness owing to the minute size of immature plant or animal life. (Strainers are effective with more mature plant or animal life.) Spring runoff in polar or subpolar areas can promote system plugging, notably by iron-feeding bacteria leached out of the muskeg during the melt. Marine growth can be controlled with varying degrees of success by several methods. Periodic mechanical cleaning of the heat exchanger, aftercooler (where it is seawater-cooled), and so on, removes accumulated growth. Chemical treatment may be employed to combat marine growth, but the chemical types and concentrations selected must not foster deterioration of the seawater cooling system components. Continuous low-concentration chemical treatment via either bulk chemical drip feed or self-generating processes may be employed, but it is mandatory that all ecological protection regulations be observed.

23.23 GAS OR AIR IN COOLING SYSTEM

To reduce the risk of air locks being formed in the cooling system, air-bleed devices are fitted in the cooling outlet manifold. When the bleed device is opened, what appears to be a formation of air may well be an accumulation of combustion gas which is leaking into the coolant. This can be confirmed and the errant cylinder identified by the method described in Sec. 23.5.

When an engine has been inactive for some time or has had its cooling medium removed and replaced, there is a chance of air locks being present in the cooling system. The cooling system must be tested for any evidence of air both before and after start-up.

It must be remembered that air or gas within the engine will not necessarily gather at the vent ready to be blown off at one time and in total. A fair amount of bleeding may be required, perhaps at intervals, to displace air or gas pockets. If the presence of air or gas persists, formal investigations must commence. Wishfulness will never cure a gas leak.

23.24 THERMOSTATS (REGULATORS)

Thermostats are also known as *regulators*, but for the purposes of this discussion only the term *thermostat* will be used. A thermostat is a form of automatic valve. Temperature changes cause it to open and close and thereby regulate the coolant flow into desired directions. The purpose of the thermostat is to maintain the engine at an optimum operating temperature by rationing the dispersal of waste heat. Therefore, the use of the term *regulator* in any coolant context is discouraged at all times.

Virtually all liquid-cooled engines have thermostat-regulated cooling systems, and many larger engines also have thermostat-regulated lubricating oil cooling systems. In either case the mode and purpose of the thermostat is the same (Figs. 23.3 and 23.4). The thermostat is an essential component of an engine's cooling system; without the installation of a correct and properly working thermostat, the automatic control of the cooling system will be absent. The absence of a thermostat from a running engine's cooling system can only have a deleterious effect.

Any thermostat can be readily tested by suspending it in a container of water and then heating the whole. When the water temperature rises to the operating temperature of the thermostat, the thermostat will start to open. This will nor-

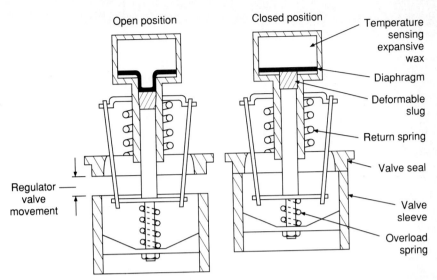

FIGURE 23.3 General arrangement of a thermostat.

mally occur between 71 and 82°C (160 and 180°F), but the ideal temperature is dependent upon the operating requirements of the engine. (Thermostats are almost invariably marked with the operating temperature range. Be sure that the thermostat being tested is in fact the correct item as specified in the OEM manual.) As the temperature continues to rise toward the boiling range, the thermostat will open to its maximum fully open position. Witness the heating and cooling process and see that both the opening and closing are faultless and match the marked temperatures.

The coolant thermostats are generally most reliable; they either work properly or they do not work at all; there is no "halfway" performance. If they are replaced, they must always be replaced with thermostats having the same operating temperature range as those originally supplied. Thermostat breakdown is indeed rare, but when thermostat trouble is encountered, it will almost certainly be brought about because of solid matter getting under the thermostat seat. If an engine has recently been worked upon and thermostat problems are suspected, dirt on the thermostat housing counterbores or misplacement of the thermostat in the counterbore should be considered (Fig. 23.5).

It is important that the thermostat be correctly positioned in the housing. Quite often the thermostat is fitted with bleed holes, vent grooves, or individual features that are very relevant to its performance. These features must be recognized and respected. There is no good reason why an engine should be operated without a thermostat in position. To do so can only adversely affect the engine's performance.

23.25 WEAR IN THERMOSTATS

In an engine cooling system, wear of the thermostatic valve will occur on the valve's seats and seating faces but may be found in other areas in the path of

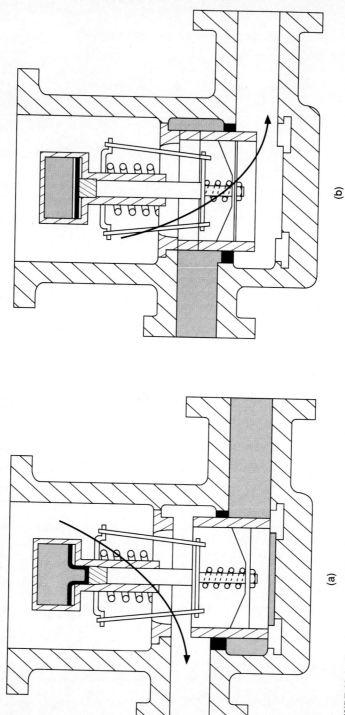

FIGURE 23.4 Regulator coolant flow: (*a*) Hot coolant flow: direction past thermostat (thermostat in open position). (*b*) Cold coolant flow: direction past thermostat (thermostat in closed position).

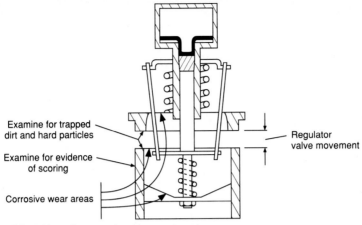

FIGURE 23.5 Common damage areas in a thermostat.

coolant flow. Wear will be accentuated if the pH of the coolant falls below 7, which will indicate the need for a coolant conditioner. Any defective thermostats must be discarded and removed from the power plant. No suspect thermostats are to be retained.

23.26 CHANGING THE THERMOSTAT OPERATING RANGE

A great deal of careful thought must be given to any proposal to change out the thermostats for ones of a higher-temperature range. If it is felt necessary to take such a step, the OEM representative must be consulted first. If the jacket operating temperatures are raised by the installation of new thermostats, there will be an effect on the lubricating oil and its life, particularly in engines fitted with jacket water–cooled lubricating oil coolers. The lubricating oil temperature should normally run about 5°C (10°F) lower than the coolant temperature, and any oil temperature rise above 82°C (180°F) will result in a foreshortening of the oil life. Distortion of the engine and unnatural wear or glazing of the cylinder liners will result in increased overhaul costs.

23.27 DIRECTIONAL TEMPERATURE CONTROL VALVES

Directional temperature control valves are commonly known by the generic title of AMOT after the well-known manufacturer of such valves. The valve casing is provided with three ports, and it contains two or more thermostatic elements similar to the normal thermostat or regulator installed in the engine's coolant outlet chamber (Fig. 23.6).

The directional temperature control valve is usually found mounted adjacent to the engine and in the discharge pipe system of an engine's jacket water and/or lubricating oil cooling system. These valves are quite simple in construction, and

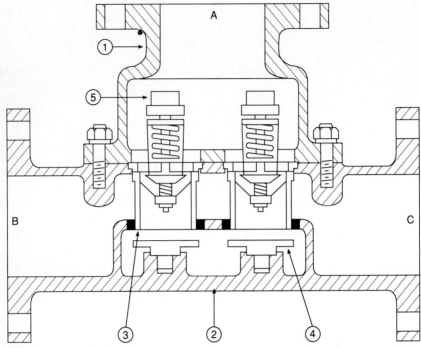

FIGURE 23.6 Multidirectional thermostatic valve. Two thermostatic elements are shown; complete units may contain as many as nine elements. Ports A, B, and C are marked on the flange exterior. Other components are as follows: 1, entry housing; 2, discharge housing; 3, valve sleeve; 4, valve seat; 5, element assembly.

they give years of trouble-free service. In a cooling system that has previously operated satisfactorily and then starts to give temperature control trouble, it will be pointless to change to a lower-range thermostatic element. The existing elements can be tested for performance, and if they do not operate properly, they should be replaced. The temperature range will be found marked on the limiting strap. An operating life of the thermostatic elements between 50,000 and 100,000 h can be expected provided that excessive temperatures, chemical electrolysis, or cavitation conditions can be kept in check. Improper coolant additives or lubricating oil in the coolant will have an aging effect on the O-ring elastomeric seals, which will become brittle.

23.28 TEMPERATURE CONTROL VALVE TROUBLESHOOTING GUIDE

Problem: System Temperature Too Cold

Symptoms

- Insufficient heat being passed to the coolant to maintain the desired temperature.
- Too much cooling system capacity being utilized for the amount of heat available for rejection.
- Following its repair or overhaul, the valve may have been installed backwards and is forcing all of the coolant through the cooler.
- Badly worn seal on sliding portion of the element.
- Debris jamming elements.
- Pressure imbalance between ports (greater than 25 psi).

Problem: System Temperature Too Hot

Symptoms

- Insufficient capacity of the heat rejection system open to circulation (particularly in the case of multielement coolers).
- After repair or maintenance, the valve has been improperly reinstalled and is reducing the flow to the cooler.
- Bypass is not closing fully because of damaged seats, sliding valves, or seals.
- Thermostatic element(s) are plugged or obstructed.
- Excessive pressure difference between ports.

23.29 JACKET WATER HEATERS

To maintain an engine at a reasonable starting temperature, a jacket coolant heater is supplied. The jacket water heater is fitted with a thermostatic control that permits the engine jacket water to be maintained at 15 to 30°C (60 to 90°F). This is an adequate temperature for a standing unit in a ready-to-run condition. Any higher temperature level will be unnecessary, uneconomic, and damaging to the engine in several respects.

It is not necessary to maintain the jacket temperature on any standing engine at anywhere near normal operating temperature. Heating engine jackets above the recommended level can, over a period of time, be extremely costly in terms of wasted power. The ideal capacity of a jacket heater for an engine standing in temperatures of $-29°C$ ($-22°F$) is 3 W/in^3 of displacement (1 Wt/0.55 cm^3). It is necessary to heat the jacket of an available engine above 38°C (100°F), and to attempt to do so becomes very costly. For example, a Caterpillar D399 with a displacement of 64,500 cm^3 (3928 in^3) will require roughly 12 kW/h.

The principal problem encountered with block heaters is that of debris precipitation. Being located in the lowest part of the jacket cooling system, it is not uncommon for quantities of sludge, garbage, scale, and other solids to accumulate around the heater element as a sediment. The poor heat transfer qualities of the sediment leads to heater element overheating and subsequent burnout. It is recommended that the jacket heaters be cleaned out at approximately 5000-h intervals. The hoses fitted in the jacket heater system are frequently of substandard quality. They should regularly be inspected, and if any signs of blistering, lamination, or bulging is observed, they must be replaced.

23.30 STRAINERS AND FILTERS

Strainers protect the cooling system from physical damage that is due to circulating abrasive materials and the plugging that will occur when large pieces of foreign materials enter the system. Water drawn from areas having abundant marine life or shallow water wells are most benefited by strainers that prevent marine life, stones, waterlogged driftwood, and so on, from being ingested.

Full-flow strainers of the duplex type are commonly used. The strainer screens are made of a noncorrosive material and are generally about 0.125-in (3.2-mm) mesh for use in seawater circuits and 0.063-in. (1.6-mm) mesh for use in closed fresh water circuits. The use of a differential pressure gauge across the duplex strainers indicates the pressure drop and enables the operator to determine when the strainers need servicing.

23.31 FLEXIBLE PIPE CONNECTIONS

Flexible connections are installed in several points in the pipe systems serving diesel generation sets. Their purposes are to take care of any small amounts of misalignment between the pipes or terminals that are rigidly attached to the engine and those of the services, to permit the thermal growth of the generator set without imposing stresses on the rest of the system pipework, and to limit the transmission of vibration from the generator set to the system pipework.

Most flexible connections are located as close as possible to the generator set (or the source of movement). They will be inserted in both the coolant delivery and discharge lines, in the low pressure fuel delivery and return lines, and in the lubricating oil lines on engines that feature remote lubricating oil coolers.

The type of flexible connection utilized in pipework is commonly of wire- or fiber-reinforced synthetic rubber construction with a woven wire exterior armor. The connection is often fitted with one fixed and one floating flange. When replacements are fitted, it is important to see that the flexible material is suitable for the intended service (e.g., neoprene liners for lubricating oil service).

Flexible connections can be expected to fail occasionally because they are continuously working and do become fatigued. Nevertheless, many flexible connections fail prematurely because of a gross misalignment of the interconnected pipes or because of improper installation. The misalignment may occur after commissioning of the equipment, after work has been done on the engine or related equipment, or when there has been an abrupt or substantial change in subsurface conditions. Any misalignment must be rectified before a replacement flexible connection is put in place. In the event that a fuel oil or lubricating oil leak develops in a flexible connection, the connection must be repaired forthwith because it will represent an enhanced fire hazard.

23.32 FLEXIBLE HOSE INSPECTION AND REPLACEMENT

When flexible hoses are replaced, the replacement hose must be suitable for the service intended. Check the length, bore, and pressure capacity. New hoses must be blown out or pulled through to ensure there is no foreign matter in them. Do not use hoses that have been crushed or excessively bent or any hose that has been used previously.

Check the hose interior lining for medium compatibility. In some cases the hose lining will not be suitable for lubricating oil, glycol, fuel oil, elevated or freezing temperatures, or other conditions, and special care must be taken for the individual applications in this respect. Hoses in the installed position should be clipped or otherwise tied back in an engine contour–following mode, and they should be arranged in such a way that they are not able to rub against anything when the engine is running.

Do not overtighten hose clips and clamps. The surface of the hose may be indented, but the surface cover material must not be cut through. No attempt should be made to tighten hose clamps on the engine cooling system to stop leaks when the engine is running. Only gear clamps should be used. Nut and bolt clips, spring clips, and the like are not suitable.

Rubber hoses or flex connections may not be painted, but they should be washed or wiped occasionally to keep them free from oil. If ethylene glycol is used in the system, careful selection of the hose-type coupling is important so that compatible material is utilized.

The operator should inspect all of the flexible connections on the engines and their attendant systems at least once during each shift. Typically, flexible connectors will be found between the main fuel storage tanks and the pipeline to the fuel transfer pump, and there may be several flexible connections in other parts of the pipeline. There will probably be one between the fuel transfer pump and the fuel day tank; there will also be one or more between the fuel day tank and the engine. There will be connectors between the raw water pumps and the raw water pipe system. In the cooling system there will be flexible connectors adjacent to the engine-driven cooling water pump and also close to the radiator inlets and outlets. The vent lines to the header tank and the crankcase breather vent line are other places where flexible connectors will be situated.

On smaller engines the flexible connectors are often simple rubber- and fabric-type sleeves secured with gear clamps. These are subject to deterioration through the effect of antifreeze, fuel oil, lubricating oil, and temperature. Although flexible pipe connectors readily facilitate the connection of slightly misaligned pipes, the plant operating personnel must take notice of any developing distortions. A certain amount of movement can be tolerated and accommodated following installation, but a complete rupture will occur if the damaging movement continues without remedy. If at the time of overhaul the rubber pipe appears to be developing layer separation, it must be discarded and replaced. For most practical purposes, the flexible connections can be regarded from a structural point of view as being the weakest part of any system.

23.33 RADIATOR RETURN FLEX CONNECTION COLLAPSE

The operator is advised to inspect the flexible connection on the coolant suction line for any indication of inward collapse. Such a collapse, particularly in freezing weather, may indicate that the coolant is not returning from the radiator and that the coolant pump is pulling a partial vacuum. The most probable cause of such a condition will be gelling of the coolant in the radiator through overrunning the fans or because of incorrect coolant mix ratios. When there is evidence of return-line negative pressure, there will likely be noise associated with pump cavitation and the engine will not carry full load without overheating.

23.34 FLEXIBLE HOSE PROBLEMS

Several common troubles are encountered on flexible fabric, metal, or rubber hoses that apply to both coolant and lubricating oil hoses.

1. *Hose hardening* results from excessive heat in the vicinity of the engine, and the rubber becomes "overcured." Heat-hardened hose will fail without warning. Also, when the hose becomes hardened and loses its resilience, it will transmit vibrations from the engine, and radiator damage may be induced.
2. *Cover deterioration* is visible evidence of such things as hardening or overcuring in progress, fraying out of the protective braid work, and abrasion caused by vibration. External damage will progress toward the interior of the hose, and failure will gradually occur.
3. *Core flaking* is common in coolant hoses but not in lubricating oil hoses. Rubber particles gradually break away and tend to plug radiator cores, aftercoolers, and lubricating oil coolers, and they may impede the proper working of thermostatic regulators, butterfly valves, and shutdown devices.
4. *Exposure to oil and grease* will decay the exterior of a flexhose. The rubber will soften and swell and will lose its resilience. Some paints will affect the outer layer of hoses in a similar manner.
5. *Replacement hose or flexible connectors* must always be utilized when any replacements are fitted. New sections must be thoroughly inspected before they are installed. The new sections must not be twisted during installation, nor may the specified bend radius be exceeded. Sections that have been flattened are bulged or have their external armor damaged must be rejected. Route new hoses so that they do not impede other work.

23.35 RADIATORS AND PARASITIC LOADS

A parasitic load is the load imposed by auxiliary machinery or equipment necessary to operate the prime mover. The parasitic load requirement may be partly intermittent. In this category may be placed prime-mover-driven battery-charging alternators, cooling fans, compressors, and similar equipment that does not necessarily have to be driven continuously.

To overcome the inertia of any moving equipment, power must be supplied, and the provision of that power detracts from the prime mover's gross power output. To obtain the maximum power capability of an engine, many generating units are so arranged that a static battery charger rather than a charging alternator is used. Similarly, remote radiators with thermostatically controlled, motor-driven fans can supplant the continuously driven fans on engine-mounted radiators with great economic advantage. Depending on the load being carried and the prevailing ambient temperature, the power losses of an engine using thermostatically controlled fans are substantially less than the conventional constant-load engine-driven fan. In a well-ordered plant where the fans are set to trip in and out at correct moments, the efficiency of the plant may be as much as 10 percent higher than that of a plant where the fans run continuously regardless of load or temperature.

23.36 RADIATORS

The typical radiator cooling system is made up of a fan-blown radiator in which the coolant liquid can cool, a pump which circulates the coolant around the system, a regulator that directs the flow of coolant according to its temperature, a radiator cap that acts as a safety valve and thereby regulates the pressure within the coolant system, the engine jacket in which the coolant collects the heat being rejected by the engine, and all the pipe work connecting the system components together.

The radiator is a complex of thin-walled copper tubes. (Steel tubes are utilized in those radiators which may be subject to abnormal abrasion.) Fins attached to the tubes provide a greatly increased surface over which the fan-blown air must pass and absorb heat from the coolant as it goes. A light coating of dust or oil on the tubes or fins will severely impair the conductivity of the radiator and its cooling ability. Therefore, it is important that the radiator core (the assembly of fins and tubes) be kept very clean at all times.

23.37 RADIATOR CAPS

The radiator cap incorporates a two-way valve that regulates the pressure in the cooling system. The cap seals the cooling system until the pressure developed by the thermally expanded coolant exceeds the cap relief pressure. The valve then modulates the pressure in the system by venting coolant to a drain. When the pressure drops after the engine is stopped, the secondary section of the valve opens to prevent the vacuum which would be developed through the thermal contraction of the coolant (Fig. 23.7).

Usually the radiator drain is so arranged that any excess coolant driven off through thermal expansion is siphoned back into the system at cool-down. By this means, entry of air into the cooling system is minimized and the loss of expensive coolant is minimized. Every effort must be made to keep air out of the engine's cooling system because the oxygen component in air contributes greatly to cooling system corrosion no matter what coolant medium is being used.

At no time should the engine be operated without the radiator cap firmly in place and in operable condition.

Never attempt to remove the radiator cap on an engine that is running. If it is

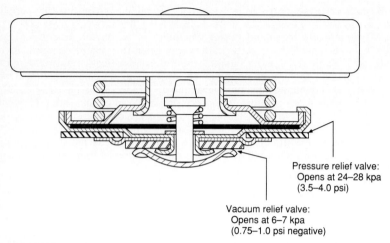

FIGURE 23.7 Radiator cap.

imperative for some reason to remove the radiator cap (or any other caps on a pressurized cooling system), the engine must be stopped and allowed to cool. If the cap is removed during engine operation, there is a great danger of personal injury.

23.38 RADIATOR FANS

The fan which blows the air through the radiator is designed to handle the amount of air necessary to reduce the full-load coolant temperature to the proper return temperature. To do that the fan must rotate at the right speed and the pitch of all the blades must be correct. In a crude sense, the fan screws a column of air through the radiator. The fan will run below its design speed if the drive belts are loose or slipping or if the bearings are in distress. If the fan blade pitch is improperly set, there may be untoward turbulence or perhaps the fan blades will not "bite" deep enough into the air to operate efficiently.

23.39 RADIATOR FAN ROTATION

It is important that all of the motor-driven fans rotate in the proper direction depending upon the arrangement of the radiator and its shrouds, plenums, and cooling fans. It is possible for one or more fans to work against the others, and the plant electrician should review the fan rotational direction at intervals. The blade pitch of all the fans in a particular unit should be the same, and the fan rotation should be checked after any radiator core or fan-motor-cleaning operation.

In the event of any rotational or air directional problems, the profile of the fan blades is often a good guide. Fan blades usually have a curvature or an aerofoil section the concave side of which forces air away from the fan and into the radiator core. If there is any doubt about the rotation of one or more fans in a radiator cooling unit, the electrician should read off the current draw of each fan as it switches in. He must also read any changes in the current draw of any fans that are already running. For example, in a four-fan radiator unit it will be necessary to read No. 1 fan, Nos. 2 and 1 fans, Nos. 3, 2, and 1 fans, and finally Nos. 4, 3, 2, and 1 fans. It is not unknown for the fans to buck against one another if the fan motors are improperly connected or if the blade pitches are incorrectly set.

23.40 CLEANING RADIATORS

The use of exotic cleaning materials for radiators is economical only when the radiator core is dismantled and immersed in a bath and soaked clean. To use expensive solvents in a high pressure spray gun is wasteful and uneconomic. Radiator cleaning may be carried out only when the fan is stationary. When the fan is belt-driven off an engine, the engine must be at rest. Motor-driven fans must be immobilized at the motor control center, and a tie must be attached to one of the fan blades to prevent sudden spinning from being induced by an adjacent fan. If a high-pressure gun is used, ordinary water is probably effective as a cleaning solvent because it is the lancing action that does the bulk of the first cleaning.

23.41 RADIATOR FAN SHROUDS

For optimum efficiency, the fan must operate within a close-fitting shroud. In addition to providing good air-flow characteristics, the shroud prevents the recirculation of hot air that has already passed through the radiator. The clearance between the fan tips and the shroud should not be more than 1.5 cm (0.5 in). Any attempt to cut the shroud or otherwise alter it will seriously interfere with the radiator's cooling ability. The fan blades must never be clipped or have their pitch altered, and to do so could result in very damaging effects on the radiator.

23.42 FAN BELTS

The tension of fan belts occasionally has to be adjusted, but continuous overtensioning of fan belts can lead to the failure of the fan hub bearings through overload. To obtain the proper tension on a V-belt, it is not necessary to pull the belt excessively taut. The belt should be tightened only enough to take out slack and sag. A good method of checking for proper tension is to strike the belt with one's hand. Slack V-belts feel dead under this test whereas properly tensioned V-belts feel vibrant and alive. An alternative test is to press down on each individual belt in a multibelt system. When the top of each belt can be depressed so that it is in line with the bottom of other belts in the drive system, the correct amount of tension has been applied.

Overloaded fan hub bearings are the major cause of radiator core damage. All too often the bearing(s) fail, the fan tilts, and the blade tips come in contact with the radiator core. The only sure way to avoid this problem is to see that the belt tension is correct. A low-riding belt may bottom, thereby reducing the wedging action and resulting in slippage, overheating and reduced life. Fan belts may also slip if the fan hub bearings are failing. A smell of overheated or burning rubber may be indicative of this condition. Slipping belts will also screech at start-up.

Dirt must not be allowed to form in the bottom of the sheave V. Dirt compaction will occur and it will reduce the V depth and lead to the belt being raised out of contact with the sides of the sheave. Any dirt in the V must be gouged out at each inspection. Belts should be kept clean, free of oil, and protected from direct sunlight. Belts may be washed effectively with soap and water. Excessive amounts of oil may be removed using an electrical cleaning fluid.

Sheaves should not be allowed to become polished on the wedge sides through belt slippage. Belt dressing must never be used on a V-belt system. Only adjustment or replacement will enable the belts to work properly.

23.43 COOLING AND VENTILATION AIR HANDLING

A large number of generator sets ranging up to about 750 kW are supplied with integral radiators and fans, the radiator being so arranged that the cooling air enters a plenum. The plenum is equipped with two sets of dampers which act in concert under the control of a thermostatic switch and a modulating motor. The warm air may be diverted back into the plant or exhausted outside as the temperature demands.

All the air that is consumed by the engine(s) or expelled from the building through the radiator must be replaced, and for that reason the power plant building must have adequate ventilation. Self-opening or barometric dampers may be fitted; manually operated dampers will be found in some plants, and in larger plants induced-draft fans will pull air into the building. In many cases, the dampers on both the ventilation and the engine cooling system are modulated by thermostatically controlled motors.

The movements of the two sets of dampers on the radiator plenum is actuated through linkage connected to a modulating motor. The dampers are so arranged that they will rotate in unison through 90° from fully open to fully closed on one set while the other set goes from fully closed to fully open. The opening of one set is therefore always in inverse proportion to the closure of the other. By this means the modulating motor, which is controlled by a thermostat, alters the positions of the dampers to conform with the desired internal temperature of the plant. Many plants rely on this method only for space heating (Figs. 23.8 and 23.9). The amount of heat rejected by the radiator varies with the load. If the damper linkage is incorrectly adjusted or disconnected, it will be quite difficult to maintain the plant temperature at a consistent and comfortable level.

The plant operator must be aware of any changes in the atmospheric pressure within the plant. The entrance door of the plant is an excellent indicator of such changes. Regardless of which way it opens if the door suddenly becomes heavier to open or close or much lighter, the pressure differential between the inside of the plant and the outside atmosphere has altered and it is probable that one or the other of the dampers is at fault.

The volume of air used by an engine is considerable. For each pound of fuel consumed, approximately 15 lb of air will be used. Thus a 300-kW unit running at

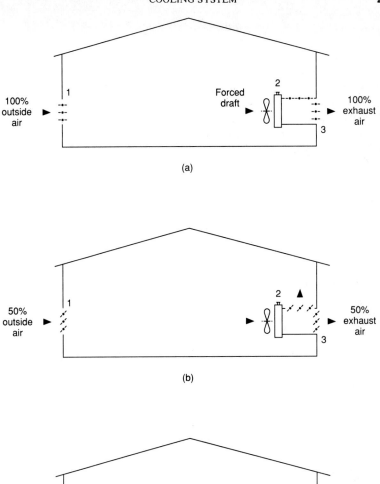

FIGURE 23.8 Louver positions for ventilation. (*a*) Zero air recirculation: damper 1 is open, 2 is closed, and 3 is open. (*b*) 50 percent air recirculation: dampers 1, 2, and 3 are 50 percent open. (*c*) 100 percent recirculation: damper 1 is closed, 2 is open, and 3 is closed.

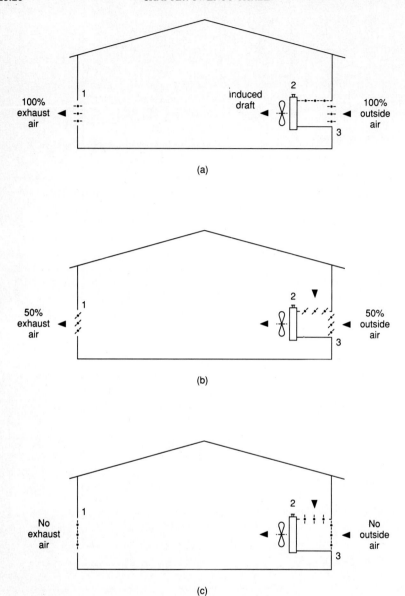

FIGURE 23.9 Louver positions for ventilation. (*a*) Zero air recirculation: damper 1 is open, 2 is closed, and 3 is open. (*b*) 50 percent air recirculation: dampers 1, 2, and 3 are 50 percent open. (*c*) 100 percent air recirculation: damper 1 is closed, 2 is open, and 3 is closed.

15 kW/gal will use 20 gal of fuel per hour, which represents 170 lb of fuel. To burn 170 lb of fuel, 2550 lb of air is required, equaling 31,598 ft^3.

The air-handling system in a diesel power plant has three prime objectives:

1. To pass sufficient quantities of aspirating air to the engine for combustion purposes without detracting from or impeding other air systems
2. To pass sufficient air through the cooling system without detracting from or impeding other air systems
3. To permit a comfortable working atmosphere within the plant for both men and machines regardless of the exterior climatic conditions

23.44 COOLANT LEAKS

Coolant leaks will be indicated by a drop in the normal level of the radiator header tank or the surge tank. Leaks will be visible when they occur at most pump glands, in radiator cores, and at flexible connections. In their early stages external coolant leaks do not harm the engine, but they must be attended to before they progress to a difficult stage. When internal coolant leaks are suspected, the engine should be stopped immediately because it is certain that the leaking coolant will contaminate the lubricating oil.

The following are the most common causes of coolant leakage.

- Aging flexible or hose connections.
- Heat, sunlight, and ethylene glycol. All three contribute to the degradation of the rubber resulting in cracking, bulging, and loss of elasticity.
- Overtightening of gear clamps. This will cause crushing or cutting of the nonmetallic fabric and accelerated breakdown of the connection.

Failed cylinder liner O-rings are a fairly common cause of internal leakage. The life of the O-ring, when subjected to ethylene glycol on one side, lubricating oil on the other side, and all operating in the vicinity of 100°C (212°F), cannot be safely taken beyond 30,000 engine hours. Early failure of O-rings can be expected if the rings have been improperly stored for some time prior to their use.

Cylinder head core plugs, otherwise known as *frost plugs*, will occasionally start leaking. Such leaks, if external to the engine, are often only cosmetic problems, but if the coolant is able to enter the lubricating oil system under the valve covers, for instance, severe problems can develop in the engine if the source of leak is not identified and rectified. This type of leak will most likely show up first in a lubricating oil analysis report, and it will require immediate attention.

The grooves at the lower part of the liner in which the O-ring seals are set sometimes erode through a form of cavitation known as pitting. A similar effect is occasionally found in the mating area of the cylinder block where the liner O-rings contact. Leakage in these areas can be detected by inspection with the engine at rest, but the first evidence of leakage will be seen in the lubricating oil sample reports. It is highly unlikely that a major leakage in this area would occur abruptly.

On some engines the cylinder liner is fabricated in such a manner that the liner has its own jacket. A classic example of this construction can be found on the EMD engine. It is not unknown for the weld on the outer shell to fail and permit coolant to enter the scavenge air spaces. Although coolant loss at the header tank may be observed, it is improbable that the lubricating oil level in the sump will rise by a proportional amount, because most of the leakage coolant will vaporize and eventually leave the engine via the exhaust. The only alternative is to change out the cylinder liner.

A leaking gland on the jacket cooling pump will permit a certain amount of coolant to escape. Minor leaks are tolerable, and if the loss of coolant is not too great, say 1 qt/day, the engine may continue to run. However, the coolant should be caught rather than being allowed to drip onto the floor. The pump gland can be repaired when it is convenient to stop the engine for this and other maintenance details. However, it must be remembered that providing a drip catcher is not a repair. (A coolant invasion of the crankcase must not be confused with a fuel oil instrusion; either one will have the effect of raising the dipstick level.)

A perforated radiator core will be readily identified by the regular shift inspection. The leakage may take place on either side of the radiator, but on a running unit, it will probably be first observed on the side opposite the fan. Radiator leakage is often caused by vibration and subsequent metal fatigue, fan impingement, or corrosion. Numerous patent radiator-sealing compounds are available to stop minor radiator leaks, but unfortunately they all create long-term problems of a very costly nature and therefore none are recommended.

The corrosion of radiator cores (and of aftercooler cores) is mainly brought about by the development of acids or oxygen entrainment in the cooling medium. For that reason, it is important that the operator pay good attention to the condition of the cooling fluid. New ethylene glycol contains various inhibitors and protective additives, but their effect fades in time and leaves just a straight ethylene glycol mix that may have an undesirable pH. There are several proprietary supplements that may be obtained to rectify any deleterious conditions that do develop in the coolant liquids. Test kits to determine the quantities of supplement needed also are available.

23.45 SEALING OF COOLANT LEAKS

All coolant leaks must be cured by fitting new seals, gaskets, or parts. The use of leak-sealing components in engine-cooling systems should not be permitted except in cases of great emergency. Although the leak may be stopped by any one of the "magic" components on the market today, a host of other heat transfer problems will be created. The cost of cleaning the system subsequently will be much more than can ever be justified by the expedient.

Small losses at pump glands will occur occasionally, but they will be remedied by routine upkeep. The system pressure will stay at an optimum level if the radiator cap, surge tank filler cap, or other accessories are kept properly closed. Unattended external leakage and uncleaned spillage of antifreeze-loaded coolant indicate poor housekeeping and represent safety hazards. Pools of antifreeze on the floor in the vicinity of an engine must be checked for source and cleaned up immediately.

23.46 COOLANT LEAK DETECTION

Coolant leaks into the engine crankcase are usually detected in the first instance through sodium showing up in the routine lubricating oil analysis. The exact source of such leaks is often very difficult to find except when the leak is pronounced and obvious. One method of leak source definition used by some engine-operating authorities involves the introduction of a trace dye into the main body of coolant. Such a dye is sodium fluorescein ($Na_2C_{20}H_{10}O_5$), also known as *uranine*. This material is a red-orange powder soluble in water. Generally, it is used as a fabric dye and also as an oceanic dye marker. When small quantities of sodium fluorescein are introduced into an engine cooling system, any leakage of coolant will show up under a black light as fluorescent green streaks. (Fuel oil leaks also will show up well under black light conditions *without* the use of any dyeing agent being added to the fuel.) Without an indicating dye, coolant leaks into the crankcase are particularly difficult to find.

So that false readings are not made at the time of inspection or shortly thereafter, it is better to drain all the coolant from the engine and replace it with plain water to which the dye has been added. After the test, the water is drained, the system is flushed, repairs are made, and the frost-protected coolant is replaced. A recommended product is 15174 Uranine produced by the Chemcenrail Corp. Several other products are as satisfactory.

23.47 COOLING SYSTEM CLEANING

Contamination of cooling systems may occur through the introduction of leak-sealing media, overfrequent change of coolant (i.e., open-circuit cooling system), lubricating oil intrusion and localized or persistent overheating. When coolant is drained from an engine, the flow should be observed for signs of sludge, corrosion particles, discoloration, oil, and so on, and care must be taken not to return anything but the cleanest mixture to the engine.

The methods and media used for cleaning out a contaminated cooling system depend to a great extent upon the origin of the contamination. If the cooling system has been contaminated by oil following the failure of a lubricating oil cooler the cleaning methods would be:

- Drain all the coolant from the system.
- Fill the system with clean water.
- Start the engine and run it until the thermostats open.
- Add nonfoaming detergent soap. (Automatic dishwasher soap is good. Do not use common detergent; it will foam too much.)
- Run the engine for approximately 20 min and check to see if the oil is breaking up. If oil patches are still evident, add more soap and continue to run another 10 min.
- Stop the engine and allow it to cool. Drain the system and refill it with clean water. If oil is still apparent, repeat the cleaning process. If the water remains clean, drain 55 percent of the water and add a 45-percent clean glycol solution.
- Start engine, circulate coolant until thoroughly mixed, and test for frost protection reading.

Radiators should be inspected regularly and often for signs of dirt accumulation. All areas of the radiator's external surfaces should be clear of dust, oil, insects, seeds, or other contaminants. A light held on one side of the core should be visible from the other over the whole area of the core.

Cleaning of the core may be accomplished by means of painted-on solvents or by a steam jet. Air by itself is not effective. The blow should always be in the direction opposite that of the normal airflow. Do not attempt to poke the dirt out of the radiator with screwdrivers, rods, sticks, or other blunt instruments. Damage to the core will be inevitable. The whole of the core must be cleaned, and particular attention must be paid to the fan blast area.

The main causes of external radiator dirt are leaking crankshaft seals, fuel line leaks, exhaust system perforations, defective crankcase breathers, and excessive cylinder blowby. Attention to these and similar defects will go a long way toward reducing external radiator soiling. If the engine or the radiator has been served with leak-stopping agents at some point in the past or if the engine is scaled with natural mineral deposits, the cleaning methods must follow the recommendations of the engine manufacturer. Those recommendations will be peculiar to the problem. In many cases the indiscriminate use of cleaning agents can have a very damaging effect on rubber, aluminum, brass, and other compositions or alloys, and the agents are usually totally ruinous to radiator cores.

Radiators should be flushed out after every 5000 h of engine operation. This will remove any sludge accumulations from the bottom header tank, but if there are signs of hard deposits, the radiator may have to be chemically cleaned. In that event, the manufacturer's recommendations must be followed most carefully because of the variety of metals, solders, and so on.

23.48 CLEANING MATERIALS FOR COOLING SYSTEMS

Numerous propriety cleaning agents that are on sale in different localities may be used for cleaning out engine cooling spaces. A common basic cleaning agent that may be utilized with satisfactory results is sodium bisulfate (Na_2SO_4). Mixed with clean fresh water in a ratio of 1 lb to 10 US gal, it will provide an adequate cleaning capability in all but the most difficult cases. It is followed by a neutralizing mix of sodium carbonate (Na_2CO_3), also known as *soda*. The sodium carbonate is used in conjunction with water in a ratio of 1:170 by weight. Most users will find these chemicals readily available at a substantially lower cost than most brand name products.

Because of the great variety of cleaning compounds available today, together with the selling pressures, the user must be very cautious in his selection. In any engine the range of metals and materials used is quite extensive, and there is always some risk of one cleaning agent or metal having an undesirable interaction with another. The user must be very selective when ordering his cleaning materials to be sure that the material chosen is totally compatible with all of the engine metallic and nonmetallic materials.

23.49 RAW WATER INTAKE INSPECTION

Under certain seasonal or climatic conditions the cooling water intake strainers of diesel generating facilities being served with river or lake water may become fouled by natural debris. Weather conditions and locations may have considerable influence on the sorts of problems that may be encountered. In some areas much vegetable matter washed down by heavy rain will be picked up by the intake screens. In other areas spring runoff can promote heavy bacterial growth on the intake pipe systems. Silt may cause some difficulties, and fish, eels, and other marine life can at times cause almost total blockage of the intake filters. A good example of marine life causing difficulty is the present plague of zebra clams now proliferating in the waters of the Great Lakes.

It is recommended that the likelihood of such difficulties be considered and the necessary countertactics planned at an early stage in the life of a new plant. A provisional plan that is not needed is much better than having no plan at all when one is needed. The first step toward the avoidance of cooling water intake difficulties would be for the shift operator to inspect the suction and discharge pipes once per shift. Any change in the water conditions such as flooding, falling levels, freezing, turbidity, or change in color which would indicate sand, mud, or other water supply alterations, should be noted in the logbook.

23.50 AIR-COOLED ENGINES

Liquid cooling is used on the majority of diesel engines, but there are many engines of less than 300-kW capacity that are air-cooled. Air cooling has its place in some applications, but the difficulties in the handling of the very large quantities of air, the parasitic load imposed by the cooling fan, and the fan noise make the use of air cooling on large engines impractical.

The main problem with air cooling is the regulation of the engine at an optimum working temperature regardless of the load. That is exceedingly difficult, and no such provision exists on the majority of air cooled engines. It can be said that the wear rates on air-cooled engines, as a consequence of this disability, are generally greater than on those of a conventional thermostatically regulated liquid-cooled engine. On the other hand, air-cooled engines are very portable and their initial cost is low.

When air cooling is employed, there are some particular points that will need the operator's attention. The flow of cooling air both to and from the engine must be unobstructed at all times. At certain times of the year it is possible that there will be quantities of airborne seeds, insects, dust, or other matter that can be borne in the cooling airstream, inducted into the cylinder-cooling fans, and contribute to overheating problems. If fuel or lubricating oil leaks on the engine are left unattended, the oil vapors or films readily combine with any airborne dirt and cause hot spots, fire hazards, and extra housekeeping work.

However, for power-generating units in the smaller capacities where there is little variation in the load, the air-cooled engine is ideal. The slightly higher wear rates and their attendant costs are more than offset by the elimination of the liquid cooling system and all of its inherent snags.

23.51 COOLANT FLUID CONTAMINANTS

The problems of either fuel or lubricating oil entering the coolant have been discussed earlier in the chapter.

1. *Metallic precipitates* are old metallic corrosion products combined with water scales. These materials derive from the use of improper coolant, air leaks will form at engine-operating temperatures, and the iron, copper, and aluminum parts will be attacked.

Salt

The salt in any makeup jacket cooling water may not exceed 200 ppm because there is danger of considerable corrosion to the engine. Radiators, heat exchangers, and fuel injection equipment are highly susceptible to salt damage, and such corrosion is an ever-present problem in tidal water areas both on- and off-shore.

Specific Conductance

The specific conductance of a coolant is its resistance to carrying an electric current between the dissimilar metals of an engine cooling system. It should not exceed 6600 $\mu\Omega$ (microohms). If it is too high, the engine and its cooling system become something akin to a wet battery cell. High conductance will be due to poor makeup or combustion gas leaks. If it is not controlled, it will result in metal pitting.

Sulfate

Sulfate can build up scale throughout the engine and can develop sulfuric acid or contaminate water. It should not reach more than 300 ppm. Sloppy cleanout after sulfuric acid cleaner use and combustion leaks are the usual causes of excessive sulfate. Sulfate will induce acid pitting or scale formations and result in metal loss or inefficient heat transfer. To preserve good working conditions in an engine cooling system, it is recommended that the coolant be sampled and analyzed at approximate intervals of 3000 h or every 6 months, whichever comes first.

Total Hardness

The elements giving hardness in a coolant fluid are mainly calcium (Ca) and magnesium (Mg). These two elements are present in untreated water and precipitate as scale when heated. It is important that they be kept in check, because scaling of the engine's cooling spaces will cause heat dissipation impairment and excessive engine wear. Proper care of the coolant or the introduction of a suitable water treatment will maintain the engine in a scale-free condition.

23.53 COOLANT CONDITION AND ITS TESTING

The successful use of a coolant conditioner is dependent upon regular testing (Fig 23.10). If there is too large a concentration of conditioner in the coolant radiator and aftercooler, tubes will plug and the wet side of heat transfer surfaces will de-

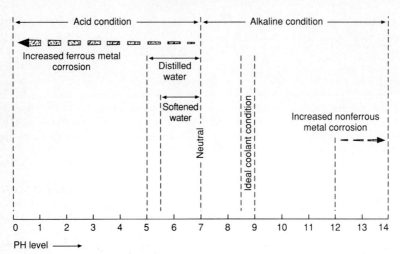

FIGURE 23.10 Coolant corrosion chart. Acid condition: blue litmus turns red. Alkaline condition: red litmus turns blue.

velop excessive deposit formations. These conditions may well lead to engine overheating, coolant pump breakdown, or cylinder head cracking.

When there is insufficient conditioner in the system, cylinder liner pitting will commence and the coolant will assume an acidity followed by decay of radiator and oil cooler solders and result in leakage. Blindly adding conditioner to the coolant will not produce satisfactory results. The coolant must be monitored. When additional material is required, adding the correct amount will restore the coolant to the ideal working level.

Testing and monitoring the coolant condition can be done successfully only by following the manufacturer's instructions for handling the material and the test apparatus. The superintendant should obtain a coolant test kit and nominate the person who is to do the test routines. It is suggested that the senior operator or the senior mechanic have this duty.

CHAPTER 24
LUBRICATING OIL AND SYSTEM

24.1 LUBRICATING OIL

The lubricating oil in an engine serves several purposes; it:

1. Lubricates by maintaining a viscous film between the moving parts and thereby minimizes friction and wear
2. Assists in cooling the engine by acting as a heat transfer medium carrying away excess heat from the underside of the pistons, bearing surfaces, cylinder walls, and other areas subject to high-temperature heating
3. Acts as a sealing component on piston ring lands, valve stems, cylinder walls and, in some cases, turbocharger shafts
4. Continuously washes the engine interior flushing contaminants out of the engine to the lubricating oil filters
5. Protects the engine components from corrosion which may be initiated by oil oxidation or products of combustion
6. Cushions components that are under high intermittent stress, such as gear teeth, cam rollers, and tappets
7. Is a hydraulic fluid medium for governors and other controls

When the lubricating oil in an engine can no longer perform 100 percent of its design functions, it must be taken out of the engine and replaced with new oil. The oil itself does not wear out, but it becomes loaded with contaminants to such an extent that it cannot properly lubricate the engine. All engine lubricating oil becomes contaminated in time and is a normal process in any engine operation.

The contaminants have their origins in the by-products of combustion, soot from partially burned fuel, acids derived from the sulfur that is present in most diesel fuel oil, sludges, varnishes, and gums resulting from the decomposition of some lubricating oil constituents, and dirt, usually in the form of silicon, ingested with the combustion air. All these contaminants have negative effects on the lubricating oil of the engine.

Lubricating oils are a combination of a base oil together with numerous additives each of which has a particular function. There are antifoam additives to inhibit any foaming tendency that the oil may have; they help prevent air bubbles from being introduced to the lubricating oil pump. Antiscuff agents are introduced to provide a chemical polishing and coating of components to lessen the natural wear rates of highly loaded bearing surfaces. Corrosion inhibitors are added to counteract acid attack. Detergent supplements keep insoluble matter in suspension and movement to the lubricating oil filter elements. Antioxidation agents promote the chemical stability of the lubricating oil. Viscosity index improvers are used to defend against the effects of high temperatures on the lubricating oil.

The viscosity of an oil is a measure of resistance to flow. Oils that meet the low-temperature $-18C°(0°F)$ requirement carry a grade designation with a W suffix. Oils that meet both low- and high-temperature requirements are referred to as multigrade or multiviscosity grade oils. Multigrade oils are formulated by adding viscosity index improver additives to retard the thinning effects that a low-viscosity base oil will experience at engine operating temperatures. The use of multigrade lubricating oil assists oil consumption control, improves engine cranking in cold conditions while maintaining lubrication at high operating temperatures, and may contribute somewhat to improved fuel consumption.

The primary criterion for selecting an oil viscosity grade is the lowest temperature that the oil will experience while in the engine oil sump. Bearing problems can be caused by the lack of lubrication during the cranking and start-up of a cold engine if the oil being used is too viscous to flow properly. It will be necessary to change to a lower-viscosity grade of oil as the temperature of the oil in the engine oil sump reaches the lower end of the viscosity range of the existing oil.

24.2 SYNTHETIC LUBRICATING OIL

Synthetic oils for use in diesel engines are primarily blended from synthesized hydrocarbons and esters. These base oils are manufactured by chemically reacting lower molecular weight materials to produce a lubricant that has planned and predictable properties.

Synthetic oil was developed for use in extreme environments where the ambient temperature could be as low as $-45°C\ 20(-50°F)$ and extremely high engine temperatures at up to $205°C\ (400°F)$. Under these extreme conditions, petroleum base stock lubricants (mineral oil) do not perform satisfactorily. Synthetic lubricating oils may be used at higher or more normal ambient temperatures provided they meet the appropriate API service categories and viscosity grades. The change intervals for synthetic lubricating oil are the same as those applied to engines using conventional petroleum-based lubricants. High cost negates the usefulness of synthetic lubricating oil for normal purposes, and it is indeed a rare situation where the use of a synthetic lubricating oil is justified in a diesel generator set.

24.3 LUBRICATING OIL SUPPLEMENTS

Caution in the selection of lubricating oil must be exercised if undesirable conditions are to be avoided. Certain lubricating oils contain zinc or molydenum dithiophosphate, which may have a harmful effect on particular engine components. Any change in the brand or type of lubricating oil originally specified by the OEM should be made only with the full approval of the engine manufacturer, not the lubricating oil supplier.

Similarly, lubricating oil supplements should not be used without a warranty statement from the supplier. This is necessary because none of the OEMs can effectively test and pronounce judgment on the great number of aftermarket additives currently offered. Most of these were developed for automobile gasoline engines, and their use in diesel engines is not necessarily beneficial.

24.4 BREAKING IN LUBRICATING OIL

The use of special oils for breaking in an engine that is either new or has just been rebuilt are not recommended and therefore should not be used. Only the standard lubricating oils specified as suitable by the OEM should be employed.

24.5 LUBRICATING OIL LIFE

The life of lubricating oil is the period of time (based on operating hours and regardless of the load carried) between changes of the oil during which the oil provides optimum lubrication.

All engine lubricating oils gradually fail through contamination and degradation. Laboratory testing can determine in fine detail the nature of any adulteration or decay of a lubricating oil sample, and the OEM will recommend the oil change intervals for any given engine. The change interval is largely set by taking into consideration the sort of service in which the engine is engaged, the volume of its sump relative to the engine's generating capacity (and fuel consumption), the class and grade of fuel used, the type of air cleaner, and the nature of the engine's working environment. Those factors, singly or in concert, have effects on lubricating oil life.

During the working of an engine, the lubricating oil becomes contaminated by matter originating in the combustion process. Acids, water, ash, semiburned fuel, soot and carbon, gums, and salts may all enter the lubricating oil from the combustion spaces via the piston rings and valve guides. Large quantities of air are consumed by the engine. Because all air carries some organic or inorganic matter, small amounts of contaminants can be expected to eventually enter the engine from the atmosphere in spite of today's very efficient air filters.

The life of lubricating oil is also limited by the degradation of the oil itself through heating, aeration, and additive depletion. Heating of the lubricating oil results in some thermal impact that produces sludges and varnishes. Aeration will result in nitration. The additive package is intended to be of a sacrificial mode, and thus its constituents are gradually lost.

All engines wear to some degree. Wear can be defined as the loss of particulate metal from its parent body through contact with other metals. Metal loss occurs when the lubricating oil film which is normally interposed between parts in motion relative to one another suffers a momentary lapse. In high-speed diesel engines, the life of a lubricating oil in the system can be roughly calculated by a method devised by Cummins Engine Co. It has been determined by laboratory and field tests that for every 255 US gal of fuel consumed in a turbocharged engine or 280 US gal of fuel consumed in a naturally aspirated engine, 1 gal of lubricating oil in the system will achieve maximum contamination.

$$P = \frac{C \times F \times L \times S}{L - 0.5(C \times F)}$$

where P = oil change period
C = a constant (255 or 280)
F = fuel consumed
L = lubricating oil consumed (makeup oil)
S = system capacity

It is not suggested or implied that this formula would apply to any particular engine or service situation, but it may form an interesting comparison with another generator set.

24.6 LUBRICATING OIL TEMPERATURES

The lubricating oil used in any given engine is specified in accordance with the temperature at which it is expected to run in service. If the engine is operated with the lubricating oil at too low a temperature, excessive sludging can be expected. This condition will, in turn, lead to a short filter element life, filter element bypassing, decreased pan capacity, and an accelerated wear rate.

The difference between the jacket coolant temperature and the lubricating oil temperature should be maintained at approximately 6°C (10°F). If the jacket outlet temperature is kept at 85°C (185°F), the lubricating oil outlet temperature should be at 80°C (175°F).

When lubricating oil heaters are used in direct contact with the lubricating oil, there is always some risk of coking. The heater skin temperature should not exceed 150°C (300°F). This temperature will probably not be detectable, but the heater should be inspected frequently for signs of carbon buildup. Excessive carbon buildup will result in the burnout of the heater element.

Typically, SAE 30 lubricating oil must be operated in the 88 to 99°C (190 to 210°F) range and SAE 40 in the 66 to 88°C (150 to 190°F) range. If the oil is operated at too high a temperature, the lubricating oil film on the bearing surfaces will be thinner, which will permit some metal-to-metal contact and attendant wear, and the cushioning effect of the oil film between adjacent parts will be less, which will result in increased wear eventually and, more immediately, increased noise. In addition, the temperatures of the coolant circulating the engine will be affected in an adverse manner.

24.7 LUBRICATING OIL CONSUMPTION

The maximum lubricating oil consumption of selected engines is given in Table 24.1.

Some small amounts of lubricating oil may pass the piston rings and be consumed during normal engine operations. The oil and its additive constituents leave carbon or ash deposits when subjected to the elevated combustion temper-

TABLE 24.1 Maximum Lubricating Oil Consumption

Engine model	Maximum gal/day at full load
Mirrlees KV16	66
Mirrlees KV12	51
Mirrlees K8	32
Mirrlees K6	13
EMD S20 645	32
Ruston RK16	32
Ruston RK12	23
Ruston 8CSV	13
MLW 251B18	32
MLW 251B16	26
MLW 251B12	19
Caterpillar D399	10
Caterpillar D398	8
Caterpillar D397	3
Caterpillar D379	5
Caterpillar D353	4
Caterpillar D334	2
Caterpillar D3306	2
Caterpillar D3406	2.5
Detroit Diesel DD8V71	3
Detroit Diesel 671	2
Detroit Diesel 471	1
Cummins KTA2300	9
Cummins KTA1100	4
Cummins NT	3

atures. The amount of deposit is dependent upon the oil consumption, the nature of the additives, the engine temperatures, and the rate of oil consumption. Other lubricating oil loss will be through the valve guides and will be confirmed by substantial deposits of hard carbon on the stem side of the valves.

24.8 RATE OF LUBRICATING OIL CONSUMPTION

The rate of use of lubricating oil used in an individual engine or plant can be predicted with some accuracy: Lubricating oil should not be consumed by the average engine at a rate exceeding to 1 percent of the fuel consumed. Other ratios which may be used are:

- 1 imp gal per 2500 BHPH, or per 1900 kWh
- 1 US gal per 2000 BHPH, or per 1500 kWh
- 1 L per 550 BHPH, or per 400 kWh

All of the above rates may vary downward if the utilization factor of the generator set is low.

The lubricating oil consumption rate should be monitored daily. Beware of the engine on which the lubricating oil consumption begins to slow down: It may be developing another sort of problem.

24.9 HIGH LUBRICATING OIL CONSUMPTION

When high lubricating oil consumption seems to have developed, it is well to consider the transfer container currently being used. Check to see how many 1-, 2-, or 3-gal (qt or L) containers are being filled from a standard barrel. It is not unknown for ten 5 imp gal buckets of makeup oil to be gotten from a 45 imp gal (50 US gal or 205 L) barrel.

High oil consumption will manifest itself in some or any of the following signals: blue exhaust smoke, pools of oil developing on the floor from leaking covers and shaft seals, formations of oil in the jacket water header tank. There may be considerable exhaust stack slobber, and the breather may precipitate a continuous dribble of oil. An internal inspection of the engine may show either or both inlet and exhaust ports coated with abnormal quantities of soft carbon; there will possibly be considerable piston blowby.

Excessive oil consumption is expensive to live with. If a typical 1000-kW generator set has a lubricating oil consumption 10 percent above normal and it is required to run continuously, the extra cost will be approximately $5/day. This does not sound like much, but $150/month has a bit more impact and $1800/year becomes a really interesting topic.

24.10 LUBRICATING OIL CONSUMPTION ESTIMATE

The total amount of lubricating oil required by an engine over an extended period will be roughly:

$$1\% \text{ of fuel volume used} + \text{capacity of sump} \times \frac{\text{operating hours}}{\text{oil change frequency}}$$

For example a 1000-kW unit running at 75 percent of capacity for 5000 h will have an approximate oil consumption of

$$\frac{1000 \times 5000}{16 \text{ kW imp gal}} \times 0.75 = 234{,}375 \text{ imp gal of fuel}$$

This will reflect as 2343 imp gal of lubricating oil.

The engine has a lubricating oil system capacity of 300 imp gal, and the oil is changed at 2000-h intervals.

$$\frac{300 \times 5000}{2000} = 750 \text{ imp gal}$$

The total annual oil requirement will therefore be

$$2343 + 750 = 3090 \text{ imp gal}$$

The maximum stock of lubricating oil for this engine will be 3093 plus 10% reserve for a gross total of 3402 imp gal.

24.11 LUBRICATING OIL STOCK QUANTITY

It is suggested that the bulk supply of lubricating oil placed at each plant be supplemented by a 10 percent reserve quantity. When the oil is supplied annually, there will be a substantial interest charge. The charge must be considered in conjunction with the increased freight and handling charges that would be incurred with more frequent deliveries.

24.12 CANNED LUBRICATING OIL

The use of canned lubricating oil has certain limited practical advantages, but there are also some considerable economic disadvantages. A small diesel generator set with a load averaging not more than 50 kW will probably run unattended for most of the time with the operator visiting it only once or twice daily. The maximum lubricating oil consumption of such a unit would certainly not be more than 1 gal/day, and it would probably be a great deal less. In this sort of situation the use of canned lubricating oil in quart or liter containers is attractive on the grounds of simplicity, cleanliness, and control.

In all cases when more than 2 qt, or 2 L has to be put into the engine at a time, the economic benefits of packaged oil begin to fall off and the advantages of drawing makeup oil from a barrel or bulk supply improve. If canned lubricating oil is provided for regular use, false consumption figures may materialize unless there is some control over the access to supplies. All too often, canned oil destined for diesel engine use is purloined for use in motor vehicles. Diesel plant operators would be horrified if it were suggested that ordinary gasoline engine lubricating oil be put into the diesel engines, but it is not uncommon to find diesel engine lubricating oil being put into private motor vehicles. The damaging effects of misapplication in either case are about the same, and the operator will do well to remind those who wish to "borrow" one or two quarts or more of that.

The formulation of lubricating oil suitable for diesel engine use is quite at variance with that designed for gasoline engine usage. The differences relate mainly to the differences in the fuels employed, the temperatures and pressures prevailing, and the end products of combustion in the two types of engines. Diesel-type lubricating oil is a notorious cause of seized valves in gasoline engines.

24.13 LUBRICATING OIL CHANGES

The lubricating oil change interval specified by the OEM must be observed. Extension of the specified change intervals cannot be recommended, and any attempt to extend it will have a detrimental effect on the life of the engine. Similarly, the lubricating oil filter elements must be changed at the intervals specified. Modern lubricating oils contain various elements which enable the oil to function very effectively and in a predictable manner. Because of the pressures and temperatures to which the oil is subjected and the contaminants that normally accumulate in the sump, all lubricating oils have a finite service life. Although the bulk of the solid contaminants, mainly carbon, can be filtered out of the lubricating oil, it is not possible to extract the water, acids, ethylene glycol, and fuel that gradually amass and mix with the lubricating oil.

At the time of a lubricating oil change, it is important that the lubricating oil filter elements be removed, cut open, and inspected for metal debris. At the same time, the crankcase doors must be removed and the engine base interior must be thoroughly cleaned out. If there is no access through the side, the pan must be dropped so that it can be thoroughly cleaned. During the cleaning process, the solid material found lying in the pan should be removed and examined for pieces of split pins, fragments of locks and tub washers, broken piston ring bits, or any other identifiable material and their points of origin checked.

Again, exacting attention must be paid to the frequency of lubricating oil filter element change. Many engines can and do suffer quite unnecessary wear because their lubricating oil filters fill up completely with solids before they are changed. The filter then bypasses, permitting unfiltered lubricating oil to circulate, and severe engine wear then commences. Few engines have instrumentation that warns of choked lubricating oil filters, and thus a rigid observance of the hours in service interval is essential. The operator must always remember that filter elements are cheaper than bearings.

The greatest proportion of all the solids picked up by the lubricating oil filters is carbon. The carbon comes from two sources: (1) products of combustion and (2) carbonized lubricating oil. As carbon is generated during the combustion process, it is washed away from the piston rings, the cylinder walls, and the valve guides. It gravitates to the sump and thence to the filters. As long as the filter elements are working effectively, the lubricating oil movement will carry the abrasive carbon particles away from wear points.

Lubricating oil also acts as a coolant and serves to cool the moving parts of the engine in much the same manner as water and ethylene glycol cool the static parts of the engine. Large quantities of oil pass through the crankshaft in both lubricant and coolant modes, and much oil is projected into the underside of the pistons as a coolant. There will be accelerated lubricating oil carbonization in an engine that is overloaded, has a jacket cooling defect, is being overheated because of aspiration problems, is incorrectly fueled, or has worn valve guides or piston rings.

There is no economic advantage in stretching the lubricating oil change intervals in an effort to save oil. The metal wear rates will accelerate, and the overhaul frequency will have to be increased to compensate for the shortened service life of wearing parts. Maintaining a proper oil change interval is a very important factor in preserving the integrity of an engine. Lubricating oil contamination is the direct result of engine operation and the load factors involved. The amount of contamination generated depends in part on the amount and quality of fuel the engine consumes.

Laboratory and field tests have determined that when using the recommended quality oils and filters, a turbocharged engine in good condition and equipped with a secondary oil filter system can consume 255 gal of fuel for each gallon of oil in the lubrication system before the maximum level of oil contamination is reached.

24.14 LUBRICATING OIL CHANGE MANNING

It is usually assumed by the diesel plant staff that lubricating oil changes can be carried out by pretty well anyone in the plant work force. To the contrary, it is recommended that any person assigned to lubricating oil change duties be thoroughly checked out by the superintendent or the overseeing journeyman to determine if he has the knowledge and training to do the work fully and competently. The person conducting the lubricating oil change must be able to select the correct new lubricating oil, recognize the conditions of removed oil, filters, and strainers, and be able to react to anything abnormal in a manner beneficial to the engine's well-being. Lubricating oil changes are not one of the most popular tasks in a diesel power plant, but they cannot be delegated to low-grade untrained labor with any hope of long-term success.

24.15 LUBRICATING OIL CHANGE PUMPS

In the larger diesel plants it is usual to find a portable pump that is used to draw the lubricating oil out of the engine when it is due for change. In many cases, the hoses are fitted with quick connections enabling them to be readily reversed on the pump so that the same pump can deliver clean oil to the engine. It is very desirable that the hoses be clean and free of old oil before the new oil is introduced. The use of the same pump and hoses for the movement of engine coolant is an evil practice and cannot be recommended.

24.16 ENGINES FITTED WITH LARGE-CAPACITY PANS

Many engines are fitted with larger-capacity oil pans so that the interval between lubricating oil changes can be prolonged. It follows quite naturally that the capacity of the lubricating oil filters also must be increased by a similar proportion so that the extra oil volume will be thoroughly filtered throughout its service life. Various supplementary filtering arrangements are provided for the large-capacity-pan engines, and they often involve a secondary oil filter detached or semidetached from the engine. These auxiliary or secondary filters must be serviced on the same frequency as the main lubricating oil filters. Lubricating oil sampling frequencies should not be extended beyond the frequencies recommended for standard engines.

24.17 EXTENDED INTERVALS BETWEEN LUBRICATING OIL CHANGES

If lubricating oil changes are extended, there is a possibility of dirty oil being recirculated through the crankshaft. The rotation of the crank throws can have a centrifugal action on the oil flow that casts out solids onto the walls of the crankshaft oil passages, which eventually close up and restrict the flow of oil to the main and bottom end bearings. The oil pressure gauge may well indicate that full pressure is being maintained, but that can be misleading and the crankshaft can be endangered. It is, therefore, again recommended that the oil change intervals be adhered to and that the oil passages of the crankshaft be thoroughly cleaned out during those overhauls, which involve the exposure of the main and bottom end bearings.

24.18 USE OF LUBRICATING OIL FROM UNSEALED BARRELS

Before any product is drawn from an unsealed barrel, the true nature and condition of the barrel's contents must be verified. Unsealed barrels may often contain spent or condemned fluids such as oil-contaminated antifreeze, fuel-diluted lubricating oil, and discarded transformer oil. The color of a barrel is not necessarily a good guide to the contents because a barrel of a particular color may be used by one company for a cleaning solvent and by another company for a lubricating oil or some quite different product. Furthermore, a company will occasionally change the barrel color of a particular product for a marketing reason.

If there is any doubt whatsoever, do not use fluids from an unsealed barrel. The consequences could be catastrophic.

24.19 ANCILLARY LUBRICATION SYSTEMS

In engines that have separate lubricating oil sumps in their fuel pumps, turbochargers, cam boxes, governors, and so on, the oil change intervals will, in all probability, be radically different from the main lubricating oil change intervals. Because of the severe thermal conditions prevailing in the turbocharger, the high fuel dilution likelihood in the fuel pump and cam boxes, the lubricating oil may become contaminated much earlier than that of the main lubricating oil supply. The operator must be aware of this sort of anomaly and ensure that his maintenance routine caters to these variations. It happens that some of these ancillary lubrication systems require a grade or type of oil different from that used in the main lubrication system.

24.20 DIPSTICKS

Close examination of the dipstick will assist in the early recognition of a lubricating oil level-related problem. The lubricating oil level dipstick carries a great deal of information to the aware operator (Fig. 24.1). Basically, the level of the crankcase fluids is shown on the dipstick. The fluid content of the lubricating system of a normally operating engine will be 100 percent lubricating oil. Other fluids such as coolant or fuel oil may be present to some degree and affect the level. The normal pattern of oil levels showing on the dipstick will be governed by the

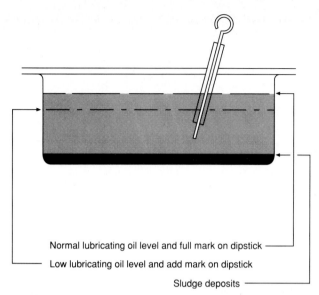

FIGURE 24.1 Typical oil pan or sump. The dipstick does not measure the quantity of oil in the pan; it measures only the level, or ullage. The reduction in the volume of good usable oil caused by sludge and solids deposits will shorten the life of the lubricating oil. *Note*: Pans must be cleaned at each overhaul.

quantities of makeup oil added in each time frame and also by the rate of burnoff, leakage, or vaporized loss. Ordinarily the operator will have a fairly good idea of the amount of lubricating oil consumed over a given period of time, and the log will confirm his opinion. If a coolant leak to the crankcase should occur, the level pattern will change and be noticed. The level will not fall at the same rate; it may even start to rise. When an engine that is cooled with an antifreeze mixture develops a coolant leak, the oil on the dipstick will be seen to be much darker and thicker, and it will have a sweetish taste. If the engine is cooled by plain water, the lubricating oil showing on the dipstick may take on the appearance of emulsification and/or have a milky tinge.

Fuel oil dilution is another common cause of lubricating oil contamination. The symptoms are rising lubricating oil levels, thinness of lubricating oil and rapid dripping off the dipstick, strong fuel odor, and occasionally vapor escape from the dipstick aperture. Fuel oil dilution is more common on lightly loaded two-cycle type engines than in any other circumstances, but in any event it is extremely detrimental to bearing life.

All dipsticks have high- and low-level marks inscribed on them by the manufacturer. Under no circumstances may those marks be altered or additional marks be added. Dipsticks are carefully designed in the modern engine. The dipstick that appears to be "splattered" and is difficult to read is symptomatic of extraordinary crankcase conditions. Overfilling, gross leakage of coolant or fuel, and excessive crankcase pressure are main contributors to the condition.

It is particularly important that several successive readings be taken following start-up after an oil change or overhaul. The operator must not forget to examine the turbocharger or fuel pump housing dipsticks (when fitted), nor must he be misled by the presence of lubricating oil in sight windows. He must be sure that the level is correct, the color looks good, and the oil is in fact fluid. Some engines that employ fuel pumps have dipsticks in their bodies; they also must be recognized and monitored.

24.21 LUBRICATING OIL LEVEL

The two most common reasons for a rising lubricating oil level are those of coolant leaks to the crankcase and unburned fuel coming down the cylinder. Either condition can develop when the engine is running or at rest. Check for leaks around the water pump. On most engines there is a drain hole between the seal for the coolant and the seal for the bearings in the water pump. When this drain hole is absent or is plugged, coolant can pass into the lubricating oil if there is a failure of the seals in the water pump.

When a rising lubricating oil level is detected on a running engine, the engine must be stopped before damage is incurred. A thorough inspection can then be carried out to determine the source of the intrusion. Glycol will quickly turn the lubricating oil into a sticky sludge that is bereft of lubricating qualities; crankshaft and bearing damage is inevitable if the condition is not recognized in good time. Fuel oil will reduce the lubricating quality of the oil and will also foster the development of an explosive atmosphere within the crankcase if subsequent overheating of the bearings takes place.

A falling oil level may be observed on an engine that has just been started after an oil change or an overhaul. If all of the oil has been drained out of the system and then new oil is introduced to the full mark of the dipstick, additional oil will be required to load all of the galleries, passages coolers, and other localities that are normally loaded with oil when the engine is idle or running. When the engine is running, these voids will be filled by the engine-driven pump, and the oil level in the sump will drop to a certain degree.

24.22 MAKEUP OIL ADDED NEEDLESSLY

Because of internal churning and splatter, it is often difficult to read an engine's dipstick when the engine is running. In those circumstances, it is not unknown for the operator to throw in an "insurance quart" of oil day after day. The overall lubricating oil consumption appears to be high in comparison with other engines of the same marque, and it is probable that the operator will be complaining of leaking crankshaft seals (Sec. 8.2, "Crankshaft Seals"). The practice of adding oil carelessly can have expensive and needless consequences. Churning is mostly confined to high-speed engines in the 1500- to 3600-rmp range.

24.23 LUBRICATING OIL FILTRATION

Maximum engine life is dependent on the proper use of both primary and secondary and perhaps bypass lubricating oil filters to protect the vital engine parts

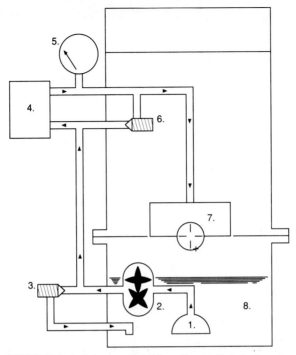

FIGURE 24.2 Lubricating oil full-flow filter: 1, Lubricating oil suction strainer; 2, lubricating oil pump; 3, lubricating oil pressure regulator; 4, lubricating oil filter; 5, lubricating oil pressure gauge; 6, lubricating oil filter bypass valve; 7, engine bearings; 8, lubricating oil sump.

from abrasive contaminants in the lubricating oil. The primary or full-flow filter system filters the lubricating oil that circulates through the engine (Fig. 24.2). Its purpose is to remove large contaminant particles (40 μm or larger) from the oil. The secondary bypass filter receives only a small proportion of the oil flow, filters the oil, and returns it to the oil sump (Fig. 24.3). This filter removes the small contaminant particles (less than 5 μm) that pass through the primary filter. Its purpose is to keep the level of abrasive contamination in the oil low enough to prevent excessive engine wear.

The contaminants have several origins. In new engines very small amounts of dirt may be left behind by the manufacturing processes. The contaminants of this class will be largely of metallic origin. Much of the contaminating matter is a product of combustion or the remains of imperfectly burned fuel. A great deal of the dirt entering an engine from external sources arrives in the aspirating air. This matter will be variable in nature, and some of it will be abrasive. The organic matter may be composed of pollen, insect debris, decomposed vegetation, and multiple other materials. The size, density, and shape of all of the particles, regardless of their sources, will differ widely. The products of combustion evolve in part from smoke, which is carbon in a soft and pure form. Accumulation of such carbon becomes the basis of sludge. The sludge accumulation is also fostered by coolant intrusion into the lubricating oil system.

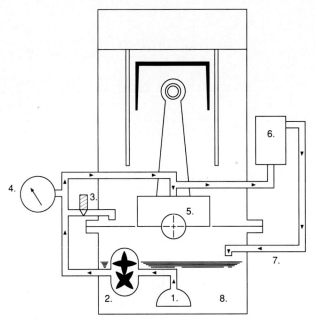

FIGURE 24.3 Lubricating oil bypass filter: 1, Lubricating oil suction strainer; 2, lubricating oil pump; 3, lubricating oil pressure regulator; 4, lubricating oil pressure gauge; 5, engine bearings; 6, lubricating oil bypass filter; 7, filtered lubricating oil return; 8, lubricating oil sump.

Whatever their beginnings, most of the lubricating oil contaminants will have detrimental effects on the engine's mechanism if they are permitted to circulate through the lubrication system. Therefore, an essential feature of an engine's design is the provision of the correct lubricating oil filter elements.

New lubricating oil filters must be stored in a dry area and must remain in their original packaging until needed for service. Each filter element inserted into the filter housing should be carefully examined for any sign of damage. If any damage is observed, the element must be discarded. The lubricating oil filter elements installed in any engine must be of OEM supply or be specifically approved by the OEM. The use of jobber filter elements may appear to be economically attractive or otherwise expedient, but the engine owner or operator must understand that the use of nonstandard elements will jeopardize the warranty position.

Each OEM filter element is carefully matched to the engine's lubricating oil flow rate and to the type of lubricating oil specified for the particular engine. Oil filter change frequency must follow the engine manufacturer's recommendations, but in the event of a bearing failure, coolant leak to the lubricating oil, or fuel oil dilution of the lubricating oil, the filter elements must be changed. Whenever the filter elements are removed from an engine, they must be cut open and scanned for metal fragments. If the engine is still on warranty, the filter must be retained for OEM inspection.

Notice should be taken of the condition of the removed elements for evidence of bursting or complete plugging. In either event, an appropriate report should be made to the superintendent so necessary action can be implemented. At the same time the lubricating oil filter elements and the lubricating oil are changed, the sump must be cleaned out. Quantities of sludge and carbon can build up and all of it must be removed.

Lubricating oil filters are designed to pick up and retain all contaminant particles of a size that will damage the engine components. As the engine ages and wears, it is quite possible that the lubricating oil filter elements will accumulate a full load of solids before the scheduled oil change is due. In such cases, it will be necessary to increase the frequency of lubricating oil filter element changes if an acceleration of engine wear is to be avoided. The alert operator will identify the changing conditions through careful and continuous regard for the condition of the lubricating oil filter elements at changeouts by taking notice of the oil pressure gauge readings and the filter differential pressure gauge readings.

24.24 LUBRICATING OIL FILTER ELEMENT DISPOSAL

Any lubricating oil filter element that is taken out of service must be destroyed and disposed of in an acceptable manner so that there is no possibility of its deliberate or inadvertent reuse. It is not unknown for a spin-on type of filter element in particular to be reused, and it is suspected that certain pleated paper elements have been sent round for a second time on occasions.

NEVER attempt to wash out and reuse a disposable filter element of any description.

24.25 LUBRICATING OIL FILTER: PLUGGING

Lubricating oil filters do not plug as long as the oil remains suitable for use in the engine. If a filter plugs, it is doing the job for which it was designed to do. It is protecting the engine by removing particulate contamination from the lubricating oil. There are several reasons for filter plugging; some of them are:

1. Excessive oil contamination indicated by a heavy, thick, loosely held sludge in the filter element. This type of contamination is caused by fuel soot, oxidation products, and products of combustion.
    a. Excessive fuel soot is generally caused by defective injectors, overfueling, a restricted air intake, a generally poor mechanical condition of the engine, or inefficient loading.
    b. Oxidation by-products are usually brought about by operating the generator set at excessively high engine temperatures, running with inadequate cooling, and again, generally poor mechanical condition of the engine or overextending the oil change interval.
    c. Combustion by-products are caused by combustion gas blowby, high sulfur content in the fuel, poor combustion, or overextending the oil change interval.
2. Coolant leakage and moisture condensation are indicated by wavy pleats in

the filter element and deterioration of the paper outer wrap. The additive package will separate from the oil and become ineffective when a small amount of coolant contamination is present in the oil.
  a. Some of the reasons for coolant leakage are:
    (1) Leaking filter head assembly
    (2) Cracked cylinder block or head
    (3) Leaks in the oil cooler
    (4) Leaks past the cylinder liner sealing rings
  b. Some of the sources of condensation include:
    (1) Excessive engine idle running
    (2) Low engine operating temperature
    (3) Inadequate crankcase ventilation
3. Gel or emulsion formation (a gel formation in the pleats of the filter) may be observed but the filter may appear to be clean. When the filter is tested, it will show a restriction if moisture is still present. This problem often occurs when oil in a bulk storage tank has been contaminated with small amounts of water (less than 0.5 percent). Filter plugging can take place in as little as 5 min after the contaminated oil has been added to a running engine. This condition is prevalent where main storage tanks are replenished at subzero temperatures.
4. Fuel dilution is indicated by a brown or red color of the filter. Fuel dilution may be a result of overfueling, incorrectly calibrated and/or wrongly adjusted injectors, restricted air intake or exhaust system back pressure, excessive engine idling, cracked injector cup, mutilated or missing injector O-ring, seized injector nozzle, or sticking valves.

Information Plates

Near or on the lubricating oil filler cap there may be a plate stating the OEM-recommended lubricating oil grades, and there may also be plates indicating oiling points or levels elsewhere on the engine. Other oil level indicators may be found on the engine cam box, turbochargers, governor, pedestal bearing, fuel pumps, air-start motors, turning engine, air filter bath, reduction gear housing, thrust block, and possibly other places.

24.26 LUBRICATING OIL STRAINER

The purpose of the lubricating oil strainer is to protect the lubricating oil pump from ingesting any metallic debris, rags, or small parts from the suction intake of the lubricating oil pump. The lubricating oil strainer is always fitted on the suction side of the lubricating oil pump. It may be situated in the bottom of the pan, or it may be located external to the pan where it is more accessible for cleaning and inspection.

In all cases, the lubricating oil suction strainer is immersed in lubricating oil when in service. The strainer on larger engines is often of dual configuration to enable one element at a time to be serviced while the engine is running, or it may be one of the self-cleaning types which demand that the operator give the element one or two manual turns per shift. A cleanout of the housing takes place at each lubricating oil change.

Whatever arrangement of lubricating oil strainer is fitted, it is imperative that the strainer element be examined for the presence and nature of any foreign material at each lubricating oil change. In extreme cases, it is possible for the lubricating oil strainer to become so obstructed with carbon, rags, sludge, or other materials that the lubricating oil pump is unable to supply sufficient oil to the engine. In such an event it is probable that the engine will shut down on a low lubricating oil pressure signal, but it is also possible that an alarm, if fitted, will have sounded earlier. The lubricating oil pump itself may have given off noises of cavitation. All of these events should be recognized and investigated.

Any rags, paper, gasket fragments, filter wrappers, or cotton waste found in the pan or the lubricating oil strainer points to careless inspections during previous overhaul or maintenance actions. Any metal found in the lubricating oil strainer will indicate that an immediate and thorough mechanical inspection must take place. The sources of the metal must all be completely identified and the necessary repairs must be carried out before the engine can be restarted. Any metal found in the lubricating oil strainers, the lubricating oil filters, or the lubricating oil pan warns of imminent catastrophe if the engine is run again without repair.

If the lubricating oil strainer is overloaded with sludge or carbon, it will be necessary to review the overall condition of the engine and consider a short-term revision in the lubricating oil change frequency and the associated lubricating oil filter changes. Any unusual lubricating oil strainer conditions give notice of incipient, expensive trouble.

24.27 CENTRIFUGAL FILTERS

Centrifugal-type filters are not a new development, but of late they have gained a great increase in use by OEMs. Because of the efficiency of the centrifugal filter, the smallest particles are separated from the fluid being filtered. In the case of lubricating oil, the result is a greatly reduced wear rate in the machine. The reduced clearances and higher loadings now being imposed on modern engines calls for the use of centrifugal filters more and more often.

The additives in the lubricating oil tend to survive better if they are not repeatedly exposed to the same containments, as happens on a normal filter element. The centrifugal filter has a limited throughput capability when compared with the ordinary full-flow filter; accordingly it is fitted in a bypass mode. These filters spin out solids as small as 1 μm which results in the formation of a dense mass or cake in the filter housing. The cake is readily removed, and when it is examined in good lighting conditions, any metallic particles are easily seen. Such debris is not so easily observed in the oily sludge of a conventional filter. If the cake is soft and spongy, it is possible that there is a water leak in the engine. A shiny black jelly indicates a glycol leak.

A reasonable amount of care in handling centrifugal filter components is requisite when the components are being cleaned. Plastic or wooden tools are recommended for cutting out the accumulated cake. Centrifugal units revolve at very high speeds and rough treatment will cause the moving elements to bind or run in an unbalanced state. To operate effectively, they require an oil pressure of not less than 30 psi.

24.28 MAGNETIC PLUGS

Magnetic plugs will be found fitted in several locations on different engines and other equipment. Their purpose is to attract ferrous metal cuttings from the passing lubricating oil stream and warn an inspector that excessive wear is taking place somewhere in the machine. Whereas most of the fine metallic debris will be caught by the full-flow filters, along with carbon soot and sludge, the magnetic plug will exhibit only iron and steel fragments, but in a very visible manner.

Magnetic plugs are to be found in turbocharger lubricating oil systems, transmission housings, and thrust blocks; they are frequently used in the dual role of oil pan drain plugs. The operator or maintainer must familiarize himself with their location on the machinery he is serving and examine them at each oil change.

24.29 CRANKCASE SLUDGE

Crankcase sludges are formed by abnormalities in the lubricating oil, and they may develop in several ways. The sludges may be directly derived from the lubricating oil or by the addition of another medium to the lubricating oil. Probably the most commonplace cause of sludge is ethylene glycol coolant intrusion into the lubricating oil following the failure of the liner O-ring seals. Water pump seal failure also may cause sludge development. As noted earlier, the sludge generated by mixing glycol and lubricating oil will be a shiny black jelly.

The development of the sludging condition can be detected by paying reasonable attention to the dipstick. It is not possible to operate an engine after ethylene glycol has entered the lubricating oil without destroying the engine bearings. Ethylene glycol may also enter an engine through a perforated lubricating oil cooler, but that will happen only when the engine is at rest. The occurrence is not common, but the problem will probably be identified when the engine is run and oil is seen to accumulate in the header tank of the cooling system.

Lubricating oil may convert to sludge when the makeup lubricating oil added is of the wrong sort. It is not unknown for hydraulic, transmission, or transformer oil to be used inadvertently. The lubricating oil specified for any engine is in accordance with the temperature at which the engine will normally operate. A gritty or granular sludge will form in the lower parts of the lubricating oil system if the engine lubricating oil is run for prolonged periods of time at abnormally low operating temperatures (Sec. 24.6).

In all cases when lubricating oil sludge develops, the condition of the lubricating oil filters should give ample warning that something is amiss. Sludge in a sump has a very negative effect on the lubricating oil heater elements, particularly in a standing engine. Crusts of carbon will build up over the lubricating oil heater element, which will probably burn out in time. Pieces of congealed carbon will break off from the crust; they can plug up oil passages, relief valves, and filters and cause bearing damage.

If sludge is found to be present, the whole lubrication system must be thoroughly cleaned out, the origin of the sludge must be identified, and the situation must be rectified before further new oil is added and the engine is restarted. Do not put the extracted sludge in oil barrels that are to be returned for demurrage recovery. Nor must contaminated oil be entered into a fuel–lubricating oil blending and burning system, because great damage will be done to the fuel injector system.

24.30 AERATION OF LUBRICATING OIL

Aeration of the lubricating oil may be caused by loose parts on the lubricating oil pump permitting air to be drawn into the suction side of the pump. However, the most common cause of aeration is overfilling the lubricating oil sump. Aeration may occur on an engine that is normally cooled by straight water if a coolant leak into the crankcase occurs. If the engine uses an antifreeze coolant, gelling, rather than foaming, of the lubricant will occur.

When aerated oil is passed to the bearing, the film separating the moving parts breaks down more readily. Similarly, the ability of the oil to carry away and dissipate heat from the bearings is impaired. Under those conditions premature failing of the bearings can be expected. To guard against foaming or aeration, the operator must check the dipstick carefully and look for rising levels or any evidence of bubbles, which may be quite minute. The dipstick may also be splattered along its length and the true level difficult to read. A drop in the lubricating oil gauge pressure may also be evident. If water has raised the oil level so that churning and subsequent aeration take place, it is probable that the oil will have an emulsified or milky appearance on the dipstick.

24.31 FUEL DILUTION OF LUBRICATING OIL

Fuel dilution of the lubricating oil is not uncommon on engines in which fuel lines are situated in compartments open to the crankcase. Lubricating oil sample analysis shows that approximately 10 percent of the samples taken from engines which are sampled once in each oil change cycle show an unacceptable level of fuel oil contamination. The value of regular and systematic oil sampling cannot be overstressed.

24.32 LUBRICATING OIL PRESSURE RELIEF VALVE

The purpose of the lubricating oil pressure relief valve is to regulate the working pressure of the lubricating oil throughout the engine and to relieve any excessive pressures that may be developed by the engine's lubricating oil pump. Many engines have oil pressure gauges at various points in the lubrication system. When the engine is running at normal operating temperatures and pressures, the valve is closed. When it has just been started and the oil has not yet reached its normal operating temperature, the oil pressure will be higher and the pressure relief valve will open to relieve the excess pressure. The valve will dump oil back into the pan.

Once the adjustment of the lubricating oil pressure relief valve is set, the valve should not require alteration at any time thereafter. If an adjustment appears to be justified, it should be carried out only by a journeyman mechanic and after very careful consideration has been given to reasons for and the consequences of making such an alteration. After the adjustment has been effected, the adjustment must be locked, and no further adjustment will be needed for many thousands of hours. The lubricating oil pressure relief valve, once set, will regulate the standard operating lubricating oil pressure. No attempt should be made to alter the valve setting in an effort to improve an abnormal lubricating oil pressure condition. The superintendent should be consulted and his permission to alter the setting and the event shall be recorded in the log book.

The reason for the lubricating oil pressure anomaly must be found elsewhere and rectified before the engine suffers serious damage through improper lubrication of one or more components. In all cases, the lubricating oil pump is of a positive-displacement type, and therefore it must be accommodated with a fully workable pressure relief valve. The pressure and temperature conditions of the lubricating oil in an engine must receive the constant attention of the operator if the engine is to give a long lasting and reliable performance.

Any deviations from the designed pressures and temperatures are signs of immediate or future problems. If premature wear, unplanned stoppages, or breakdowns are to be avoided, the early recognition of warning signs is essential.

24.33 LOW LUBRICATING OIL PRESSURE PROTECTION

Nearly all engines have some form of protection against low lubricating oil pressure. Usually the protective device for this condition is a pressure-sensitive switch which, upon loss of pressure, signals the shutdown solenoid to return the fuel pump rack to the zero-fuel position. On many engines the electrically operated protection is backed up by a mechanically operated device, commonly in the form of a pressure-supported trigger. When the lubricating oil pressure falls, a spring-loaded rack return pin pushes the fuel pump rack to the zero-fuel position. The sole purpose of the low lubricating oil pressure shutdown is to save the engine from damage that would result from loss of lubrication.

24.34 SHUTDOWN FOLLOWING A LOW LUBRICATING OIL PRESSURE SIGNAL

When an engine is shut down on a signal from the lubricating oil pressure sensor, the situation must thereafter be handled in a controlled manner if damage to the engine is to be avoided. The lubricating oil filter elements must be removed and examined for evidence of metallic debris. The lubricating oil dipstick must be inspected. The oil level may be high, in which case either fuel oil or coolant intrusion into the crankcase can be suspected. On the other hand, if the level is too low, leakage to the lubricating oil cooler should be considered. Failure of the manual or automatic lubricating oil makeup service is another common possibility.

Consider the solids retained by the filter elements, because they provide some evidence of the origin of the problem. Besides any metal, which is not always easy to find, there may be quantities of soot or sludge in the filter elements. A slight brownness of color may point to the presence of fuel in the lubricating oil. Notice should be taken of the lubricating oil on the dipstick (Sec. 24.20). Merely changing the lubricating oil and the lubricating oil filter elements following a low lubricating oil pressure shutdown will serve no useful purpose. A full and careful investigation into the cause must be made before attempting a restart.

With the fuel shut off and the indicator cocks open (where provided), the engine should be turned over by hand to see if there is any unusual stiffness or any other indications of seizure.

Do not attempt to remove any crankcase doors or covers until the engine has cooled down for at least an hour.

Under no circumstances should the engine be restarted following a low lubricating oil pressure shutdown until the steps noted above have been carried out. If there is any doubt whatsoever about the condition of any of the bearings or bearing surfaces, the engine must remain at a standstill until it has been thoroughly examined and qualified by a journeyman mechanic.

24.35 LUBRICATING OIL ANALYSIS

The basic purpose of an oil sample analysis program is to reduce overall engine upkeep costs and at the same time improve the machine's safety. Continuous monitoring of engine lubricating oil conditions and the initiation of appropriate and timely action based upon the laboratory advice assist in the avoidance of costly repairs.

24.36 LUBRICATING OIL SAMPLING

To have overall success of a lubricating oil sampling routine, the participating plant must act in a consistent and predictable manner. All the lubricating oil samples drawn should all be sent to the same testing laboratory. Test results vary considerably from one laboratory to another in scope, test methods, equipment used, reporting formats, and other aspects.

The engine user must develop and maintain a sampling schedule so that the test results can be meaningfully compared. The oil samples should be taken at set running time spans so that any untoward deterioration in the oil's condition will show up when compared with preceding analytical results. The samples should always be drawn from the same point on the engine when the engine is at operating temperature and running. Any engine user about to or already participating in an oil analysis program will find the installation of lubricating oil sample self-closing bibs to be very worthwhile. These bibs are commonly supplied as OEM items.

Oil samples may be drawn from the oil pan (sump) through the filling plug or a sample extractor. Extractors are usually supplied by the analyzing laboratory as part of the service. Samples should not be taken from the lubricating filter housings, filter elements, or filter pots. Care must be taken not to include any external dirt from areas surrounding the sample points. The sampling equipment and container must be absolutely free of any contaminant such as fuel, coolant, other lubricating oil solvents, or loose material that may obscure the test results. It must be remembered that the testing is done in terms of *parts per million* (ppm) and it does not take very much foreign material to affect the accuracy of the sample analysis.

A frequent contaminant is engine coolant. This shows up in the analysis as water, sodium, boron, or sludge. The sodium and boron elements are components of water treatments and antifreezes. Lubricating oil analysis indicates the presence in the oil of the various metals used in the engine. These metals and their sources are shown in Table 24.2. On a new engine or one that has just been overhauled, higher values may be expected until the engine has "run in."

TABLE 24.2 Elements Commonly Found in Lubricating Oil

Element	Symbol	Origin
Aluminum	AL	Pistons, bearings, blower housings
Barium	Ba	Lubricating oil detergent additive, smoke-modifying agent in fuel supplements, additive used in coolants and lubricants
Boron	B	Additive used in coolants and lubricants
Calcium	Ca	Lubricating oil detergent additive, road salt, salt derived from marine atmospheres and seawater
Chromium	Cr	Bearings, top end and rocker arm bushes, cam roller bushes or thrust bearings, thread lubricants and lubricating oil supplements
Iron	Fe	Cylinder liners, pistons, camshafts, crankshafts, gears, tappet faces, cam followers and rollers
Lead	Pb	Bearing flash, poured bearings (babbitt)
Magnesium	Mg	Lubricating oil detergent additive, aluminum components or magnesuim derived from seawater
Molybdenum	Mo	Antifriction supplements, lubricating oil additives, flash coatings on components
Nickel	Ni	Crankshafts, camshafts, gears, valves and valve seats
Silicon	Si	Dirt, atmospheric dust, grinding compound, casting sand, silicone gasket material and sealants, antifoam additives in lubricating oil
Sodium	Na	Coolant inhibitor additives, lubricating oil additives, water treatment additives, road salt
Silver	Ag	Piston flashing, bearing flashing, bearings or bushes
Zinc	Zn	Lubricating oil wear and oxidation additive, die-cast parts (valve covers, and so on)

It is important that each lubricating oil report received be compared with the preceding reports from the same engine so that a pattern of wear rates can be established and an appreciation of developing trends can be formed.

24.37 THE ORIGINS OF WEAR METALS AND OTHER ELEMENTS SHOWING IN LUBRICATING OIL ANALYSIS

Table 24.2 lists in alphabetic order the more common elements exposed in lubricating oil analysis.

The wear that normally takes place in a running engine will result in various metals or other elements showing up in the oil analysis. In all cases the testing laboratory will advise the service user of any extraordinary metal loss. However, the user must compare the metal loss rate between one sample and another and identify trends as they develop. Expected wear rate figures for the various elements cannot be quoted as examples because of the great variations in engine models, type and condition of service, locations, and so on. However, there will always be a certain amount of metal entering the oil.

There have been many cases of lubricating oil analysis showing a high copper content. That might indicate a bearing failure, but in fact it is caused by the use of an antiseize compound containing copper. It is important that the function of the lubricating oil sample be unimpaired, and it is for that reason that the use of copper-based compounds is disallowed. Several equally effective antiseize compounds are available. In any case, great caution must be exercised in their situation and application because of their apparent torque-altering propensities. It is known that bottom end bolts have been inadvertently overtorqued by 40 percent owing to the use of improper antifriction compounds. Consult the OEM manual for advice.

24.38 LUBRICATING OIL ADDITIVES

There is no such thing as the perfect diesel engine lubricating oil. Furthermore, the type and grade of oil that are best for one model of engine may not be suitable for another. The fuels used and the conditions of service in which the engine operates have a great influence on the lubricating oil selected for a given engine. Because of the multiple factors affecting the choice of lubricating oil, it is strongly recommended that no changes be made in the type of oil used without OEM approval.

Because of the complexity of formulation in today's lubricating oil, it would be foolhardy to attempt modifications of the supplied oil with any of the numerous supplements offered by aftermarket oil companies. Depending on the intended use of the finished lubricating oil, additives are blended into the feedstock to impart particular qualities to the oil. Much effort is exerted by the oil companies to incorporate the best possible blend and combination of additives in their products.

1. *Antifoam agents* are utilized to assist in the release of entrained air from the circulating lubricant. Without foam control, accelerated wear can be expected.
2. *Antiwear additives* are designed to provide a high strength film to minimize metal-to-metal contact. Antiwear additives provide a reserve alkalinity to offset potential corrosive wear. In an engine using a high-sulfur fuel, it will be necessary to employ a lubricant with a high alkalinity reserve.

3. *Corrosion inhibitors* reduce the effect of acids developed in the blowby gases. Forming a protective film on bearings, these inhibitors are necesary to combat premature bearing wear.
4. *Detergents of metallic origin* are used for their detergent and dispersing abilities and to ensure that deposits of insolubles do not accumulate but instead remain in suspension. These metallic detergents also contribute to acid corrosion control.
5. *Dispersants* with ashless characteristics help in controlling deposit formation and in keeping the engine interior clean. They are also useful in combatting the formation of low-temperature-fostered sludges.
6. *Oxidation inhibitors* minimize the attack of oxygen upon the oil. Without inhibitors being present, the oil will eventually thicken to a solid state, at least in part, and it will also become quite acidic.
7. *Pour point depressants* modify the wax formation characteristics of an oil and improve the oil's low-temperature fluidity.
8. *Rust-inhibiting* additives provide a protective film to guard against the attack of corrosive elements. However, these rust inhibitors do not neutralize the corrosive elements themselves.
9. *Viscosity index modifiers* slow down the rate of change in an oil's viscosity when the oil is subjected to heat. When such additives are employed, the oil will not thin out so quickly, and thus wear rates are minimized.

24.39 LUBRICATING OIL CONTAMINANTS

The lubricating oil analysis report will also show the presence of other contaminants or conditions which indicate the current state of degradation. Different testing laboratories use various rating figures or scales which may not compare too readily. Nearly all laboratories use an unmistakable signal when danger levels are found in the samples.

Sludge is a term used as a reportable item in lubricating oil analysis reports, but it may be a composite of contaminants. Just as there is a chemical reaction between iron and oxygen producing a corrosion of the metal, so the interaction between lubricating oil and oxygen results in a degradation of the oil. This *oxidation*, as it is known, reduces the lubricating properties of the oil; acids form and thickening of the oil is promoted. Thus, sludges composed largely of degraded oil, soot, ethylene glycol, and suspended solids develop.

Indications of lead showing up in a lubricating oil analysis may, in some cases, be attributable to the lead content of certain greases. Grease used as a thread lubricant or gasket adhesive during engine work activity may be a subsequent oil contaminant. The zinc content of some greases will produce the same effect and also be misleading.

24.40 LUBRICATING OIL PUMP

Lubricating oil pumps always work on the principle of positive displacement. They are invariably of the gear-type configuration, and they are direct-driven by the engine. In the majority of engines of 500-kW and greater capacity, auxiliary motor-driven lubricating oil priming pumps are provided.

Because of the medium they pump, lubricating oil pumps are generally trouble-free, but they do suffer some wear eventually. The wear rate of a lubricating oil pump will accelerate if the unit is required to pump dirty oil. Seals will fail for the same reason and also if there is excessive wear in the auxiliary drive gear train. All positive displacement pumps have to be fitted with some sort of relief valve, most such valves are adjustable. Worn valves, broken springs, and dirt under the relief valve seat may result in reduced pressure performance from the pump.

Failure to clean the sump regularly can lead to a substantial buildup of sludge and carbon buildup in the vicinity of the lubricating oil pump suction strainers. If the flow of lubricating oil to the lubricating oil pump is restricted, problems will develop (1) in maintaining an adequate lubricating oil working pressure at the bearings and probable shutdown of the engine by the protection devices and (2) in gross wear in the lubricating oil pump casing in way of the pump rotors.

24.41   OVERCOOLED LUBRICATING OIL

If the lubricating oil is grossly overcooled, it will not flow easily, and hence it will not pass through the lubrication system in a satisfactory manner. In the worst case, the oil flow through the cooler may cease entirely and the only circulation will be from the pan, through the lubricating oil pump, past the lubricating oil pressure relief valve, and back to the sump. In that case, a seemingly correct lubricating oil pressure may be observed but it is probable that little or no lubricating oil is being delivered to the engine's oil galleries. The condition is not likely to occur on engines in which the lubricating oil cooler is served with jacket coolant, but it can occur when the lubricating oil cooler is superimposed on a fan-cooled radiator and the air is at 0°C (32°F) or cooler.

The temperature of the lubricating oil leaving the oil cooler should be approximately 6°C (10°F) less than the temperature of the oil arriving at the lubricating oil cooler. The operator must refer to the OEM manual for each individual engine normal lubricating oil operating temperatures. The overcooling of lubricating oil may be made evident by unusually high lubricating oil pressures. No attempt should be made to adjust the lubricating oil pressure relief valve at this or any other time (Sec. 24.32).

24.42   LUBRICATING OIL COOLER THERMOSTAT

Lubricating oil cooler thermostats (regulators) are extremely reliable, but if they are suspected of malfunction, they should be removed and tested in the same manner as jacket water thermostats (regulators). Engines must not be run without the lubricating oil thermostats in place and in proper working order. When multiple thermostats are employed, they must be identical in temperature rating.

24.43   PERFORATED LUBRICATING OIL COOLER

Lubricating oil coolers will leak when they are corroded, subjected to high vibration, or overpressurized. In all cases, the pressures prevailing in the lubricating oil cooler will be higher than in the cooling liquid. When leakage occurs, the oil will flow to the coolant as long as the engine is running, but it is possible that there might be a small amount of coolant leak back into a static engine.

A leaking lubricating oil cooler will readily be detected by observing the sight glass in the coolant header tank, which will soon be occluded by oil. Shortly thereafter, a drop in the sump lubricating oil level will be noticed. As soon as a lubricating cooler oil leak is determined, the engine must be stopped, the cooling system must be drained, steps must be taken to replace or repair the cooler, and the whole of the cooling system that has been contaminated by oil must be cleaned out (Sec. 23.47).

In plants so arranged that the lubricating oil is cooled by a raw water source, the operator must look to the raw water outfall for traces of oil which would indicate a lubricating oil cooler perforation. All lubricating oil coolers will leak to the cooling water because pressure of the lubricating oil system is higher than that of the raw water circulating system. Most lubricating oil coolers have a cooling capacity greater than is needed for normal service. If a leaking tube is detected, it is quite feasible to plug the leak with plain tapered wooden or neoprene plugs until such time as a permanent repair can be effected.

24.44 DIRTY LUBRICATING OIL COOLER

A lubricating oil cooler may get dirty on either the oil or the water side. Oil coolers will agglomerate quantities of carbon inside the tubes if the lubricating oil filtration system is not being operated properly and oil bypasses the filters. Oil that has been contaminated by glycol will tend to stick to and coat the tubes and thereby reduce heat transfer.

Flushing the oil side of the cooler with a solvent-type cleaner is a very effective way to remove masses of carbon, gummy deposits, and ethylene glycol-induced sludge. Brass or bronze wire brushes are acceptable, but steel wire brushes, scrapers, or steel rods must never be used to push muck out of the cooler tubes. Caustic solutions must not be used for wash-out purposes because they will have a decaying effect on the cooler bundle metals. Lubricating oil coolers should normally be cleaned out at about 10,000-h intervals, but they must also be thoroughly cleaned following a bearing failure, the breakdown of any component resulting in metal debris falling into the lubricating oil sump, or a coolant leak. On smaller engines, the structure and design of the lubricating oil cooler is such that it is often impossible to effectively flush them after an internal engine failure and the cooler core must therefore be replaced with a new one.

24.45 OIL BARRELS

- Barrels containing new oil should be stored on their sides. Ingress of rain water, snow melt water, and so on, is thus kept to an absolute minimum. The make and type of oil stencilled on the top of the barrel is not readily obliterated by weather conditions.

- Barrels must be emptied completely. Most pumps do not draw every last drop of oil out of the barrel and it is estimated that an average of 5 percent of the

overall lubricating oil consumption in any given plant is attributable to this form of wastage.
- In cold weather, oil barrel draining will be facilitated if the new oil supply is allowed to warm up to room temperature before use.
- Both filled and empty barrels must be handled with care. Barrel demurrage and damage charges represent as much as 10 percent of lubricating oil costs, and every effort must be made to obtain maximum barrel charge recoveries.
- Barrel bungs must be kept in place and always be replaced in empty barrels.
- Oil barrels should be kept closed except when oil is being drawn off. That will prevent the entry of dirt or water.
- Lubricating oil is slightly hygroscopic. Oil in an opened barrel will absorb quantities of moisture from the atmosphere, and the moisture will be transferred to the engine when makeup oil is added.

24.46 OIL BARREL HANDLING

Barrels must be treated carefully at all times during transit, movement from storage, and on return to the empty barrel park. Carelessly handled barrels will leak quite easily, and barrels delivered from the supplier in a leaking condition should either be refused or shown immediately to the supplier's representative for claim. The supplier will not hesitate to refuse refund on any distorted, leaking or corroded barrels when returned empty, and in many cases, freight companies will not carry barrels with apparent leaks. In 1989, the demurrage charge of each oil barrel was $30 U.S. (An engine generating 3,000,000 kWh annually will consume approximately 1608 U.S. gal of oil, or 35 bl, which is equivalent to $1072 U.S. in barrel charges.)

24.47 FLUSHING THE LUBRICATING OIL SYSTEM

Flushing of the lubricating oil system of an engine will be required when there has been a part failure resulting in the detachment of metal debris, the engine has been unused for a considerable time, the lubricating oil has been contaminated with an excessive amount of ethylene glycol or water, or the engine has become grossly dirty internally.

Flushing is best carried out by using a supply of the standard flushing oil heated to 65 to 90°C (150 to 200°F). Flushing oil that has been used previously may be used provided that it has been thoroughly filtered and cleaned. Any flushing oil contaminated with ethylene glycol or water must be discarded. Lubricating oil system flushing has its hazards, and so it should be done only by an OEM respresentative.

If it is necessary to flush the lubricating oil system of an engine that has been heavily contaminated with ethylene glycol, it may well be expedient to use Butyl Cellosolve, which is a product of the Union Carbide Co. This product will clean the system effectively, but it has its hazards. Under no circumstances can the material be used without the approval of the OEM. If and when that approval is obtained, all of the personnel involved in the operation must read the product instructions, the OEM notes, and the safety notes before the work commences. Failure to follow the instructions will endanger the crankshaft and other moving parts of the engine.

24.48 TURBOCHARGER LUBRICATING OIL

The turbocharger bearings are fed with lubricating oil by the engine lubricating oil pump or by a lubricating oil pump built into the turbocharger. In the first case, crankcase oil is circulated to the turbocharger bearings and drains back to the crankcase; in the second case, the turbochargers have their own exclusive lubricating oil supply. The grade or type of oil used in this system may well differ from that used in the main engine lubricating oil supply, and the reservoirs will have their own sight glasses and/or dipsticks and lubricating oil coolers. Filters may also be fitted on the larger-engine turbochargers.

24.49 GREASE APPLICATION GUIDELINES

- Reapply grease to normally greased components soon after cleaning to inhibit corrosion of balls, rollers, and races.
- If it is necessary to use a different grease for some reason, the greased component should be dismantled and thoroughly cleaned before the new grease is applied.
- Keep grease containers, grease guns, and grease supplies clean.
- Grease types and grades must not be mixed. If there is doubt about the correct grease, *ask*.
- Antifriction bearings that have been cleaned but not yet regreased must not be hand-spun; they must first be lubricated.
- Bearings must not be overpacked because there is often a danger of seal breakdown, particularly in a ventless housing. Do not force the grease gun.

At weekly or monthly intervals one of the plant operators will be sent on a fixed tour to grease and oil the overhead door mechanism, the ventilation louvers and their linkages, the raw water pump, valve spindles, various motors and pump bearings, and any other point requiring regular attention at extended intervals. Because this area of activity is normally very badly neglected, a grease point checklist and diagram should be posted in the control room for the operator's reference.

If the frequency is a bit flexible, it does not matter too much; what does matter is that all of the points needing some form of lubrication are identified and listed so that there is no excuse for individual points being continuously overlooked or ignored.

24.50 GLOSSARY OF LUBRICATION TERMS

Additive. Any material added to a lubricating oil at the refinery to improve its suitability for service. It may improve a property already possessed by the lubricant or give it properties not naturally possessed. Typical examples are antioxidants and corrosion inhibitors.

Anhydrous. Devoid of water. Specifically, a finished lubricating grease in which there is substantially no water.

Antioxidant. An additive, usually incorporated in a relatively small proportion, to retard oxidation of lubricants, including greases and gear lubricants.

Base Number. See *Neutralization Number*.

Cavitation. In a hydraulic system, cavitation in the oil that is due to the sudden collapse on the pressure side of the pump of air bubbles formed on the suction side of the pump. The bubble collapse causes gradual and severe damage to the metal surface.

Corrosion. A gradual destruction and/or pitting of a metal surface that is due to chemical attack.

Detergency. Characteristics imparted to an oil by suitable additives, called *detergents*, that suspend harmful sludge, varnish, and carbon in the oil and prevent their deposition on cylinder walls and bearing surfaces. Detergents act as cleaners to remove existing varnish, lacquer, or carbon deposits only in a secondary role.

Emulsion. A mixture of water and oily material in which either very small drops of water are suspended in oil or small drops of oil are suspended in water. The whole is stabilized by a third component, such as soap, called an *emulsifying agent*. Emulsibility is desirable for some products such as soluble cutting fluids.

Film Strength. The ability of a film of lubricant to resist rupture due to load, speed, and temperature.

Flash Point. The temperature to which an oil must be heated to give off sufficient vapor to form a flammable mixture with air under the conditions of the test. The vapor will flash but will not support combustion.

Foaming. The effect of active agitation in the presence of air. It may occur if the oil level is too high. Use of antifoam additives helps to break the foam bubbles. Contaminants, especially water, aggravate foaming.

Fretting. A form of wear resulting from an oscillating or vibratory motion of limited amplitude; very finely divided particles are removed from rubbing surfaces. In the case of ferrous metals in air, the wear particles oxidize to a reddish, abrasive iron oxide (alpha-Fe_2O_3) which has led to the name *fretting corrosion*. The phenomenon has also been called *false brinelling* and *friction oxidation* when it occurs in rolling contact bearings.

Hydrophilic. Having an affinity for water; capable of uniting with or dissolving in water.

Hydrophobic. Having antagonism for water; not capable of uniting or mixing with water.

Hydrostatic Lubrication. That state of lubrication in which the lubricant is pumped to a plain bearing under sufficient external pressure to separate the opposing surfaces by a continuous lubricant film. Most commonly used prior to start-up of heavy engines with plain bearings.

Inhibited Oils. Petroleum oils which are fortified by additives to impart resistance to rust, oxidation, and foaming.

Multigrade Oil. Oil manufactured to have the viscosity-temperature relationship necessary to place it in more than one SAE grade classification. Thus, if an oil is within the viscosity bracket of an SAE 10W oil at −18°C (0°F) and also that of an SAE 30W at 99°C (210°F), it can be referred to as multigrade (i.e., SAE 10W/30). This reference is to viscosity only. Many other characteristics are necessary for satisfactory service. Multigrade oils are not generally recommended for continuous diesel generator service.

Oxidation. The combining of constituents of a lubricating oil with oxygen from the air to form harmful acids. Oxidation is accelerated by heat and by copper.

Pour Point. The lowest temperature at which a lubricant will pour or flow under specified conditions.

Sludge. An insoluble material formed resulting from deterioration reactions in an oil, or contamination of an oil, or both.

Specific Gravity. The ratio of the weight of a material, such as oil, to the weight of an equal volume of water at the same temperature. It is usually determined by the weight of a known volume of the material, but it is rarely used for oils lighter than heavy residual road oils.

Straight Mineral Oil. A petroleum oil which contains no additives or fatty oils. More correctly called a *mineral oil*.

Viscosity. A measure of resistance to flow. It is determined by measuring the time taken for the lubricant to flow through a standard orifice at 38°C (100°F) and at 99°C (210°F). The result is expressed in saybolt seconds or kinematic centistokes.

Wear. The removal of materials from surfaces in relative motion.

Wear (Abrasive). The removal of material from surfaces in relative motion by a cutting or abrasive action of a hard particle (usually a contaminant).

Wear (Adhesive). The removal of materials from surfaces in relative motion as a result of surface contact. Galling and scuffing are extreme forms.

Wear (Corrosive). The removal of materials by chemical action.

CHAPTER 25
INSTRUMENTATION

25.1 REMOTE AND DIRECT-READING INSTRUMENTS

By using extended wires or tubes, many instruments can be arranged for remote reading. Often the gauge that is used is the one that would normally be monitored directly on the engine. Remote locations for the gauges may be on a panel or console adjacent to the engine or some considerable distance away in a central control room. In the largest plants it is common practice to have remote-reading gauges backed by direct-reading gauges at all points.

25.2 ENGINE INSTRUMENTS

Ammeter

An engine fitted with auxiliary alternators for battery-charging purposes also has an ammeter to show the rate of battery charge or discharge. The dial or face of the gauge will be marked +0−, meaning zero at center, or a −0+, meaning the charge or discharge status of the battery.

Air Filter Differential Pressure Gauge

As the air filter element becomes loaded with solids, the imbalance between the atmospheric pressures on the two sides of the air filter element becomes more pronounced. The air filter differential pressure gauge serves to indicate the serviceability of the air filter element. Usually the indicator is in the form of a spring-loaded plunger in a transparent cylinder, but it may also be a diaphragm-type vacuum gauge. The gauge is usually fitted between the air filter element and the pressure charger, but it may be arranged for remote reading.

Ambient Air Temperature

Although the ambient temperature cannot be controlled, it is important that it be recorded, because the temperature variations have a marked effect on the generation demand and hence the fuel consumption. The ambient temperature record is an important element in future load forecasting. The thermometer will be sited outside in the shade and be of the DAY MAX/MIN type.

Boost Air Pressure Gauge

Boost air pressure is often referred to as turbocharger air pressure or aspirating air pressure. The gauge is mounted after the turbocharger outlet, and it may be either before or after the aftercooler. The gauge or gauges will be of the Bourdon tube type either directly or remotely connected to the engine air manifold(s), and it will read in kilograms per square centimeter, pounds per square inch, or kilopascals. On V-engines one may expect both air pressure gauges to normally read a very similar pressure when the turbochargers are giving balanced performances.

Boost Air Temperature Gauge

The boost air—turbocharger air or aspirating air temperature—gauge reads the temperature of the aspirating air being delivered to the cylinders after the air has passed through the turbocharger or blower and the aftercooler. On V-engines, the temperatures of the air passing through both air manifolds should be very similar. Direct-reading instruments in this application are of the bimetal stem type, and remote thermocouple-connected instruments are common.

Coolant Temperature Gauge

The coolant temperature gauge is sometimes known as the jacket temperature gauge or cooling water temperature gauge. It indicates the temperature of the coolant at the point where the coolant leaves the engine. The temperature of the coolant may vary with the load carried, but it should never exceed the boiling point of the cooling medium relative to the standard operating pressure and altitude of the cooling system. Occasionally, colored bar-type gauges are utilized, but most gauges are marked in either Celsius (°C) or Fahrenheit (°F) increments. These gauges may be of the bimetal or the vapor tension type. In the latter a sealed bulb and capillary tube are utilized. The capillary tube is rather fragile, and it should be inspected frequently for signs of impending damage. The tube must never be folded or kinked or have any weight placed on it.

Dipstick

The dipstick is the original measuring device for a liquid level; its origins are prebiblical. It indicates the level of lubricating oil present in the sump, or pan, of an engine.

Never start an engine without first reading the dipstick even if the engine is started and stopped many times daily. On engines with more than one lubricating oil system, it is probable that there will be more than one dipstick to be monitored. Look for the dipstick on the turbocharger, cam box, governor, governor drive casing, fuel pumps, pedestal bearing, and any other place where there is a separate lubricating oil system (Sec. 24.20).

Engine Room Temperature

The ideal temperature of any engine room is in the 20 to 30°C (70 to 85°F) range, and the temperature can be moderated by adjustment of ventilation to a certain extent. The temperature is usually read from a mercury type thermometer often of the DAY MAX/MIN type located in a position where there is little draft or localized heat emanation.

Exhaust Temperature Gauges

There are several differing methods of monitoring exhaust temperatures. Generally, all exhaust temperature gauges are pyrometers, which work on thermoelectric principles. Temperature variations are determined by the measurement of an electric voltage that is generated by the action of heat on the junction of two dissimilar metals (a thermocouple). Other pyrometers work on the variation of electric conductivity that is due to changes in temperature of the sensing metal. These pyrometers have either dials or scales marked in Celsius or Fahrenheit degrees. In most installations extreme care and expert knowledge are required to replace damaged thermocouples, connections, and cables.

Fuel Gauge

The fuel gauge is fitted directly to the instrument panel of only a few smaller engines; it indicates the fuel level in the service tank. The gauge is electrically operated and indicates the fuel level only when the engine is switched on and running. Fuel gauges are generally marked E (empty) and F (full). Large engines generally do not have fuel gauges; instead, the fuel tank itself is fitted with a gauge, generally of float mechanical type, with a large dial that is readable from approximately 20 ft away. The markings on the gauge may indicate the quantity in tons, liters, gallons, or percentages.

Fuel Pressure Gauge

The fuel pressure gauge indicates the pressure of the fuel in the manifold between the fuel filter elements and the injection pump. If the fuel filter elements become plugged with fuel-borne debris, the fuel pressure will drop. On some engines, the fuel pressure gauge face is marked in colored graduations; on others, units of pressure are used (i.e., kg, #cm, psi, and kPa). Only gauges showing recordable figures should be fitted. The colored bar type of gauge gives only approximate readings and is quite useless for gauging recordable figures.

Hour Meter

The hour meter is a most useful instrument for regulating the planned generation unit routine activities. Almost every installation has its oil change frequencies, overhaul work, and utilization regulated by the hours of operation as displayed on the hour meter.

The most favored types of hour meters are those which are mounted remotely from the engine. Usually they are found on the engine control panel, and ideally they have an ability to read to 99999 h, which amounts to the lifetime of most engines. Most are without resetting facilities, but in any event the hours should never be altered. These instruments are essentially electric clocks that receive their motive power only when the engine-driven generator is producing power. Some engines are fitted with mechanical hour meters mounted directly on the engines. However, these instruments are less reliable, and they are often quite difficult to read. It is recommended that the mechanical types never be replaced when they fail but instead be supplemented by electric remote meters.

Jacket Water Expansion Tank Level Gauge

This gauge is fitted either in the top tank of the radiator, in an engine-mounted radiator cooling system, or in the separate header or surge tank of common rail or intercooled engines. Usually they are of a float-operated mechanical design, and sometimes they have a low-level alarm-sounding capacity. On smaller installations sight glasses are occasionally inserted in the header or surge tank, but they have a marked tendency to become occluded, and thus they must be cleaned quite frequently.

Lubricating Oil Filter Differential Pressure Gauge

The lubricating oil differential pressure gauge is provided to read the lubricating oil pressure both before and after the filter elements. The gauge has two pressure-sensing mechanisms built into it and two needles that indicate the two pressures sensed on a common dial or scale. When the difference between the two readings reaches a certain point as recommended by the OEM, the lubricating oil filters will require servicing. These gauges will only be found on the larger engines of, say, 2.0 MW capacity and up.

Lubricating Oil Level Sight Glasses

This is another instrument without any moving parts. Quite commonly it will be found fitted on larger turbochargers at both turbine and compressor ends, on governors, and on the sumps of smaller engines. It is also used to observe piston cooling oil returns on certain large engines.

Lubricating Oil Pressure

The lubricating oil pressure gauge is the *most important gauge on the engine*. The pressure indicated will be at its highest point immediately after start-up when the lubricating oil is still relatively cold. As the oil warms up, the indicated pressure will drop slightly and then remain steady. Some oil pressure gauges are marked with colored bars indicating GOOD, CAUTION, or DANGER but they are not recommended and should be replaced by standard numerically marked gauges.

Most engines have oil pressure gauges marked in units of pressure (i.e., kg, #cm, psi, and kPa). Large-face gauges are better than small-face gauges because they are generally more robust and are visible from a greater distance. Never replace a defective oil pressure gauge with a gauge that has been used before. Liquid-filled gauges can be recommended in this instance.

Lubricating Oil Temperature Gauges

Lubricating oil temperature gauges will be found at the inlet and outlet of the lubricating oil cooler. On larger engines they are also inserted into each of the piston-cooling oil return outlets. These instruments are of the bimetal stem type except when thermocouples give remote readings.

Raw Water Pressure Gauge[1]

This instrument is fitted between the raw water circulating pump and the intercooler; it is a standard Bourdon tube type of pressure gauge. The face will read in pounds per square inch, kilograms per square centimeter, or kilopascals.

Raw Water Temperature Gauge[2]

In systems equipped with an intercooler it is probable that there will be two raw water temperature gauges. One will be fitted at the raw water inlet to the intercooler and the other at the outlet. The instruments are usually of the bimetal stem type of thermometer in those locations, but occasionally thermocouples are utilized for remote reading. The gauges will read in either or both of the commonly used temperature scales.

Tachometer

Tachometers indicate the rotational speed of the engine in revolutions per minute. These instruments may be of mechanical drive or of the electronic speed-sensing type. Tachometers are used on large engines, particularly for monitoring the turbocharger speeds, and on engines that power loads other than electricity generation.

25.3 DATA PLATES

Numerous data plates are situated throughout the power plant complex. Stamped or otherwise permanently inscribed with certain information essential to the operation or upkeep of the equipment, these plates must not be obscured, defaced,

---

[1]For the purposes of the remarks concerning the raw water pressure and raw water temperature gauges, the term *raw water* is taken to mean any water that has been taken directly from the sea, a river, lake or pond without any treatment before use.
[2]Ibid.

altered, or removed. They may be either of metal or synthetic material, and they will be permanently fastened. They must not be cleaned by means of emery cloth or other abrasive materials.

The instrument gauges are often accompanied by small plates identifying individual functions, and many of the switchgear panels are similarly marked. In some facilities individual globe valves, gate valves, or other flow controls also will have a function-identifying plate.

25.4 GENERATING PLANT INSTRUMENTS

Instruments and controls for the generator function are mounted on the plant switchboard, which is divided into two basic parts. One part is the power section, and the other is the metering and control section. The power section contains the circuit breaker, which may be manually operated. Today, however, most circuit breakers are power-operated. In the metering and control section some or all of the following instruments will be found.

Alternator Winding Thermometer. This instrument monitors the temperature of the alternator stator windings.

DC Ammeter. The dc ammeter reads the dc current of the exciter field windings.

Indicator Lights. There will be various indicator lights on the plant switchboard. Industry-wide, there is little uniformity in what the lights are saying. The lights may be white, amber, green, red, or blue, and the operator must make a careful study of them to learn what status they indicate within the individual plant.

Frequency Meter. The frequency meter shows the generator output frequency. It may read in hertz (HZ) or cycles per second (cps).

Hour Meter. Based on an electric clock principle, this instrument counts the hours of individual generator operation.

Kilowatt-Hour Meter. A kilowatt-hour meter records the total output of the generator in terms of kilowatt hours. The plant gross production is the sum of all of the generator's kilowatt-hour output as shown on the kilowatt-hour meters over a given time span, usually 24 h.

Power Factor Meter. The power factor meter indicates the power factor of the generator output circuit. It may read in PF (power factor) as either a percentage or a decimal. When running generators in parallel, all PF readings must be the same.

Wattmeter. The wattmeter is normally marked in kilowatts indicating the power output of the generator.

25.5 FEEDER PANEL INSTRUMENTS

Another group of panels which are usually sited alongside the generator control and power panel are the feeder panels. Generated power is fed through the generator breaker onto the main bus and hence to the feeder breaker so the sections of the distribution system can be supplied or isolated individually. Each breaker has three instantaneous overcurrent relays, one for each phase. In the event of a gross defect in the distribution system (e.g., a conductor on the ground), the overcurrent relay will trip the feeder off the bus and thus protect the generating equipment.

Each feeder panel will also contain a kilowatt-hour meter, a kilowatt meter, and an ammeter. The ammeter will have a transfer switch so that the current on each phase can be read. The kilowatt-hour meter, from a revenue point of view, is an exceedingly important instrument because it reflects the net plant production and hence the revenue earned.

On some instruments a mirror is mounted behind the scale to permit very accurate readings: By lining up the needle and its image in the mirror, a true reading is obtained. On other instruments the pointer is a blade so the gauge can be read more accurately than is possible with the more common pointer.

No gauge that has an adjustment facility should be interfered with unless the means of properly testing and calibrating the gauge are at hand. This recommendation applies particularly to the instruments mounted on the switchgear panels.

25.6 PRESSURE GAUGES

Pressure gauges usually fail for one of three reasons:

1. *Vibration caused by improper support or transmitted vibration from another source.* A gauge that is poorly supported or located in a position in which it will be subjected to vibration will fail prematurely. If a gauge should begin to vibrate when it previously did not, the underlying reason for the vibration should be determined so remedial steps can be taken.

2. *Flutter of the needle and internal mechanism.* It is not normal for a pressure gauge to flutter (the indicator needle moves backwards and forwards at high speed). An instrument that flutters will be practically useless as an accurate gauge, and it will wear out very quickly. If the pressure being measured causes the pressure gauge to flutter, a snubber—an orifice disk or a small ball valve—must be inserted between the pressure source and the gauge. However, if the gauge once read steadily but now flutters, the problem is probably not with the gauge but instead somewhere in the system.

    Gauge flutter occurs most frequently in a fluid-handling system in which the pressures are developed by a positive-displacement type of pump. In such cases, it is normal practice to insert a snubber or throttling valve between the gauge and the pressure source. When flutter develops, the snubber may be stuck or the setting of the throttling valve may have been altered.

3. *Inability of the gauge to handle pressures imposed on it.* Sometimes a pressure gauge cannot handle certain intermittent pressures imposed on it. This type of failure is not uncommon in fluid-handling systems. That is especially true of low-temperature start-ups, when very high initial pressures are sometimes experienced. The pressure gauges get strained under such circumstances, and subsequently the needle fails to return to zero when the system is shut down. Because the gauge no longer reads accurately, it must be replaced.

25.7 CAUSES OF INSTRUMENT FAILURE

Some common causes of instrument failure are:

- Fatigue of internal mechanism through vibration or pulsation
- Casing, glass, or connection damage through abusive work practices
- Overloading by way of improper operating practice or replacement instrument selection
- Wear of thermocouple cables through abrasion, crushing, or strain
- Poor protection, strapping, and routing of pressure pipes, cables, and so on
- Unauthorized adjustment or repair

Gauges sited in positions subject to an unacceptable level of vibration, pulsation, shock, or surges will have a short life. If the conditions leading to early gauge failure cannot be moderated, the use of glycerine-loaded gauges may be considered. Generally these gauges give a good performance, but are very expensive. It is not good practice to install such a gauge if the gauge it is replacing worked well until the conditions which destroyed it developed. Better to rectify the conditions.

Never ever attempt to change out or adjust the pressure gauge on a pressurized vessel of any description. The pressure must be dropped to zero and the vessel isolated before any work is attempted.

Never tap a gauge to make the needle move to a desired or expected point.

25.8 DEFECTIVE GAUGE DISPOSAL

Any pressure or temperature gauge removed from an engine because it is faulty must be destroyed. No useful purpose will be served by keeping a defective instrument around, and to do so could be the cause of a catastrophe at a later date. A newcomer could quite reasonably assume that the apparent spare gauge lying on the shelf is a good, usable item and subsequently find out in a calamitous manner that it was no good at all. Further, no attempt to repair any gauge or instrument in the field should be considered.

25.9 INSTRUMENT CALIBRATION AND ALTERATION

Nobody should ever alter instrument calibration. Many instruments such as voltmeters, kilowatt meters, ammeters, and pyrometers have external calibration devices. Job-related pressures or faulty education may induce a person to adjust the calibration of an instrument, but the knowing operator or journeyman will not do such a thing without having the means and knowledge to recalibrate the instrument properly and accurately.

Altering the calibrated setting of an instrument without a legitimate and comparisonable measure is quite unethical and grossly misleading. Such a practice can result in considerable damage to engines, generators, switchgear, or transmission equipment. If an instrument is suspected of giving an inaccurate reading, it must be reported, tested, and recalibrated if necessary. Such work must be carried out by a suitably qualified journeyman in a properly equipped instrument repair shop.

If the calibration of an instrument is altered improperly or without authority, it is quite probable that nobody other than the person making the alteration will know what has happened. Thus, any subsequent reading made from that particular instrument will be incorrect. When that is realized, all of the other alterable instruments in the plant will be regarded with mistrust. Proper tuning and adjustment of the engine generator and switchgear equipment will be impossible until the whole instrumentation package has been thoroughly and expensively checked out.

Each plant should have a set of test gauges and adapters so that the true pressures can be verified. Those test instruments must not be allowed to get into general service, because they are very costly compared with ordinary good-quality gauges.

25.10 THERMOCOUPLE LEADS

It is common practice to arrange for the engine temperatures and pressures to be monitored at some remote point, usually at a control console which may be adjacent to the engine or in a control room. The temperature seen at the reading point depends upon a constantly transmitted signal being sent by the sensor on the engine. The lead conveying the signal between the read instrument and its sensor is known as the *thermocouple lead*.

Frequently, thermocouple leads are much abused and their care is neglected. If accurate temperature readings are to be regularly obtained, the complete thermocouple must be in good condition. Damage to the thermocouple occurs when the lead is improperly tied, routed, or supported and thus becomes frayed or abraded. Crushing is another common cause of damage; it is brought about by repeated pedestrian traffic or by trapping the lead between engine components, pipework, or floor plates.

When thermocouple leads are renewed, it is important that the connections be made with regard for the correct polarity. To facilitate that, the leads are color-coded. If for some reason it is necessary to extend or replace the leads, the new ones must be of regular thermocouple wire of a material matching that which is already installed. In diesel applications this is invariably iron-constantine (Type J), but other types may have been inadvertently installed if there are any gas turbines or other very high temperature machinery at the same site. Whenever a lead is replaced, it must be properly resistance-calibrated at the meter.

25.11 COUNTING METERS AND DEVICES

Several counting devices will be found in any power plant. Depending upon the size of the power plant complex, some or all of the devices described below will be installed. The *engine hour meter*, sometimes known as the engine service meter, may be mechanically or electrically powered. A mechanical meter may be mounted on the engine and driven directly off the engine timing system; it may be remotely driven by a flexible shaft in a manner similar to that of an automotive odometer; or it may be installed on the engine frame and be motivated by vibration. The electrically powered engine hour meter may be mounted either on the engine or at a remote location. Depending on the arrangement, the meter may be powered from the engine-mounted battery-charging alternator or from the main generator output.

Engine running hours must be recorded in the engine log at the end of each shift. To eliminate any possibility of error, no minutes, fractions, or decimal parts of an hour should be either read or recorded. Any attempt to do so will, in the long term, result in confusion and erroneous records. Mechanical meters do not generally have a very long life. Whenever they fail, they should, if possible, be replaced by electrically powered meters with the maximum capability of 99999. They should preferably be mounted on the control console or switchpanel and be remote from any vibration source.

The engine hour meter is a most valuable instrument because it provides the all-important time element of the engine's history. At no time should the meter be altered, turned back, or otherwise interfered with.

25.12 BREAKER TRIP COUNTER

On more modern installations, breaker trip counters will be fitted on main generator breakers, station service breakers, and feeder breakers. The counters cumulatively record each breaker closure cycle. The OEM manuals for the circuit breakers will recommend that certain adjustments, maintenance, and overhaul activities be carried out after a given number of operational cycles. The number of operations recorded by the counters is therefore a useful guide for the plant maintenance staff in its work-planning activities.

The operator should record the breaker count at least daily and preferably each shift. That may not be necessary in a power plant serving a residential community, but in a plant serving a mine or other industrial complex with wide load swings and frequent electrical fault trips, reading the breaker count will be beneficial.

In the normal course of events the number of closures made by the feeder breakers will be much smaller than that made by the generator breakers. Alterations in the frequency of closures may indicate that problems are developing in the distribution system. The breaker trip counter is usually a multidigit type similar to the bicycle-type distance meter.

25.13 FUEL METER

The number of fuel meters in a diesel power plant is variable; it is largely dependent upon the size and complexity of the facility. A fuel meter can be expected at the point where fuel deliveries to the plant are received. The reading of this meter should be recorded at least daily and certainly before and after each delivery from the fuel supplier. It is essential that an accurate record of all fuel delivered to the plant be continuously maintained.

So that each engine's fuel consumption can be calculated on a daily or even hourly basis, fuel meters may be found in the lines supplying each of the daily service fuel tanks. In some cases the meters may be placed between the daily service tank and the engine, but that will not be true of an engine with a fuel return line that is led back to the daily service tank.

It is most important that an accurate accounting of the fuel consumed by each engine be kept. The fuel consumption of any engine has the greatest economic impact of any of the expense factors in a diesel power plant, and any fall-off in the fuel performance of a given engine must be attended to without delay. Similarly, the unit most appropriate to the load must be operated in the interests of fuel efficiency, and the fuel meter will provide part of the all-important guiding information.

25.14 ENGINE START COUNTER

An engine start counter is sometimes installed on larger engines. This instrument is mainly of interest to the engine upkeep planner, who will wish to compare the number of starts made by similar engines over a period of time with reference to the wear observations and other conditions of deterioration. The cumulative number of starts should be recorded in the log sheet on a shift or daily basis. At no time should the start count meter ever be returned to zero. If it is replaced, a permanent sign or notice must be placed adjacent to it to give the date of changeout and the reading at that time so that the history does not get distorted. The provision of engine start counters is important to any application of load allocation and equipment utilization.

25.15 LUBRICATING OIL METER

Lubricating oil meters are fitted to engines set up with automatic lubricating oil level controls. They may also be installed in the manually controlled feed line when there is a direct manually operated connection between the makeup lubricating oil supply and the engine.

The reading of the lubricating oil meter must be recorded on every shift, and it must be regularly compared with previous readings to see that there has been no departure from the norm. Under ordinary circumstances, the lubricating oil consumption of a diesel generating set should not be more than 1 imp gal/1860 kW (1 US gal/1550 kW, 1 L/492 kW).

CHAPTER 26
FOUNDATIONS, MOUNTING, AND ALIGNMENT

26.1 HOLDDOWN BOLTS

Among the extended interval inspection details that must receive attention are the equipment hold-down bolts. They must be checked for tightness as any looseness will lead to alignment problems, damage to fasteners, fretting, concrete floor tramping, cracking of connected pipework, deterioration of electrical connections, and similar difficulties.

It is suggested that at approximately 90-day intervals, a complete tap-test inspection be made of all the hold-down bolts of the engines, generators, auxiliary machinery, and any areas of structure framework that are subjected to transmitted vibration. Any loose fasteners discovered must be retorqued to their correct values in conjunction with subsequent alignment inspections.

26.2 ENGINE AND GENERATOR ALIGNMENT

Before commencing any realignment procedures or making an alignment check, it is advisable to refer to the relevant workbook and the OEM manuals to determine if the readings should be taken hot or cold. Ideally, the workbook will show sets of deflection readings that were taken both hot and cold.

The drive and driven shafts of an engine-generator combination must rotate precisely about the same axis. A full-alignment routine must take place at the time of installation and must be redone at each overhaul. When there is any disturbance of the generator, frame, floor, or engine, the alignments must be reset. If the unit's vibration characteristics change from those noted at the time of installation or trend toward the intolerable, the alignment must be reinspected as a first step toward understanding and resolving the problem (Sec. 27.8). Failure to maintain a proper alignment will induce fretting of the generator stub shaft spigot and the coupling. The conductor metals will become fatigued, the bearings will be overloaded; and the insulating materials will fail.

The air gap between the rotor and the stator should be equal all around within 10 percent, but any effort made to adjust the gap clearance may affect the alignment. It follows that the alignment process and the air gap setting are part and parcel of the same exercise.

Alignment Procedure (Single-Bearing Generators)

Remove all paint, burrs, and dirt from the contact surfaces of the flywheel, coupling disks, bell housings, adapters, feet, and underframe. Clean the flywheel bore and the rotor shaft spigot.

Use a dial gauge to read off the axial float of the crankshaft-flywheel. (Do not attempt to insert a lever behind the torsional vibration damper.) Record the float measurement. With the dial gauge mounted on the flywheel housing, check the axial runout of the flywheel (Fig. 26.1a). Record the reading. Still with the dial

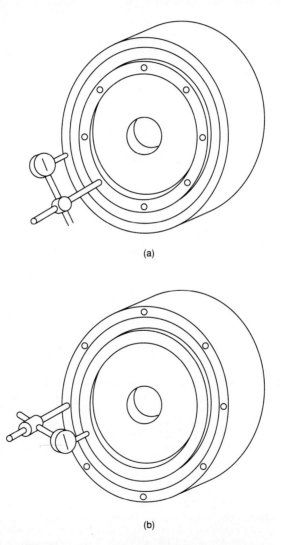

FIGURE 26.1 Flywheel and flywheel housing runout. (*a*) Checking face runout of the flywheel. (*b*) Checking face runout of the flywheel housing.

gauge, this time mounted on the flywheel, read off the runout of the flywheel housing face (Fig. 26.1*b*). (For this check the weight of the engine must not be resting on the flywheel housing because some distortion may occur.) Compare the readings taken with the tolerances indicated in the engine manual.

Place the coupling disk pack in the bore of the flywheel and check for all-around clearance. If the clearance is good, bolt the disk coupling to the generator rotor shaft. Mate the generator to the engine and install the bolts that fasten the coupling disk to the flywheel. Do not turn the locks or tabwashers up at this point.

Install the bolts fastening the alternator housing to the engine flywheel housing. Use the dial gauge to read the axial float of the crankshaft-flywheel. Compare the readings with the readings taken earlier. If the readings differ, remove the generator and adjust the shim thickness between the coupling disk and the generator rotor shaft to obtain the original crankshaft float reading (Fig. 26.2). Reassemble the generator to the engine. If the float is correct, torque the fasteners to specification and turn up the locks or tabwashers.

The engine now being coupled to the generator, is necessary to fasten the generator set to the base. In the case of a unit that is without supports at the flywheel housing, loosely fit the bolts that fasten the engine and generator supports to the base or underframe while ensuring they have adequate clearance.

Loosen the base-to-foundation fasteners. Inspect the clearance between the base and its foundation at all fastening points. Insert shims as necessary to prevent deflection of the base when the base-to-foundation fasteners are tightened. When the engine is mounted on spring vibration isolators, it may be necessary to insert shims under individual isolators to equalize their deflections to within 0.5 cm (0.25 in) of each other. When the base-to-foundations or isolator deflections are satisfactory, tighten the fasteners to their final torque.

Now check the clearances between the engine-generator supports and the base. The entire length of support must be checked because some generating equipment has supports running along almost the full length of the machine. Use shims as needed to prevent any deflection of the supports of the base as the hold-down bolts are tightened (Fig. 26.3).

Place the dial gauge on the base with the gauge finger adjacent to a hold-down bolt. Tighten the bolt to its final torque and check the dial gauge. If it shows more than a 0.005-in (0.15-mm) movement, the shim thickness below the bolt must be corrected. Repeat this routine at all of the hold-down bolts so that all the bolts are tightened to the specified torque without deflection (Fig. 26.4).

26.3 VIBRATION ISOLATORS

Many diesel–engine driven generator sets are mounted on vibration isolators. The purpose of the isolators is to limit the transmission of vibration from the generator set to the surrounding structures and equipment. The isolators are so placed that each carries a designed share of the generator set weight. Most isolators utilize coiled springs, but some employ rubber blocks as damping media. When loaded, the deflection of each isolator should be approximately the same (0.25 in, 65 mm). Spring collapse, rubber decay, or floor movement will cause an alteration in individual isolator deflections, and it is suggested that all isolators be examined and measured at semiannual intervals whether or not the unit runs during the period.

in the cooling system, and defective electrical grounding of the generator set. Insofar as these particles are abrasive, premature failure of pump seals will occur. There may be rapid metal loss in aluminum or copper parts, and some fretting of the lower end of liners may take place where sludge accumulates.

2. *Soluble metal* in the sample is an indicator of the rate of metal loss. Coolant samples regularly showing an iron loss of 2 ppm are predicting an eventual component loss. Similarly, a copper corrosion level of 0.5 ppm is above the level of acceptability. Aluminum in excess of 1.0 ppm also will suggest an impending failure. Undue amounts of soluble metal in the coolant will be brought about by using incorrect coolant, air leaks into the coolant, defective ground systems, broken or disconnected ground strap, or engine wiring system in poor condition.

23.52 WATER TREATMENT CHEMICALS

The conditioning treatment used in a straight water-cooled system will probably not be the same as that normally used in an engine with ethylene glycol frost protection. Individual users should be very cautious in this respect. If the test standard readings are too low, there may be a conditioner loss through leaks in the system which uses straight water, there may be a defective relief valve, or cold water may be used for makeup, which promptly expands and overflows and has a diluting effect. All these situations will result in a reduction of the water treatment strength. The test level will be too high if makeup quantities of treatment are not carefully calculated and measured out.

Dissolved Solids

The level of dissolved solids should not exceed 5000 ppm. The solids generally can be attributed to contaminated water, but combustion gas leaks can also have an effect. When an excess of solids is present in the coolant, foaming will occur, heat transfer will be restricted, and aeration pitting of the liner metal will commence.

Free Carbon Dioxide

Free CO_2 should not be more than 150 ppm. If air or combustion gases enter the coolant, CO_2 will develop. When heated to engine operating temperatures, it converts to carbonic acid (H_2CO_3), and again the engine components are attacked. It is said that the rate of metal pitting is in direct proportion to the quantity of air leaking into the system.

pH (Acidity-Alkalinity)

The alkalinity level in any cooling system should be in the 8.5 to 10 pH range. If the pH is below 8.5, there will be a risk of corrosive damage to the system. Acid

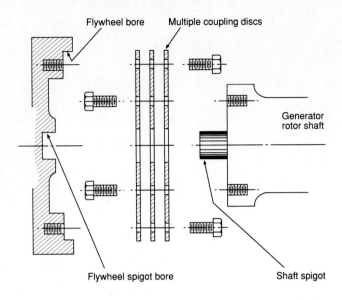

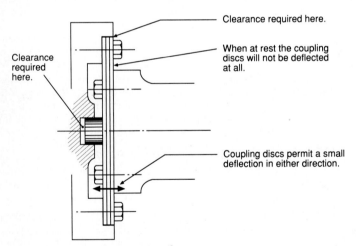

FIGURE 26.2 Rotor shaft and disk coupling.

A simple test for overdeflection of the springs can be carried out by using a wedge approximately the thickness of a coin. If the isolators are well adjusted, the average gap between the spring coils will be between one-third and two-thirds the thickness of the coil itself. It is useful to have a reference diagram drawn in the related engine workbook so the observed spring deflections can be noted on it for future comparison. Movement of the floor or foundation will often be the cause of one isolator becoming overloaded.

At no time should any unit be operated with the isolator adjusting screws under load. These screws are intended to be used only to facilitate the insertion of

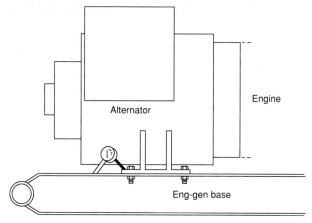

FIGURE 26.3 Deflection readings of alternator hold-down bolts.

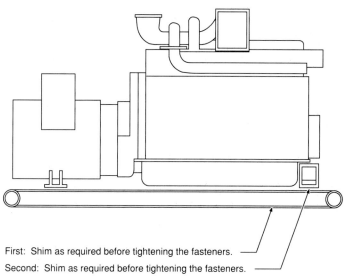

First: Shim as required before tightening the fasteners.
Second: Shim as required before tightening the fasteners.

FIGURE 26.4 Fastening sequence of hold-down bolts.

alignment shims. If the unit is run with the screws carrying the weight of the engine that is meant to be resting on the isolator alone, the threads will rapidly tear out and be rendered useless. Running the unit with the isolators unevenly loaded or deflected will probably cause undue vibration and alignment problems (Fig. 26.5).

During the semiannual inspection of the isolators, attention should also be given to the skid or underframe. Examine it for evidence of cracking, particularly in the vicinity of welds. The screws positioned on the side of the isolator housing must not be so tight that they restrict the movement of the isolator top plate. If

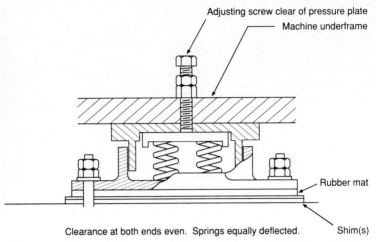

FIGURE 26.5 Correct isolator setup. Spring-type isolators should be inspected at approximately 30-day intervals.

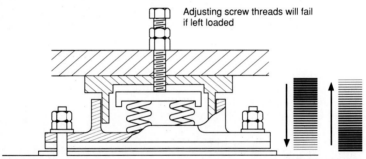

FIGURE 26.6 Adjusting screw in wrong mode. Vibration isolators are intended to limit the transmission of vibration both from and to dynamic machines. A standby unit can be damaged when at rest by vibration transmitted from another unit if some or all isolators are defective. Vibration isolators when improperly loaded may result in machine misalignment. If it is found expedient to adjust the vibration isolators, the alignment of the unit must also be checked.

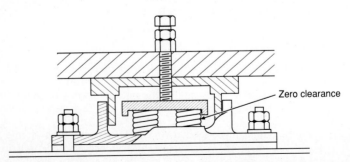

FIGURE 26.7 Improper isolator loading. Spring collapse through overload, misalignment, fatigue, or floor movement. Vibration isolators in this condition cannot function. The vibration readings will not reflect the true condition of the machine.

FOUNDATIONS, MOUNTING, AND ALIGNMENT 26.7

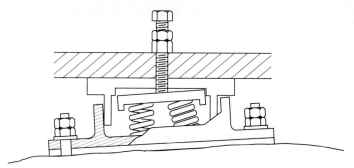

FIGURE 26.8 Misaligned isolator. Bad shimming causes uneven floor and spring imbalance, jambing of frames, and excessive wear. Vibration isolators left in this condition cannot function and the readings will be misleading.

the isolator is not free to work, vibration will be transmitted through the isolator housing to surrounding equipment and structures and the floor on which they stand. The tramping action of the isolator housing will cause fretting of the concrete floor, and machine alignments will gradually be affected (Figs. 26.6, 26.7, and 26.8).

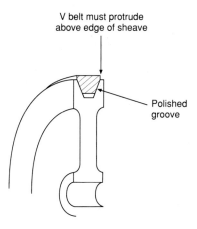

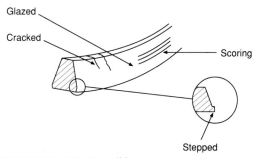

FIGURE 26.9 V-belt conditions.

26.4 V-BELT SELECTION AND INSTALLATION

V-belts have differing cross-sectional angles. Automotive application belts are of a configuration different from that of industrial or domestic belts. Any belt intended for a given application must be carefully checked for suitability before it is installed. Replacement belts must be ordered according to the OEM parts book.

The angle of the belt wedge must match that of the sheave groove. The profile of the groove should occasionally be checked against a test gauge. When new, all belts and sheave grooves are straight-sided and without either glazed or polished contact surfaces (Fig. 26.9). V-belt defects and their causes are given in Table 26.1.

TABLE 26.1 V-belt Defects

Defects	Cause
Slipping belt	Polished sheave groove Insufficient tension Overload Oil or water splash
Squeaking belts	Overload Insufficient arc of contact Starting load too great
Overturned belt(s)	Improper installation Broken internal cords Impulse loads Misaligned pulleys Worn grooves in sheaves
Flapping belts	Old and new belts used together Uneven wear on sheave grooves Misaligned driver and driven shafts Insufficient tension
Broken belts	Shock loads Heavy starting loads Bad installation practice Foreign object in drive
Rapid belt wear	Worn grooves in sheave Mismatched belts Overload Slippage Misalignment Overheating
Fabric decay	Oil or grease Overheating
Ply separation	Belt too large for the sheave diameter

26.5 FLEX DISK BALANCE WEIGHTS

Single-bearing generators are often trim-balanced by the use of washers of various weights under the flex disk coupling bolts (Fig. 26.10). If the unit is dismantled, there is a possibility of the trim washers being incorrectly assembled which will lead to unbalance and high vibration. All the washers will be of the same diameter, but the thicknesses may vary considerably and the weight may vary from 0.5 oz to 0.5 lb.

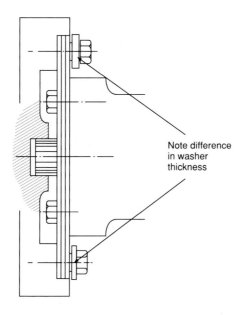

FIGURE 26.10. Coupling disk fastener balance.

By observation and careful notation, the generator can be separated from the engine and then reassembled so that it operates without increased vibration. It should be noted that the use of one or more replacement bolts can disturb the balance. Look for differences in the bolt head thickness and stem length (Fig. 26.11).

26.6 GENERATOR SET BALANCE

When supplied new, all generator sets can be expected to have uniform bolts fastening the flex disks to the flywheel and the generator shaft. There is a possibility that after one or more work routines involving the disconnection of the generator from the engine that bolts will have been renewed. This creates a problem of bal-

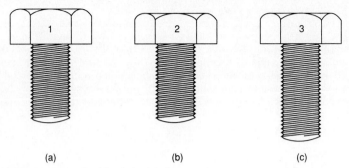

FIGURE 26.11 Coupling bolt vibrations. (*a*) Bolt head volume and weight 25 percent greater than those of bolt 2. (*b*) Standard bolt. (*c*) Bolt stem volume and weight 25 percent greater than those of bolt 2.

ance if the replacement bolt is not exactly the same as the one discarded. A difference of ounces in the static bolt weight can produce an unbalanced force of pounds when the unit runs at full speed. The result will be increased vibration level. It is best if any replacement fastener used is weighed in comparison with the reject bolt (Fig. 26.11).

26.7 FLOOR AND FOUNDATION VIBRATION LEVELS

The ranges and classes of vibrations, actions to be taken, and the methods used to report the actions are given in Table 26.2.

Point floor 11 right and floor 11 left are normally located on the floor at 90° to and 48 in away from point 11 on the engine. These points are read to determine the level of vibration transmitted through the floor and foundations (Fig. 26.12).

TABLE 26.2 Vibration Levels

Reading range, in/s	Class	Recommended action	Reporting method
0.05–0.11	Normal vibration	No action is required.	
0.11–0.22	High to rough	Correction is required but is not urgent.	Advise head office if above 0.16 in/s.
0.22–0.44	Rough to unsafe vibration	Correction is urgent.	Advise head office; response and advice are required.
Over 0.44	Ultra-unsafe vibration	Shut down.	Advise head office; response and advice are required.

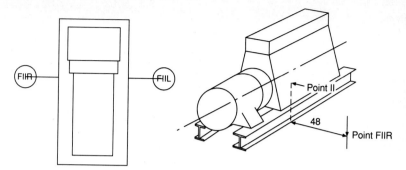

FIGURE 26.12 Floor vibration.

CHAPTER 27
VIBRATION

27.1 ENGINE CONDITION MONITORING

The main duty of the diesel power plant operator is to watch over the machinery while logging and interpreting the many details of indicated data so that the equipment is able to operate continuously, reliably, and efficiently. Besides the logging of data and the frequent walk-around inspections, more and more plants are adopting the routines of vibration monitoring and lubricating oil sample analysis. These two activities have tremendous value in the early detection of undesirable conditions and avoidance of very costly breakdowns, and they are now regarded as a standard practice in any well-run diesel facility. Lubricating oil sample analysis is discussed in the topics dealing with lubrication. (See, for example, Sec. 24.35.)

27.2 VIBRATION MONITORING

Vibration is a form of energy dissipation. All dynamic machines and the structures that house them vibrate to a certain extent, so the vibration acceptable to various classes of machine or structure has been or can be established.

Vibration monitoring on a continuing basis must be recognized as an expected routine in a diesel power plant. The benefits from regular vibration monitoring are far too valuable to be ignored or avoided. The process is simple and should be routine. Reading a vibration meter is no more difficult than reading any other meter, and obtaining a set of routine readings does not require more than 5 min of a person's time even on the largest machines. The activity is normally carried out by a person of journeyman status, but it can be adequately carried out by someone with little related skill. Only a minimal amount of training is necessary. When vibration is regularly measured, the mechanical health of the machine generating the vibration can be assured.

In plants that are continuously manned, a vibration meter should be available to the operating personnel. The operators will, of course, be instructed in its use and in the basic interpretation of the obtained readings. It is expected that a set of vibration readings will be taken each shift and inserted in the log sheet. It will also be found worthwhile to take regular readings from any other rotating equipment within the plant environs—compressors, bulk fuel transfer pump, fan motors, centrifuges, and so on—at weekly intervals or thereabouts.

The station mechanic(s) and the plant superintendent will compare the daily vibration readings with those recorded previously and take into account the loads being carried when the readings were taken. Their interpretation of the readings, coupled with their experience, will dictate what immediate or future actions are indicated.

In other plants the duty of vibration monitoring may be assigned to one of the maintenance staff or any other person who has been suitably instructed and delegated by the superintendent. Just who does the vibrator monitoring does not really matter so long as it gets done accurately and frequently.

27.3 VIBRATION METER USAGE

Most models of vibration meters are quite easy to read. In all cases, however, the person doing the vibration monitoring must refer to the manufacturer's manual if accurate readings are to be obtained. The monitor must take care to read from the correct scale. Depleted batteries in the vibration meter will give false readings, and so the batteries must be tested each time the monitor is used.

The probe of the signal generator must be pressed firmly onto the monitoring point, and it should be so held that it is as near as possible to 90° of the axis of the main shaft for all horizontal and vertical plane readings. For axial readings, the probe must be so held that it is in line with and parallel to the main shaft axis. It may be found convenient to use probes of various lengths for reading different points. The use of probes exceeding a length of 12 in (30 cm) is discouraged.

The monitoring of vibrations should not be confined to just the axis of the basic engine generator assembly, although that is the most important area of attention. It is quite reasonable to monitor the vibration of the individual turbochargers, each main bearing, the foundations or any other part of the complex that may supply meaningful or significant vibration readings.

Never attempt to take vibration readings from sheet-metal surfaces such as filter housings, alternator end covers, large-diameter cooling water pipes, and radiator casings. Never attempt to place the probe of a vibration meter on a moving part or on a surface behind a guard.

27.4 MEASUREMENT OF VIBRATION

An individual may, through long experience, be able to sense good or bad vibration conditions in a particular machine. This ability cannot readily be interpreted or communicated, nor can it be quantified. However, through the use of vibration reading instruments, consistent measuring methods can be carried out. The basic devices used are transducers, which convert mechanical motion into an alternating-voltage signal. The signal is converted into a readable vibration term and level and is shown on the dial of the vibration monitoring instrument. Most monitoring instruments are hand-held, and they provide both velocity and displacement readings depending on the switch settings. The instruments are sufficient for day-to-day comparative purposes.

On hand-held instruments the probe and transducer also are hand-held, but some monitoring instruments may be panel-mounted and connected to a transducer fastened to a fixed point on the machine. Such an arrangement may be connected to an audible alarm. The latter mode is more common on the larger machines.

There are three characteristics to consider when monitoring or analyzing vibration: amplitude, frequency, and phase.

1. *Amplitude* is defined as the maximum absolute value attained by a wave disturbance or by any quantity that varies periodically. In simple terms, amplitude is the amount of vibration, and it can be measured as displacement, velocity, and acceleration.

Displacement is the distance about a fixed point that the vibrating body moves. It is often referred to as peak-to-peak vibration, and it is measured in mils. (1 mil = 0.001 in-0.0004 mm.) Velocity is, in effect, the speed of the vibrating body. All vibrating bodies are moving from one point of rest to another in a reciprocating motion. From that it follows that a point of maximum velocity must occur midway between the two zero-velocity points, at which the displacement is zero.

Velocity is considered in vibration monitoring and analysis because it is directly related to the energy imparted to the vibration. *Acceleration* is the rate of change in velocity with respect to time; it is measured in g's (1 g = 32 ft/s^2 = 980.6 cm/s^2). The accelerating force can be measured only if the accelerating mass is known. In a multiple-component unit, it would hardly be practical to field-calculate the total and true mass.

2. *Frequency* is one cycle of vibration. It is defined as the movement of the vibrating body from a given starting point through the positive- and negative-peak positions and back to the original position within a given period of time. For the purposes of vibration measurement, frequency is expressed in cycles per second (cps) or cycles per minute (cpm) (Fig. 27.1).

3. *Phase* is the vibratory relationship of one vibrating body with another. Two bodies oscillating or moving in concert are said to be in phase. If the two bodies move in a manner directly opposed to each other, they are 180° out of phase. If they move in another relationship, they will be proportionally out of phase. Phase relationships are of great importance when balancing or isolating problems in rotating machines are addressed.

27.5 ENGINE VIBRATION TROUBLESHOOTING

All machinery vibrates to a certain extent when it is operating. The level of vibration in a given type or class of machine is an indicator of the machine's condition, and the tolerances may vary considerably from one type of machine to another. For example, the acceptable vibration of a rock crusher would not do in a typewriter. In diesel engines, as a class of machine, the vibration tolerances are very much the same, be the engine large or small, fast- or slow-turning. Acceptable levels of vibration have been established for diesel engines, and the condition of an individual engine can be gauged by them.

The vibration level of an engine-driven generator set is an excellent guide to the unit's condition. The vibration is caused by numerous slightly out of balance forces acting on both the static and rotating assemblies of the entire machine. The imbalances can also be caused by looseness, misalignment, and unequal loading

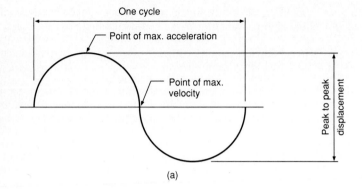

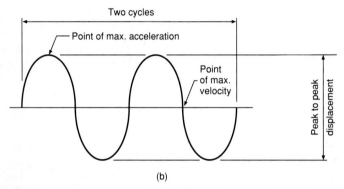

FIGURE 27.1 Vibration cycle. The displacement in (*a*) is the same as in (*b*). The maximum velocity in (*b*) is twice that in (*a*).

of cylinders. To a certain limit, a degree of vibration is tolerable and expected; provided it remains so, the engine will give good, reliable service. If the limits are neglected or grossly exceeded, damage to the unit is inevitable. Some components, having been stressed beyond their designed capabilities, will eventually give way. Generator winding failures, cooler tube breakage, radiator leakage, and many similar failures usually have their origins in excessive vibration. Standards of vibration have been developed to identify the vibrations that are acceptable and those that are not. The continued use of an engine that has a condition of excessive vibration can eventually lead to a failure of one or more major components such as crankshafts, bottom ends, and alternator windings.

Most troublesome engine vibrations are linear, and they can be countered in a relatively easy manner. Linear vibrations are in-line: They occur as an object moves along a straight line, that is, from point A to point B and return in a reciprocal motion. When the vibration level at one or more monitored points begins to show a progression of small increments, there is usually sufficient time for a planned investigation to be organized.

On the other hand, if the vibration level on a unit that has operated satisfactorily in the past abruptly increases, it is advisable to stop the unit forthwith and proceed with an investigation before major damage takes place. (Refer to the table of vibration levels and standards in Sec. 27.06.) If the vibration level reading exceeds 0.77 in/s or has altered upward by 25 percent or more since the last reading, the unit has a very serious problem.

Vibrations are generally caused by one or more of the following:

- Imbalance of rotating parts
- Imbalance of reciprocating parts
- Imbalance of combustion forces
- Misalignment of drive and driven equipment
- Torque reaction (torsional vibration)

The first step in identifying the source of a vibration problem is to determine if the vibration increases with speed at no load or as load is applied.

Although vibration monitoring routines are usually carried out at intervals that are too far apart, it must be stressed that the more frequently the readings are taken the more useful they become. In any continuously manned plant there is no reason why the generator sets in use should not be monitored during each shift. In daily attended plants there should be a vibration check on each running unit each day. Whatever the plant's status, the more checks the better.

The easy-to-use vibration monitoring instruments commonly in use permit the determination of the dominant frequency. This determination is by no means a perfect exercise, but in the absence of vibration analyzing equipment it is an excellent guide and a primary step toward problem resolution. The dominant frequency will be a multiple of the unit's rotational speed expressed in revolutions per minute.

When the dominant frequency is 1 × rpm, the trouble will likely be imbalance and/or misalignment. Phase imbalance of the generator must also be considered. If the dominant frequency when read in an axial plane is half or less of the vertical or horizontal readings, there may be a bent shaft or misalignment of bearings or couplings. On two-cycle engines there may be problem with the fueling of one cylinder.

If the dominant frequency is 2 × rpm, there may be a misalignment condition; on a four-cycle engine there may be one weak cylinder. Again the axial measurement should be used to check. When the dominant frequency is several or many times the rpm, there may be trouble with the timing gears or antifriction bearings.

To determine the dominant frequency, use the formula

$$\text{Cycles per minute} = \frac{\text{velocity}}{\text{displacement}} \times 19{,}120$$

It is not thought that phase imbalance will contribute greatly to the vibration characteristics of a generator unit, but if this condition is suspected, the generator set can be tested several times in the same day. The phase loadings will change relative to each other in the early morning, noon, and late evening, and such changes must be taken into consideration if phase imbalance is suspected.

Alignment variations will take place often through foundation or floor movement. Running the unit at incorrect operating temperatures also will create conditions of misalignment. Many alternators are assembled without the benefit of locating dowels on their end bells. It is recommended that the dowels be fitted whenever possible so that in subsequent reassemblies the alternator will be properly lined up.

Equipment manufacturers sometimes state the levels of vibration that are acceptable or can be expected in their machines. Such statements form a good starting point for gauging the vibration performance of an individual unit, but the history recorded over a period of time is the best guide to a machine's state of health.

27.6 VIBRATION LEVELS IN DIESEL GENERATOR SETS

For monitoring purposes, vibration magnitude is best measured by measuring the vibration velocity. A vibrating object is not only moving, it is changing direction in an oscillating manner: its speed is continually changing. At the end of its motion cycle, the speed of the object is zero; as the object passes through its neutral position, its speed or velocity is greatest. This change in velocity is an important characteristic of vibration. The peak velocity is read by the monitor; the reading is expressed in inches or millimeters per second.

A table of vibration levels for a diesel engine illustrates the levels at which remedial action should be taken. When vibration readings are recorded frequently, however, there is a good opportunity to perceive figures which are useful in early detection of developing trends or problems. In a systematic vibration monitoring scheme, there will be certain fixed points on the machine at which readings are taken. They are often identified by a center punch mark highlighted by a prominent color and perhaps numbered. By using the marks, a person can take readings from an engine without his individual touch affecting the readings.

TABLE 27.1 Vibration Readings, Status, and Recommended Action

Vibrations in/s	Status	Recommended action
0.00–0.33	Normal	No action is required.
0.33–0.55	Normal to rough	Monitor for trends, check tune, and make a careful examination.
0.55–0.77	Rough to unsafe	Immediate corrective action is required.
0.77 and above	Ultra unsafe	Shut the unit down immediately.

27.7 TYPICAL ENGINE VIBRATION MONITORING POINTS

For monitoring purposes, the vibration readings should be taken on the horizontal and vertical axis of the first and last main bearings in the prime mover and at

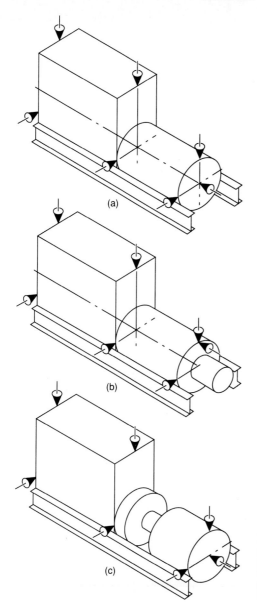

FIGURE 27.2 Vibration monitoring points. (*a*) Engine with single-bearing generator and integral exciter. (*b*) Engine with a single-bearing generator and an outboard exciter. (*c*) Engine with a two-bearing generator and an integral exciter.

the last or outermost bearing carrying the generator shaft. It is only necessary to take one axial reading (usually at the generator). No readings should be taken from sheet metal, fiberglass, or other thin-section resonant surfaces. All designated monitoring points should be clearly marked so that all personnel read at the same position (Fig. 27.2 and 27.3).

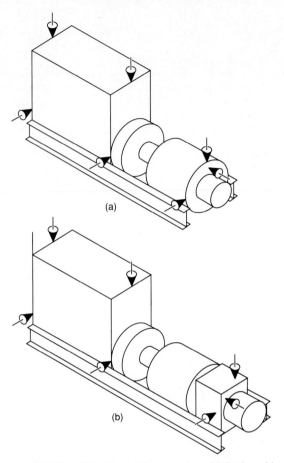

FIGURE 27.3 Vibration monitoring points. (*a*) Engine with a two-bearing generator and an outboard exciter. (*b*) Engine with a single-pedestal-bearing generator and an outboard exciter.

27.8 COMMON VIBRATION CAUSES

The level of vibration in a diesel generator set should not be more than 0.55 in/s velocity under a full range of steady loads. Through regular monitoring of the vibration levels and comparing the present with past readings, any trend toward a higher or unacceptable level of vibration can be recognized. If a trend is seen, corrective measures must be instituted. In most cases, the causes are found fairly easily, but a certain amount of journeyman time and patience must be given to the problem-solving effort.

Diesel generator vibration conditions can be classified as those that:

1. Gradually develop

2. Suddenly develop
3. Change over a period of time

The causes of change in the first two classes are quite different and are summarized below:

1. Gradual change
    - Loss of compression on one or more cylinders
    - One or more cams wearing
    - Burned valve seat(s) or valve seat pounding
    - One weak turbocharger on a V-engine
    - Dirty air filter on a V-engine
    - Cracked flexible disk coupling
    - Dirty alternator rotor windings
    - Weak vibration isolators

2. Sudden changes in vibration levels
    - An overspeed incident
    - Seized fuel pump
    - Seized injector
    - Water in fuel
    - Broken or loosened part. e.g., bottom end, tappet
    - Failed bearing
    - Broken timing tooth

There are many other less common causes of vibration, but the examples given are typical. In any event, when there is a sudden change in the monitored vibration pattern, the unit should be stopped immediately.

27.9 ALTERNATOR VIBRATION

Defective rotor windings may contribute to vibration at the alternator. Although there may be no apparent defect as indicated by the switch panel instrumentation, a fault may exist. After the basic inspection of the engine and generator has been completed, the fueling of the individual cylinders has been checked, and all other obvious areas have been explored, a voltage drop test of each pole coil should be carried out. A close visual inspection of the pole brick fastener will be in order at this time.

27.10 VIBRATION ANALYSIS

Vibration analysis is aimed at identifying the nature and cause of unacceptable vibration after the monitoring process has indicated a problem. It is a complex activity that demands a high level of technical knowledge. Analysis is necessary only when the normal remedial work that would be done following an anomalous monitored reading fails to rectify the situation.

Vibration analysis involves the use of fairly expensive equipment by a technical specialist with considerable expertise; the actual analytical work is performed by specialists. However, the specialist will need the assistance of the operator and the station mechanic for the starting and stopping activities and for the adjustments and changes that will possibly be necessary in the analyzing procedure.

Vibration analyzing units employ tunable filters which permit the reading of the salient vibration frequencies. This feature may be linked to a strobe light to allow stop-motion analysis. All components vibrating at the timed frequency appear to be at a standstill and the phase is understood. Similarly, tuning filters will provide data on vibration amplitudes.

An analyzing unit may be used for balancing procedures, and with it all of the characteristics of vibration—displacement, velocity, acceleration, frequency and phase—can be measured and evaluated and their relationships determined.

27.11 MACHINE SIGNATURE

When a program of vibration monitoring is embarked upon, it is useful to prepare a sketch showing the general layout of the machine shaft centerlines, bearing positions, and other prominent points from which vibration readings may be required at some time in the future. The machine is then operated at its normal speed and loads while vibration readings are taken from all the marked points. The readings, together with speeds and loads, are then noted on the diagram pertaining. The diagram and its figures are known as the *machine vibration signature*. It is most important that a machine vibration signature be obtained when the generator set is newly installed and is still under warranty (Fig. 27.4).

The normal monitoring routine may call for only two or three points to be monitored regularly. Whenever an excursion is noted during the monitoring routine, however, a full follow-up reading of the signature points is most helpful in pinpointing the origin of any aberrant vibration.

27.12 VIBRATION RECORD

It is all very well for a plant to have vibration reading equipment available, but from a supervision point of view any vibration monitoring effort is useless without a record of previous readings (Fig. 27.5). The superintendent must exert a continuous pressure on the shift operator and the maintenance presonnel charged with equipment health-monitoring duties to write up their observations as and when they are taken. Without a credible history of vibration to which reference can be made, it will not be possible to establish trends, strike comparisons, or determine standards of performance.

If a routine of systematic vibration monitoring and recording is introduced, a workable method of predictive and preventive maintenance can be established. Observance of trend developments in any of the vibration characteristics will show what is happening to the machine, and simple projection will indicate if, and possibly when, failure may occur. The use of predictive data will assist both the operational and the maintenance staff in their efforts to provide a high level of reliability and efficiency with an absolutely minimal interruption of service to the consumer.

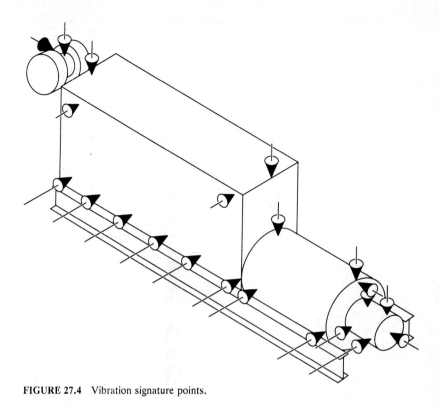

FIGURE 27.4 Vibration signature points.

Machine # _____   Date _____
Load   _____   Speed _____

Monitor	Monitor point	Horizontal	Vertical	Axial
Velocity				
Displacement				
Dominant Frequency				
Velocity				
Displacement				
Dominant Frequency				
Velocity				
Displacement				
Dominant Frequency				

Remarks (state rectifying work done):

FIGURE 27.5 Vibration report form.

CHAPTER 28
ENGINE PRESERVATION AND STORAGE

28.1 PRESERVATION, STORAGE, AND STANDBY UNITS

Engines that run infrequently require a certain amount of protection peculiar to their limited utilization. If a diesel generating plant is to be placed in a standby mode of operation, the owner, the superintendent, and the plant staff must have a common understanding of what is meant by the term standby and the implications of that understanding. Without such knowledge, there is a risk of the plant not being in the state of readiness expected when there is a need to run, and an unfamiliarity with the necessary operating procedures may have developed.

A standby plant may be one in which one or more pieces of generating equipment is kept in a state of constant readiness. In this plant the engine jackets will be kept warm, the lubricating system will be primed, the starting energy source will be topped off, and all of the equipment will be available for instant service upon command.

Another version of a standby plant is one that is idle but is available for service following certain preparatory work such as loading the cooling system, priming the injection system, removing shutters from radiators and ventilation apertures, and inserting the main breakers in their cubicles. Diesel generating plants that have been laid up with all of the fluids drained from the systems, the air and exhaust openings sealed off, the batteries disconnected, and the generating units sheeted over also are often referred to as standby plants.

From the three examples described, it is obvious that the scope of the term standby is quite broad and open to many interpretations in detail. It is suggested, therefore, that when a plant or piece of generating equipment therein is placed in a standby mode, a written statement be prepared and passed by the plant superintendent to the owner. It should outline the status of the equipment, the duties and routines that will be carried out by the plant staff, and the limits of what may or may not be done with the standby units. Immediately after the equipment or plant is placed on the standby status, a step-by-step start-up and operating guide should also be prepared by the superintendent so that future personnel can recommence operations without difficulty or equipment damage. The owner will require to be fully informed of the standby condition of each unit.

28.2 INSPECTION ROUTINE

It is recommended that any plant that has one or more engines which are planned to stand idle for a prolonged period adopt a systematic routine of engine care. At about 30-day intervals, the following exercises should be carried out:

1. Bar the engine around at least one full turn if the fluid systems have not been drained.
2. Check the sump for level, glycol, and/or fuel if the fluid systems have not been drained.
3. Monitor the vibration level of the adjacent floor and engine underframe. (Fig.26.12)
4. Inspect the unit for missing parts. (Out-of-service engines sometimes get cannibalized.)
5. Inspect the unit for obstructions. (Out-of-service engines often get used as places to leave tools, materials, parts, outdoor clothing, and so on.) Prepare a statement for the owner regarding immediate and future intentions if parts are missing, have been dismantled, or are damaged.
6. Check the air intakes, exhaust outlets, alternator housings, and so on, where rodents or birds may rest.
7. Check the battery voltage and electrolyte level if the battery is still connected.
8. Check for signs of corrosion, dust entry, or anything that will be detrimental to the engine.
9. Ensure that the diesel-driven compressor has a full fuel tank and that the sump is at the correct level if the standby unit is air started.

28.3 LUBRICATION

The level of the lubricating oil sump must be checked at frequent intervals if the engine remains loaded with coolant. The decaying effect of ethylene glycol on the liner seal rings continues whether the engine runs or not, and it is not unknown for a substantial leakage to take place with resultant sump flooding. The lubricating oil system must be primed at regular intervals so that all bearing surfaces have their lubricating oil films renewed. To reduce the possibility of circulating leaked antifreeze, this must be done after the dipstick has been examined.

28.4 BATTERIES

The batteries of an engine that is not to be run again for an extended period of time should be disconnected from the engine and put into service elsewhere if possible. If the batteries are not utilized, they should be disconnected from the load but left connected to the charger. It will be necessary to inspect them at intervals for electrolyte level and voltage.

28.5 VIBRATION MONITORING OF STANDING ENGINE

It often happens that an engine will not be required to work for an extended period of time because other engines in the same plant are better suited to the prevailing loads. Special care is needed in such cases to see that the still engine is not affected by vibrations transmitted from running engines. For that reason, it is important the standing engine be monitored for vibration at the same time intervals as the other engines in the plant. If vibration is detected, it is suggested random main and bottom end bearings be examined before the unused engine is returned to service. It is possible for a standing engine to suffer a complete crankshaft failure through transmitted vibration. A complete vibration inspection will include a reading from the floor adjacent to the laid-up engine.

28.6 BARRING OVER

Engines that run infrequently should be barred over at regular intervals so that flooding of the cylinders by coolant, fuel oil, or lubricating oil can be detected in good time. The engine should be rotated by hand, and the indicator cocks, where present, must be open. An electric starter, an air starter or the turning gear must not be used.

28.7 LAID-UP ENGINES

All work done to lay up a generator set must be recorded in detail in the relevant workbook as a guide to the unit's recommissioning at some time in the future. For engines that will definitely not be used in the foreseeable future, proofing against weather effects is required. The engine must be protected from snow, rain, heat, frost, wind, dust, wildlife, and any other natural damaging agents. All apertures giving access to the generator set interior should be closed against the intrusion of dirt, dust, water, oil, fuel, or other foreign material. All exposed machined surfaces must be protected against corrosion.

An inventory of all the normally fitted parts, including switch keys and the like, that have been removed or are otherwise absent from the unit should be taken. A copy of the inventory should be placed in the subject unit and another copy in the record file pertaining to the engine. Care must be taken that all spaces have been thoroughly drained of coolant on engines that are to be stored through a winter. Particular attention must be paid to block heaters and other coolant-carrying spaces on the lower areas of the subject unit. It is recommended that the block heaters be disconnected electrically to protect them from burnout.

If an engine is to be transported during the winter to or from a site, care must be taken to see that either the coolant is thoroughly drained or is proofed against the worst-case temperatures likely to be encountered. Similarly, a generator set that is to be moved to another site must be totally proofed against road dust. Any engine in transit must carry a prominent notice such as the following ones:

THIS UNIT CONTAINS COOLANT
THIS UNIT IS DRAINED OF ALL FLUIDS
THIS UNIT HAS NO OIL IN IT

A similar notice is needed for any other extraordinary condition. The signs must be multiple and robust. They should be placed in eye-catching positions.

28.8 CAUTION: ENGINE DRAINING

It cannot be recommended that the coolant and the lubricating oil be drained from any engine unless the starting systems are fully disconnected and/or the starting circuit fuses are removed.

28.9 FUEL SUPPLY

The fuel supply to any engine which is likely to remain idle for more than a few days at a time should be shut off at the valve before the engine. A card indicating that the fuel is shut off should then be attached to the start lever or switch. If the fuel remains open to the engine for a prolonged idle period, there is a faint possibility it will leak into a cylinder and dilute the lubricating oil or cause hydraulic lock upon start-up.

28.10 CLEANING AND PAINTING

It is suggested that each engine that is to be laid up be thoroughly cleaned and painted. At this time the alternator windings also should be cleaned. All exposed bright ferrous surfaces should be either painted or coated with a paste-type anti-corrosive coating. Do not use common lubricating oils or greases because they do not perform well in a corrosion preventive role.

The lubricating oil should be drained from the engine, and the crankcase, sump, and oil coolers should be thoroughly cleaned. The lubricating oil filters also should be changed. The engine must be marked with signs stating that there is no oil in it. The governor should be brought down off the engine, flushed out, and returned to the engine with a fresh charge of oil.

28.11 RETURNING THE ENGINE TO SERVICE

Before an engine can be run safely after it has been out of service for a prolonged period of time, there are a number of things to be done. The workbook relating to the generator set should be consulted to determine what was done at lay-up and what must be done now.

The cooling system must be flushed out to remove all of the trash that will have resulted from drying out the coolant spaces of the unit. After flushing, the engine should be recharged with coolant of the correct mix. This should be done first so a close inspection for leakage at the liner skirts, circulating pump, coolers, and radiators can be carried out before the lubricating oil is added. The following are the steps to be taken next:

1. If it was not done at the time of lay-up, the governor must be removed from the engine, flushed out, and refilled with clean mineral oil. It will be necessary to bleed the governor to eliminate any air from the governor mechanism after reinstallation.
2. After the lubricating oil has been put in the engine and the lubricating oil system has been primed, the engine should be barred one full turn by hand.
3. The covers should be taken off the exhaust outlet, breather outlet, air intakes, alternator, and so on.
4. In the case of an electric start engine, the batteries must be reconnected and brought up to optimum charge. The safety shut-downs should be checked for electrical function at this time.
5. A very thorough eyeball inspection of the entire generator unit must be made to see that there are no obstructions, no parts are missing, and there is no other abnormality that will preclude a safe and effective start-up.
6. All grease and oil or nonsetting anticorrosive media applied to the engine when it was laid up must be removed as a safety precaution.
7. The fuel injection system will require priming before starting is attempted.

28.12 PRECAUTION AFTER RESTART

When an engine or generating set has been restored to service after a prolonged rehabilitation or lay-up, it is prudent for the operator to pay close and continuous attention to the engine and the surrounding structure immediately after the engine has warmed up to operating temperature. He should look for evidence of overheating in the exhaust system or in the structure immediately surrounding the exhaust outlet.

CHAPTER 29
SEALS AND GASKETS

29.1 SHAFT SEALS

Many seals are made of a plastic, neoprene, or other synthetic flexible material. These seals are frequently affected by exposure to unusual temperatures. (At −40°C, the material may crack; at 150°C, the material may deform permanently.) Most of these seals are not reusable if, after they have been removed from their housing, they distort readily. These seals are common on fast-turning automotive and mass-produced engines.

The spring ring seal is well known. It takes the form of an iron ring similar to an ordinary piston ring set into a housing which is clamped to the shaft, and the whole seal revolves within a baffle that is bolted to the crankcase. When they are in good condition, seals of this type are very effective, but dirt or rust can spoil them quite easily. They are frequently changed and the old ones discarded, but thorough inspection and cleanup will often restore them to first-class serviceability. They are expensive, and they should not be thrown away just for cosmetic reasons. The question that must be asked is: "How much is an oil leak worth?"

The metallic labryinth seal is frequently used on high-speed machinery in which actual contact between two surfaces is undesirable. Large turbo blowers and gas turbines use this method of seal. Turbocharger labryinth seals are often air-pressurized to keep exhaust gases, aspirating air, and lubricating oils all in their respective zones.

29.2 SHAFT SEAL MATERIALS

Attention must be given to the type of material employed in shaft seals when the seals are installed. Seals with Teflon glands should be fitted in a dry state *without any lubricant* of any description being applied to either the Teflon gland or the shaft. When the machine is started after seal renewal, a certain amount of Teflon will migrate from the gland to the shaft and a good seal will be achieved. If any foreign fluid is present, the migration will be inhibited and a less than perfect seal will result.

Seals fitted with elastomer glands (i.e., neoprene or other synthetic rubbers) should be lubricated between the gland and the shaft at the time of installation. Metallic fabric or ceramic seals should also be so serviced. The performance of many of the materials used in seal casings or glands is often limited by the operating temperature ranges. Correct selection of the seal will therefore greatly contribute to a machine's reliability.

The gland material—its composition and its compatibility with the sealed medium—and its installation positioning are some of the points basic to successful seal renewal. Rubber seals and diaphragms employed in a crude oil system should be of a nitrite or viton rubber. Buna or neoprene rubber is subject to decay under the influence of the elements present in crude oil. Although buna cannot be used in conjunction with lubricating oil, neoprene is generally quite satisfactory.

29.3 ENGINE LEAKS

In any engine five media are in movement whenever the engine is running: aspirating air, fuel oil, lubricating oil, coolant, and exhaust gas. All the media systems operate under pressure that can be positive or negative relative to the adjacent medium or atmosphere. If there is a pressure imbalance between media, there will be leakage from the higher-pressure zone to the lower-pressure zone whenever the opportunity is presented.

At points where one engine component interfaces with another, a joint sealing method is used. It may take the form of a fiber, elastomer, viscous, or metallic gasket. The purpose of the gasket is to seal any imperfection in the mating surfaces through which a gas or liquid may pass. The gasket also closes any variation of the interfaces that may be caused by uneven torquing of the fasteners.

When a leak becomes apparent, the natural reaction of the operator or mechanic is to retighten the fasteners. Certainly the torque of the fasteners must be checked, and in some cases tightening may produce the desired effect. At no time, however, should the manufacturer's torque specifications be exceeded. It is better to tolerate a leaking gasket until time as a proper repair can be effected than to run the risk of threads stripped, flanges cracked, or components warped by torque overload.

Overtightening fasteners probably causes as many if not more leaks to develop as undertightening. The lightness of the torque that is sufficient to provide a permanent and practical contact between sealing surfaces is often a source of surprise, particularly in the case of such pressed-steel components as valve covers, crankcase doors, and pans of smaller engines. [For example, a 16GA cover (0.0625 in) with 9/16-in screws at 2.5-in intervals will have the screws torqued to 8 lb · ft and be without leak or warp.]

29.4 REASONS FOR LEAKS

When a leak begins to show, it is as well to consider the real reason. Leaks often occur not because of loose fittings, but because the pressure behind the leak has increased or the pressure in front of the leak has decreased. For example, if the lubricating oil of an engine is greatly overcooled, the system pressure will be abnormally high. That will cause leakage at the lubricating oil filter housing gaskets. If the crankcase breather becomes plugged, it is probable that the crankshaft seals will begin to leak. Possibly the two leaks will occur simultaneously. All leaks and their causes should be carefully considered before thoughtless wrench-bending commences.

29.5 SEAL DRESSING, GASKET CEMENTS, AND O-RING LUBRICANTS

Many varieties of lubricating or adhesive compounds can be applied to O-rings, gaskets, and joints when they are installed in an engine. Incorrect application may result in various problems if certain basic guidelines are not followed. All cements, lubricants, sealers, or other compounds that may be used must be compatible with any of the other materials they may subsequently contact.

Glycerine, industrial hand cleaner, dishwasher liquid, soft soap, and so on, have their places, but they can be allowed only when the OEM manual permits them. For example, common lubricating oil should not be used as an O-ring-inserting lubricant, and RTV cements can be applied only sparingly if oil passages, cooling jets and fine clearances are to remain free of any blockage risk. If a new seal of the correct part number is fitted where previously there had been no problems, the use of supplementary sealing compounds will not improve the subsequent performance of the seal.

In most instances, the fluid being sealed against may be used as the seal lubricant at the time of installation. Alternatively, it is recommended that glycerine be used as a lubricant. As far as is known, glycerine has no deleterious effect on any seal material.

29.6 O-RINGS

The O-ring seal is in common use in all classes and types of machinery. As technology advances and improved machines are marketed, the continuous development of O-ring materials is necessary. Not only is the O-ring required to provide a perfect and long-lasting seal; it must be proof against ever-increasing temperatures, pressures, and other stresses. The immediate environment touching each side of the seal may have significant effects on the seal. If the correct part number is used, no trouble will be experienced. Otherwise, the installer must be certain that the material is compatible with its use.

29.7 O-RING MATERIALS

The material used in an O-ring must be in accordance with the ring's application. Just about all O-rings have an elastic quality and thus are capable of sustaining a limited deformation without suffering a permanent loss of size or shape.

Depending as it does on the service temperature, the fluids or gases contacting the seal surfaces, the pressures imposed on the ring, and the motions of the mating components, the specification of the O-ring material is critical. O-rings have many applications, and several different materials are used to give optimum performance under various conditions of pressure temperature, and exposure.

Elastomeric materials used in engine and machinery applications are usually synthetic rubbers. The following are the rubbers most frequently encountered and their applications.

1. *Buna-N* is mainly used in conjuction with petroleum products such as fuel oil and lubricating oil working in temperature ranges of −20°C up to 110°C (−4°F up to 230°F). Buna-N material is black in color.

2. *Ethylene propylene terpolymer (EPDM)* is used in cylinder liner, air manifolds, and cooling system applications. In contact with fuel or lubricants, it will swell. Seals made from EPDM should not be used in conjunction with oil; they may be used only for water service. If a lubricant is required at the time of installation, glycerine or soft soap can be used.

3. *Fluoroelastomer (FKM)* is supplied as green, white, or brown rubber. It is used in the high-temperature, high-pressure conditions found in an increasing number of late designs. Although an O-ring can be partly identified by its color, it is essential that it be positively identified by its part number. The incorrect use of O-rings can lead to early leakage or other unsatisfactory performance.

4. *Nitrile butadiene rubber (NBR)* is another high-performance synthetic rubber now coming into frequent use. Although costly, it is very effective for use with crude oil, diesel fuel, and lubricating oil. All used O-rings *must be discarded*. After prolonged use, any O-ring loses its original cross-sectional shape. With such distortion, it is not possible to obtain a 100 percent fit all around.

5. *Neoprene* is the elastomer most often used in internal-combustion engines, where it is highly resistant to attack by fuel oil or lubricating oil. It was the first of the synthetic rubbers, and it is quite costly as compared with the rubbers used in the automotive industry.

6. *Silicone rubber* is used in O-rings working in high-temperature, low-pressure situations in connection with lubricating and hydraulic oil. It cannot be used in fuel services. Silicone O-rings may be supplied in yellow, white, or red. Silicone oils or greases must not be used when these O-rings are installed, but soft soap or glycerine can be employed as a lubricant.

7. *Urethane rubber* O-rings have good abrasion resistance and are used where there may be some movement between components. Urethane O-rings are black in color.

29.8 O-RING MATCHING

Certain manufacturers supply O-rings in different colors; the colors denote the type of service for which the ring is intended and are also indicative of the different materials used in their manufacture. When any O-ring is to be changed for a new one, the related parts book must be consulted to see that the correct ring is used. It must never be assumed that because an O-ring of a particular size and color was taken off an item of equipment, it must be replaced by one of similar size and color.

29.9 O-RING PROFILE

Not all O-rings are of perfectly round cross section. Some liner O-rings are more of a D-shaped cross section with the flat toward the inside of the ring. Care will be necessary in the selection and installation to ensure that the ring is properly positioned in its groove.

In the review of O-ring spares stock, any rings that appear to be of D-section must be examined carefully to see if they are in fact new and usable or are old heat-deformed rings that should have been discarded long ago.

All O-rings taken out of service should be destroyed at the point of discard.

29.10 RUBBER PARTS STORAGE

Rubber O-rings should always be kept in a cool storage space and not be exposed to daylight. Ideally, they should be stored in a container filled with talc or french chalk. They should never be hung in a vertical position on pegs or nails. It is recommended that all stock O-rings be tagged with identifying part numbers as and when they are put into stock. Not all O-rings are proof against oil, water, ethylene glycol, elevated or low temperatures, or other conditions even if they look alike, and so a certain amount of caution should be exercised in the selection of an O-ring for use.

Before use, each O-ring must be examined for any sign of radial cracking. Discard any O-ring that has softened, deformed or decayed, or has otherwise departed from its original state. Never retain a used O-ring for later reuse.

29.11 GASKET STORAGE

Whenever possible, all paper and fiber gaskets should be stored in a horizontal position. Hanging gaskets on nails or hooks cannot be recommended. No gasket should ever be folded to save storage space. When gaskets of more than one pattern are to be placed on a shelf, cardboard separator sheets should be used so the different gaskets are kept apart and repeated search handling, which is so damaging to stored gaskets, is minimal. Fiber-based gaskets that have been in stock for a prolonged period should be saturated with the medium to which they will be exposed in service well before they are installed.

CHAPTER 30
PIPES AND VALVES

30.1 PIPE AND VALVE LOCATIONS

The operator must be very familiar with all the many valves within the generating plant complex and must know which of the valves are normally open and which are normally closed. That knowledge must be developed early in the operator's career at any given power plant. In an outage event, the operator's ability to identify and manipulate any particular valve quickly and accurately will be of great value.

30.2 OPENING AND CLOSING HAND-OPERATED VALVES

1. All operating personnel must learn to leave hand-operated globe, gate, nonreturn, and screw-lift valves in correct modes after use.
2. Any person wishing to open or close a hand-operated valve must be able to easily ascertain if the valve is in an open or closed position without having to apply undue force to the handwheel.
3. When a valve is closed, the spindle is screwed down to the point at which no further movement takes place. (Applying excessive force or leverage to a valve handwheel does not close the valve further; the spindle only gets strained.) The spindle is then backed off half a turn; the crush on the valve is released; but the seal is still maintained and the valve position is readily determined by a following person.
4. Similarly, when the valve is opened fully, the spindle is screwed back to a stop position and then reversed half a turn.
5. All hand-operated valves should be operable by hand without the need for any levering tools.
6. The packing in the spindle glands will last for many years if the valve is handled with a reasonable amount of care. The glands should never be so tight that the valve cannot be opened or closed by hand alone. The packing in the gland must be of the correct type for the fluid being passed.
7. Pipe wrenches must not be used on valve spindles because both the spindle and the gland packing will be quickly ruined. Nor should pipe wrenches be

used as keys for valve handwheels; that sort of abuse will render the pipe wrench useless. Using adjustable wrenches instead of a handwheel on the spindle square will soon lead to spindle damage.

Summary

Proper opening and closing techniques and good glands give prolonged and trouble-free life to all hand-operated valves. See Figures 30.1 and 30.2 for cross sections of gate and globe valves and the points at which damage can be caused by improper techniques.

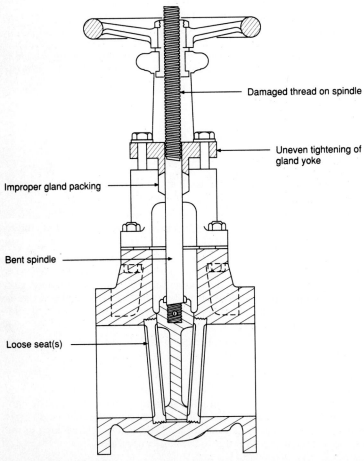

FIGURE 30.1 Gate valve.

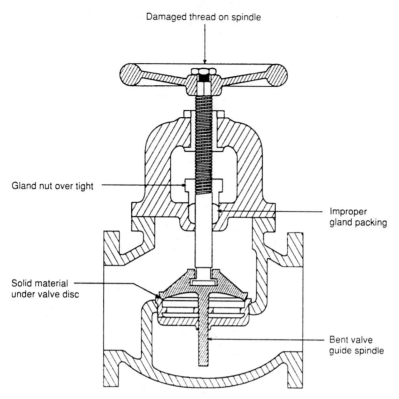

FIGURE 30.2 Globe valve.

30.3 PIPEWORK AND PIPE SYSTEM

Rarely are two diesel generating plants equipped with exactly similar pipework. The great variety of fittings inserted in the different system designs are not subject to any universal standard, and thus the drawings can be most confusing. The symbols used in the drawings and schematic diagrams are not linked to any standardization system at this time.

The plant superintendent must request a full set of drawings of any installed pipe system from the designer or installer, and the drawings must be complete with a table of the symbols used in the drawings. As part of the learning process of the plant staff, each one should make an isometric diagram of the individual pipe systems and understand it (Fig. 30.3). Artistic ability is not necessary, and elements do not have to be drawn to scale. The sketch should be rendered in pencil and finalized in ink. The operator should try to incorporate as many details as possible in the sketch. A highly detailed sketch suggests a competent operator. All of the diagrams should be compared and corrected at the next personnel and equipment safety meeting or periodic training session.

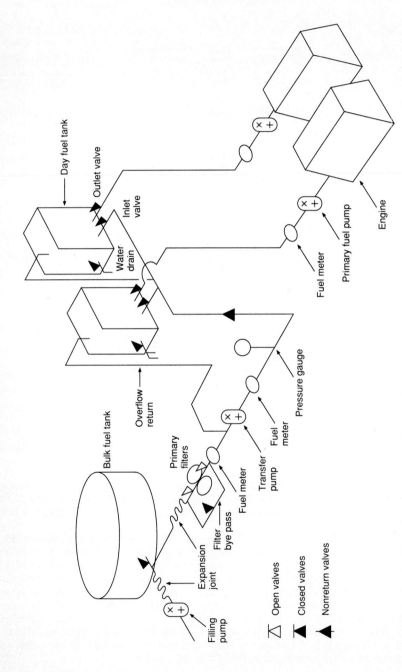

FIGURE 30.3 Typical operator's system learning sketch.

30.4 ABBREVIATIONS

The following are the abbreviations commonly used in piping drawings and schematic diagrams. They are provided by the Mechanical Contractors Association of America, Inc.

A	Air
ABS	Absolute
ANSI	American National Standards Institute
API	American Petroleum Institute
ASTM	American Society for Testing and Materials
AWS	American Welding Society
BBL	Barrel
BC	Bolt circle
BCD	Bolt circle diameter
BLE	Beveled large end
BLK	Black
BOP	Bottom of pipe; used for pipe support location
BS	British standard
BSP	British standard pipe thread
BTU	British thermal unit
BOS	Bottom of steel
C	Centigrade or Celsius
CENT	Centigrade
CFM	Cubic feet per minute
CI	Cast iron
CL	Centerline
CM	Centimeter
CS	Carbon steel
CU	Cubic
DIA	Diameter
DWG	Drawing
EFW	Electric fusion welded
ELL	Elbow
ERW	Electric resistance welded
F	Fahrenheit
FAHR	Fahrenheit
FF	Flat face(d), full face(d), flange face(d)

FLG	Flange
FLGD	Flanged
FS	Forged steel
FW	Field weld
G	Gas
GAL	Gallon
GALV	Galvanized
GPH	Gallons per hour
GPM	Gallons per minute
HEX	Hexagonal
Hg	Mercury
HPT	Hose pipe thread
HR	Hour
ID	Inside diameter
IMP	Imperial (British unit)
ISO	Isometric drawing
K	Kilo, times one thousand, × 1000
KG	Kilogram
LB,lb	Pound weight
LR	Long radius (of elbow)
M	Meter, mega, times one million
MATL	Material
MAX	Maximum
MCC	Motor control center
M/C	Machine
MFR	Manufacturer
MI	Malleable iron
MIN	Minimum, minute (of time)
MM	Millimeter
N	North
NC	Normally closed
NEMA	National Electrical Manufacturer's Assn.
NO	Normally open
NPT	National pipe thread
NRS	Nonrising stem (of valve)
O	Oil

OD	Outside diameter
OS	Outside screw
OS & Y	Outside screw and yoke
P & ID	Piping and instrumentation diagram
PBE	Plain both ends (swage, etc.)
PCD	Pitch circle diameter
PE	Plain end, pipe, etc.
POE	Plain one end (nipple, etc.)
PS	Pipe support: anchor, guide, or shoe of items combined to form the support
PSI	Pound (weight) per square inch (pressure)
PSIA	Pound per square inch absolute
PSIG	Pound per square inch gauge
RED	Reducing
RF	Raised face
RJ	Ring joint
RPM	Revolutions per minute
RS	Rising stem (of valve)
S	Steam
SAE	Society of Automotive Engineers
SCH	Schedule (of pipe)
SCRD	Screwed
SKT	Socket
SMLS	Seamless
SO	Slip on
SR	Short radius (of elbow)
SS	Stainless steel
STD	Standard
STR	Straight
SW	Socket welding
SWC	Swage
SWG NIPP	Swaged nipple
T	Temperature
TOE	Threaded one end (nipple of swage)
TOS	Top of steel
TPI	Threads per inch
TSE	Threaded small end
TYP	Typical
UNC	Unified national coarse
UNF	Unified national fine

WP	Working pressure
WT	Weight
XS	Extra strong

CHAPTER 31
MANUALS, PARTS BOOKS, WORKBOOKS, AND SERVICE BULLETINS

31.1 GENERATOR SET BOOKS

Each generating set must have its own workbook in which all work activity should be recorded. Only in that way can new or visiting mechanics or electricians obtain an appreciation of what problems are peculiar to the individual units or judge when a particular activity will become due. Memories of other personnel at the plant are not different from those of people elsewhere: They must always be regarded as short, frail, and incomplete. Thus, the workbook entries are essential if any sort of communication continuity is to be maintained.

Preferably the workbook will be a standard size 8½ by 11 in hardback. The first few pages should contain all the basic engine and generator data (make, model, serial number, governor details, injector nozzle number, fan belt size, filter number, and similar details) together with any specific remarks concerning undersize bearings, chromed liners, and so on. A large amount of information can be written in this section. Any manufacturer's recommended modifications must be noted, together with the date the modification was made and the identifying number or title of the governing bulletin. If good attention is given to this requirement, a great deal of potential future trouble will be eliminated and future power plant staff will find the data immensely useful.

The main part of the book should be written up in diary form; it will contain much detail of the incidents, activities, and work performed on the unit. It is important that both the date and the engine hours always be written in against an entry. Also, all the parts, part numbers, and quantities consumed should be noted as far as possible. Some of the repair jobs done will negate the need for the same sort of work in a closely following planned overhaul.

As the workbook becomes developed, it is possible to read it with a certain sense of perspective. Some of the events will be seen to form a pattern, and the incidence rate of particular types of failures will be identified. The book can be compared with another workbook from a similar engine, and again useful and often startling details become apparent. Many mechanics are reluctant to write down what they have done; they feel that such a chore smacks of clerical work. The fact remains that the mechanic is the person who is directly involved in the work performed. Only he has the related training and experience necessary to write coherent and meaningful reports of engine work activity, and it is his successor who will rely on the reports for essential background information.

The work information is needed to fill several basic needs. It must be possible to correlate events on one or more engines to determine if a particular engine model has a specific defect or perhaps to see if the overhaul schedules are totally satisfactory as designed. Failure patterns must be identified and analyzed so that future preventive action can be built into revised overhaul schedules. An appreciation of manpower needs to handle the maintenance overhaul and breakdown work on individual engines or plants can be obtained, and an assessment of the extra manpower needed to accelerate the work can be formed.

None of the above information can be obtained just by relying on a person's memory, nor can formal presentations be developed to explain to the owner the all-too-often high-cost improvements that appear to be needed. The information must originate on the power plant floor.

31.2   CLEAN COPY OF THE WORKBOOK

It is suggested that the superintendent maintain a clean copy of the generator set workbooks by using the plant workbooks as they are written up by the journeymen. The superintendent should transfer all possible relevant information to a clean office copy and perhaps add a limited amount of meaningful comment without, however, editing the rough original. By that means, the workbook will maintain its continuity and legibility long after a well-used plant workbook has become partly obliterated with grime. Furthermore, the superintendent can study the recorded events more readily and identify patterns, trends, or repetitious incidents at a later time and in an improved perspective.

31.3   USE OF SHOP MANUALS

An adequate supply of service manuals must be provided for the use of the mechanic and the electrician. That supply must be sufficient to cater to training purposes and storekeeper usage as well as the normal day-to-day plant use. All journeymen are expected to refer to the manuals for all stages of generator set work, and the superintendent must check to see that the manuals are in fact consulted. The shop manuals may be seen as a form of quality control tool for the work that is to be done. By following the manuals in a careful and disciplined manner, the journeyman will conduct the work in a fashion approved by both the owner and the OEM.

31.4   PARTS BOOKS

It is essential to have parts books on hand in each power plant, and they must be available for the use of all mechanics, electricians, storekeepers, and superintendents. No part number in any parts book should ever be crossed out or obliterated, but a blank sheet of paper can usefully be pasted in the back of the parts book on which to write all the part number changes as they occur. A printed part number can be starred to indicate that a new number is listed in the back of the book.

Many engines have variant part numbers to accommodate the many purposes to which they are put—tractor, fire pump, motor boat, and so on—and it is important that the build numbers of each unit in the plant be read off and marked on the front cover of the parts book. The build number must then be used to find the correct page in the master index for a given model. If that routine is not followed, there can easily be an incorrect selection of parts.

31.5   SERVICE BULLETINS

Most engine manufacturers periodically issue various service bulletins, part number revisions or updates, modification advices, or other written material having relevance to one or more of their engine models. When such notices are received, they must be handled systematically. Ideally, at least one copy will be placed in the master file and copies will be distributed to the mechanic, the electrician, the operators, and the storekeeper for them to insert in their manuals. Depending on the nature of the advice given, all the existing parts books, service guides, and shop manuals should be neatly, clearly, and appropriately marked up on receipt of each notice.

Some of the bulletins issued by the OEM state which page or previous publication has been superseded, and it will be necessary to formally discard all the obsolete material held by the plant personnel.

31.6   BLUEPRINTS

Regardless of the color, the terms *drawing*, *print*, and *blueprint* mean the same thing in reference to copies of engineering drawings. Blueprints provide the details of the size and shape description, the tolerances and allowable variations, the grade of finish, the general and detail layout, or any other information necessary to attain a given objective. The ability to interpret blueprints is an important part of a diesel power plant employee's skills, and it should be practiced at every opportunity.

Blueprints are records of what is to be or has been installed. With good care, they can have a long and useful life, and they are a most valuable training aid. Particularly in electrical work, they are essential for system training, fault-finding, and troubleshooting.

31.7   CARE OF BLUEPRINTS

Blueprints are regarded as part of the owner's record inventory, and they merit the respect that is given to the other books, manuals and similar reference material. Nobody should be permitted to write on or alter a blueprint without due authority. If such an alteration is made, all of the existing copies and the master drawing must be similarly revised.

All blueprints must be kept clean and dry, and all tears and breaks must be repaired promptly. The prints should be folded with care by using the same fold lines on each occasion. That should leave the folded print with its identification data exposed. The filed prints are best kept in numerical order and in darkness, because many blueprints will fade when continuously exposed to either natural or artificial light.

When there is any doubt concerning the accuracy of the data on a blueprint, a new copy must be obtained and compared with the issue and revision dates. If there are obvious differences between the actual setup and the latest blueprint, the journeyman must refer to the plant superintendent or designer for clarification before commencing work. Blueprints are notorious for their discrepancies, which are natural characteristics of postconstruction blueprint use in any industry.

31.8 SCHEMATIC DIAGRAMS

Schematic diagrams indicate the scheme or plan according to which equipment is connected for a specific purpose. Diagrams are not drawn to scale, and they do not show the shapes of devices. Straight lines joining the symbols representing equipment indicate the routing of the connections. Schematic diagrams may be elementary block diagrams or complete diagrams of the entire system.

31.9 BLOCK DIAGRAMS

Elementary block diagrams use rectangular or square blocks to represent the basic elements of an installation. A straight line joining the blocks indicates which elements are connected. A block diagram is useful for indicating the overall function of a system. The names of the devices or their abbreviations are placed in the blocks to indicate the functions of the individual elements. The connecting lines usually carry arrowheads to show the direction of the flow of power or media.

31.10 SINGLE-LINE DIAGRAMS

Single-line diagrams are used to illustrate the layout or connections of piping or electrical systems. The single-line diagram does not always show all of the details of the system, so that an uncluttered idea of the system is presented. In the case of a piping system single-line diagram, the different pipe diameters may be shown by different thicknesses of lines and the diagram may be drawn from an isometric or three-quarters view. In electrical single-line diagrams, it is normal to show only one phase for the purpose of better clarity.

In schematic diagrams, symbols are used to represent electrical devices and their connections are shown by straight lines. According to the number of lines between the symbols, the diagrams are classified as single-line diagrams or three-line or complete diagrams. Regardless of the actual number of wires connecting the components of a circuit, the single-line diagram uses a single line between the device symbols. The complete diagram shows the actual number of wires between any two parts of the circuit. If three wires are actually used between many elements, the diagram is known as a three-line diagram (Fig. 34.1).

The symbols used in electrical diagrams are the generally accepted standard symbols for the individual devices. Diagrams may, however, make use of symbols used by individual manufacturers or industries which will clearly indicate the intended meaning of the diagrams. Electrical diagrams are always shown in a deenergized or power-off condition. For example, a relay shown in the diagram in a closed position would normally be open when the circuit is energized.

In diagrams of pipe systems, the devices are shown in their normal operating position, but variations may be encountered. Again, "caution" is the watchword.

31.11 SCHEMATIC DIAGRAM SYMBOLS

All components or devices shown in an electrical schematic diagram are numbered. A table of standard devices and functions used in switchgear and electrical control gear is given in Sec. 36.2 and a similar table of pipe symbols is given in Sec. 30.3.

31.12 WIRING DIAGRAMS

Wiring diagrams also are in wide use, but they are not strictly schematic diagrams. In addition to symbols and straight lines, wiring diagrams show simplified drawings of the devices and indicate physical locations and actual connections of wires and equipment. They are sometimes referred to as *connection diagrams*, *equipment diagrams*, or *panel diagrams* according to their application.

31.13 AUDIT OF DRAWINGS AND MANUALS

The superintendent will find it very well worthwhile to catalog all of the drawings, blueprints, manuals, instruction books, and other printed matter supplied to the plant for guidance and training purposes by vendors and constructors. After the catalog has been compiled, the actual presence of the material must be verified. If necessary, supplement duplicates or replacements will be obtained. Thereafter, the entire information stack should be maintained thereafter in a referrable condition accessible to all plant staff on an as-needed basis. The widest possible use should be encouraged, and every assistance should be given to personnel who indicate an interest in improving their knowledge or who need information to perform their jobs.

The catalog should be devised to show mechanical and electrical sections. The manuals should be listed by OEM, and the drawings should be in numerical order by section of interest: raw water pipe system, plant building, structural steel work, feeder breaker wiring diagram, and so on. All too often, it happens that a key drawing is missing when it is needed most. Similarly, it can be said that it is not possible to carry out successful overhaul work on any modern machinery without the OEM manuals being at hand. At approximate annual intervals the catalog and the inventory should be compared, and both should be brought up to date. The opportunity to acquire new material must not be ignored; in recent times much excellent-quality training material has been produced by various OEMs, and much of it can be obtained without charge.

To protect the printed pages of frequently used manuals, transparent covers in standard page sizes are available from office supply stores. Their use will go a long way toward protecting the pages from the inevitable oily finger marks. Soiling of the written material is a natural hazard in any industrial use, but it must not be an excuse for withholding the data from people on the operating floor. A thousand dollars provides a lot of manuals, instructions, drawings, and copy services, which are far cheaper than the average breakdown so often brought about through ignorance of the plant and its equipment.

CHAPTER 32
ENGINE OVERHAUL AND MAINTENANCE ROUTINES

32.1 INTRODUCTION

Within the diesel power plant owner's organization, there has to be provision for the care of all the equipment. In many cases, this care is the responsibility of the individual plant superintendent; in others, there is a central bureau for work planning and implementation. The technology of today impels a trend toward computer-assisted upkeep programming, but many programs are still handwritten. It matters not which method is used for programming a maintenance and overhaul plan; the basic elements that contribute to a successful plan remain the same. However, a well-prepared computerized scheme will be able to handle and process a great deal more information and feedback than a handwritten system can with the same amount of effort.

If a maintenance and overhaul program is to be effective in all respects, certain guiding principles must be recognized and applied. The principles as written are applicable to any upkeep program, be it for garbage trucks, rock crushers, candy wrapping machines, or diesel engines, and the avoidance or degradation of any one of the principles outlined will have a stultifying effect upon the others.

A maintenance scheme that has a clarity of purpose will have only two basic activities:

1. Preventive maintenance
2. Planned overhaul

Many so-called maintenance planners include "breakdown maintenance" in their schemes. Since breakdowns are quite unplanned, it can be seen that such efforts are illogical and not subject to any form of control.

The whole purpose of a planned maintenance scheme is to initiate and control the routines of preventive maintenance and scheduled overhaul work for a given roster of machinery. In an ideal plan that is backed by a first-class labor force and is fully financed and free of all snags, all breakdowns and unscheduled events are eliminated. In a less than 100 percent ideal scheme, the amount of breakdown or repair work is in inverse proportion to the effectiveness of the applied plan. A 100 percent performance is virtually unattainable, but it is the only possible target at which to aim.

32.2 PLANNED MAINTENANCE DEFINITION

The term *planned maintenance* can be defined as a pattern of upkeep work on equipment and facilities within a controlled system performed by a suitably organized work force. The principal objective of such organization is to prevent unscheduled incidents involving the plant machinery from occurring. It follows that if preventive maintenance work is to be carried out successfully and with continuity, scope, timing, and cost accounting must be carefully planned.

The use of the terms of *planned maintenance and preventive maintenance* is acceptable provided the two are written in full and that their definitions are clearly understood. Use of the initials PM must be avoided at all times if confusion is to be avoided. To illustrate the point: if a PM (planned maintenance) program is not adequate, the PM (preventive maintenance) work will not forestall an incident, and a PM (post mortem) discussion will be necessary later in the PM (afternoon) at which PM (phase modulation), the PM (permanent magnet), or any other PM (pure magic) can be considered. By using the proper definition, planned maintenance can be discussed as an activity, and its governing principles can be considered.

32.3 PRINCIPLES OF PLANNED MAINTENANCE

Principle 1 There Must Be Clarity of Purpose

The cardinal principle of a maintenance scheme must be clarity of purpose. Without clear knowledge and full understanding at all levels and in all sectors of the organization of what is to be done and who is to do it, there can be no predictable outcome. It must be possible to plot progress in meaningful terms; it must be possible to forecast the attainability of the target with considerable accuracy. That accuracy will be in terms of manpower, material, money, methods, and time. An essential part of the clear knowledge and understanding requirement is acceptance and use of a universal terminology. Maintenance schemes in different industries may adopt different terms for similar activities, components, tools, or materials. In any one industry, however, there must be consistent terminology (Chap. 42).

To ensure the work gets done according to plan, the scope of work included in the maintenance or overhaul schedule must always be defined in detail. The parts to be used or changed, the materials necessary, the special tools and equipment, the type, class, and number of the work force, and the time available for the job must be stated in terms that are fully comprehensible to the man on the spot. It matters not how the terms were interpreted by the person who has just left or did the job last. There is an essential need for clearcut guidance toward implementation of policy and procedure. If that is provided without variance, work will progress smoothly, the development of personnel attitude defects will be inhibited, and workplace morale will be high.

Principle 2 There Must Be Authority Delegation

Authority delegation implies the distribution of directive responsibility to one or more persons for particular segments of the work that is to be undertaken. The delegatory instructions must be in clear terms; the delegator must be quite satisfied that his delegate can carry out the intended task; and all the peripheral resources to complete the work must be available.

In creating a maintenance schedule that is to be applied to a plant's equipment, a great deal of maintenance work will be formally defined. The maintenance instructions must be sufficiently clear that they will not require interpretation to benefit both the headquarters and the plant organization. It is essential that the workload contained in the maintenance schedule be recognized as necessary by the headquarters, and in the delegation to the plant staff there must be no doubt as to how the work will be handled and who will be in charge. There is always the likelihood that some of the work requirement has not been detailed in the maintenance schedule, so someone, preferably the plant superintendent, must be authorized to proceed with the work by using his own discretion.

In such cases, the schedules should thereafter be reconsidered and perhaps modified in detail. The "do as you think fit" type of directive can never be a useful part of a permanent system, nor can the information feedback from a situation based on such a directive be used for future forecasting purposes. The "do as you see fit" directive can be used only in extraordinary circumstances. In ordinary circumstances, it makes the superintendent responsible for maintenance planning, which is probably not his job.

When more than one delegate is employed in the attainment of the planned objective, all the delegates must be aware of the scope of each other's activities and the boundaries of such individually assigned activities must be respected in a cooperative spirit by all the participants. By that means, much confusion through duplication of command or effort will be avoided.

Principle 3   There Must Be Work Motivation

Work motivation must be present before an individual will do anything useful and in conformity to a given system. Good work motivation is created through the exercise of first-class leadership, delegation of authority as appropriate, clarity of instruction, positive recognition of efforts, and presentation of attainable targets. If those factors are observed, morale will be good. Work motivation cannot exist in an atmosphere of poor morale.

The motives to work are numerous; the behaviorist will insist that the money paid to the worker is not the first and foremost reason for a person working well. Many times in the past it has been shown that time off, free meals, bonus payments, extra overtime, and other one-shot incentives are gimmicks that have no permanency or continuity and make a zero contribution to productivity. A good comparable wage, clean working conditions, communicative supervisors, trained ability to perform the work, and logical objectivity of task assignment, together with recognition of efforts, contribute strongly to an ongoing motivation.

Principle 4   There Must Be an Economy of Effort

In this particular context, the unit of economy is manpower. In a properly organized scheme, the right type of person must be utilized either to do more work in less time or produce a superior quality of work within the circumstances of the job in hand. Economy of effort cannot necessarily be obtained just by assigning the personnel already employed to the tasks they do best. The demands of the work to be done must be continuously reviewed, and the strongest efforts must be made to recruit the personnel who can not only do what is needed but also effect improvement in quality, quantity, and method.

In the generally accepted sense, on-the-job training schemes are good only for situations in which the scope of ability needed is very limited and the really desirable type of labor cannot be hired. Labor trained for limited purposes will come from the ranks of people with limited natural and vocational abilities, and their efforts will have only limited productivity and economy. It is most important that the people introduced to a maintenance scheme bring with them superior abilities and contribute to an enhancement of the established system.

The actual work of preventive maintenance is very much the same as that of defect rectification. It is carried out by the same kind of person and, all too often, competes with other task demands calling for time, money, effort, and skill. It is quite reasonable that maintenance and repair work should generally be handled by the same personnel, but there may be very valid circumstances in which such an arrangement can be varied. Those circumstances must be stated before the occasion arrives so the work can proceed smoothly.

The job supervisor must identify the segments of the intended work that are best suited to the individual worker and assign the various tasks accordingly. Bearing in mind the attitude of each worker, the supervisor must at all times keep the priorities of the planned work intact.

Principle 5   There Must Be Provision of Information

The provision of information is of paramount importance in a maintenance system. The information demanded by the planner must be forthcoming, and the planner must ensure that the maintainer is properly advised of what is to be done. Verbal information is fragile: Any word that is to have any permanency must be written down.

The information flow within a maintenance system must be multidirectional, but the channels used must be constant, consistent, and without by-pass. The information flow in any direction must have purpose, and in every possible case there must be no duplication of information. At the same time, any information passed must be believable.

The information demands of the planner should be explained to all personnel and the reasons why the information is needed should be given. Otherwise, there will be a reluctance to provide the desired data and the information will be less than complete. The bulk of the plant information feedback is of a comparative nature; being so, it lends itself to the plotting of trends which in turn become viable forecasts. The maintenance system cannot be operated without information. The information mainly becomes history, and history becomes the instrument for measuring progress, the forecaster of events, and the teacher of what the system must be.

In the beginning, what is to be included in the information flow needed by the planner can be limited to simply providing a starting point for a developable scheme. The information banking and recording methods must be so set up that they will not inhibit future expansion in the scope of work if necessary. Maintenance schedules are devised from the obtained information, and such schedules state the when, where, what, and who of preventable work for the future.

The provision of adequate and proper information is a major contribution to the success of any maintenance scheme, but the provided information must be accepted by those to whom it is directed. That acceptance is an important component of the commitment outlined in Principle 6. If the provided information is not readily acceptable, the reasons for the resistance to its acceptance must be investigated and recognized. Any necessary rectification must be done before the overall scheme can work effectively. Causes of rejection are frequently found to be rooted in conflicting instructions or demands from more than one information collector.

A review of the annual overhaul costs and the unplanned downtime and repair costs will show the well-managed plant in a better light than the plant in which there is a lack of policy and procedure enforcement. When before-and-after comparisons are carried out in plants that at one time were lax in policy and procedure adherence, it can be shown that, when reorganized planned maintenance workloads are easily absorbed, most improvemental activity can be implemented without difficulty or an increase in the work force. All this by a work force that was formerly fully occupied in scuttling from one emergency to another as a normal course of events.

Principle 6   There Must Be Commitment

For an equipment upkeep program to be successful, the commitment of all the people involved has to be established. All the people involved in a maintenance scheme must be committed to achieving the targets and requirements of the established scheme. They must be pledged to learn, practice, and enforce the accepted principles and methods. There must be no variance from the governing policies and procedures as set out by the owner and his planners.

The commitment will begin when the engine is first selected and the selectors take into account all the upkeep and operational ramifications attendant on their choice. The chosen equipment may justify the continuous presence of operating staff; on the other hand, it may be suitable for long periods of unattended operation. Its overhaul needs may require highly skilled and specialized mechanics, or it may be of a type demanding less skill in its care. The man-hour needs can be very variable from one make of equipment to another. In broad terms, it is fair to say that the most fuel efficient engines require a greater input of skill and man-hours than the less efficient types.

Whatever engine is installed, the replacement parts position must be examined by the owner and the plant superintendent. If it is found necessary to stock a range of parts because of normally extended delivery, as is often the case with larger engines, there must be fiscal commitments on a continuing basis to that end. There must also be a plan and a commitment to the monetary demands of the long-term overhaul schedules.

There must be a commitment on the part of the station personnel to be on hand to participate in the overhaul process according to the plan devised by the owner and his superintendent. If the commitments are apparent to all, the work planning will be successful and the targets will be achievable. Without the commitments being visible, it is unlikely that the plant will have an economic, successful, and continuous work program.

A maintenance scheme that permits individual variations will be ill thought out, poorly controlled, and headed for trouble. If a deviation appears to be appropriate, the situation must be examined to see if the deviation should be developed as a modification within the entire system. If it is found to be so, the whole system must be advised of the change along with the reasons, but the commitment remains unaltered.

32.4 GENERATOR SET UPKEEP RECORD KEEPING

A continuous record-keeping system and standard informational flow advises all concerned parties of the status and progress of machines and events. The extent of the records kept and the actions engendered by the records will generally be in inverse proportion to the equipment failure rates. All of the information recorded must be set down in a manner that will permit it to be used in a comparative and cumulative mode, and it must also be so arranged that the information is easily retrievable. The information stack developed will be so arranged that segments of it can be perused on an as-needed basis by the superintendent or the planner. Any trends will be observed and analyzed as they become evident. Failure origins will be detected and suitable remedies will be developed. Subsequent adjustments to the planned upkeep procedures will be made if necessary.

With adequately comprehensive records, costs can be monitored, regulated, or estimated, replacement equipment or parts can be accurately selected, and the planner will have no difficulty in preparing an engine care program that is defensible. There cannot be an economic and effective equipment maintenance scheme without complete records to back up the management planning and execution of an upkeep regime.

The data-logging efforts of the operating and maintaining personnel are essential to the overall scheme of upkeep. The operator who writes the log and inserts comment is in fact providing vital information for the maintenance planner. The information he provides will be a mixture of statistical, circumstantial, and requesting notations. The journeymen who record their activities in their respective workbooks are creating the written work history upon which the planner bases his schemes and their modifications.

A large amount of clerical work is involved in the processing of data related to a comprehensive engine care system. It is recommended that the planner and the superintendent impose only the minimum amount of paperwork on the journeymen while ensuring that all of the reports required of the journeymen are actually completed and reviewed in good time.

All of the information recorded must be set down in a format that will permit its use in either a comparative or cumulative manner, and it must also be retrievable so that trends can be perceived. Cost records will allow estimates to be developed and expenditures regulated after the work has commenced. Records of the equipment component changeouts such as governors, fuel injection gears, and turbochargers will enable spares and replacements to be obtained quickly and accurately.

The person who has overall control of the generating machinery operation may fill an individual position, or he may be the lead operator, the plant chief, or the superintendent, depending on the size and complexity of the plant and its owning organization. He may wish to dispatch or stand down a particular generating unit in accordance with demand, work plans, or other schedules. His operational decisions as to the best equipment to use or to service are based on plans developed from the basic data sent up from the operating floor.

By using the data provided in the logs and the work records, it is perfectly feasible to devise running and dispatch schedules for the various machines so that individual units become due for planned attention and downtime at convenient times and without hindering the plant in its efforts to maintain a first-class service to its consumers.

The superintendent must make an unceasing effort to get the operating and maintenance staff to record any difficulties that they may have in carrying out their routines so as to be able to identify the real reasons why a particular thing is not being done. Often, the neglect or avoidance of work details originates in the lack of training or direction given to the individual. Whatever the cause of the problem, the superintendent must have the facility to interpret and rectify the situation.

Traditionally, machine-operating personnel view the collection and recording of plant data of any description as an onerous burden, and they often form the impression that such requirements are meant to police their activities and gauge their competence. The superintendent and the planner must dispel those ideas and explain to the operators the value and essential nature of their work. The well-informed operator will always provide better reports and figures than the poor misinformed operator who is just prodded into the collection of a bunch of meaningless numbers without any obvious reason.

It is realized that a minor amount of clerical work is demanded of the operator, but that is part of the job. It is also recognized that the maintenance journeyman must do some written work, and it is recommended that the superintendent and the planner impose only the minimum amount of paperwork. At the same time, they must ensure that the required reportage is fully completed and reviewed on time. In the last decade of the twentieth century, all journeymen must be expected to do a certain amount of reading and writing as part of their calling.

When a conscientious and ongoing effort is made to collect and correlate a mass of necessary data during the life cycles of a group of similar machines, it has been proved possible to develop a viable overhaul plan and schedule. Such a schedule will be economic, and after some adjustment it will permit the machines to operate without any discernible pattern of failure. That may sound a bit farfetched, but numerous owners have found it not only to be the case but also to be greatly to their benefit. The organization that has many pieces of similar-purpose equipment will find that the great quantities of data available will provide a comprehensive upkeep plan sooner than in a small organization. On the other hand, the small company will find that it has no monopoly of equipment difficulties and that its problems are not much different from those of other companies in the same line of business, given that the operators and maintainers are of similar caliber. In developing the overhaul and maintenance plans, the planner will find the experience of other users to be both a guide and yardstick.

A service record of each piece of equipment should be kept. This record will enable notations to be made at predetermined intervals (usually weekly or monthly), and it will encompass such details as hours operated, extraordinary parts (other than overhaul parts) used, overhauls performed, repairs carried out, modifications made, and related information. A study of the failures recorded over a period of time will often reveal a pattern, and that pattern can be eradicated. It may apply to only a particular plant, an individual unit, or one or more people who handle the machines. In such cases, the remedy is frequently self-evident. Often, however, there are several facets to the problem, and more than one remedial step may be needed to eliminate the true causes of difficulty.

Other records will show the fuel efficiency of individual units and plants, lubricating oil consumption, vibration levels, lubricating oil analysis and similar figures as the performance numbers for a given period of time and also as comparable numbers for trend identification purposes. The work record will show details of all maintenance overhaul and repair work, and it will contain notes on any abnormal parts usage. Included in the work record will be wear readings obtained during the overhaul process. These figures must be set down in a comparable mode so that wear rates can be established (Sec. 46.8).

It will be useful if the man-hours used in all of the engine care activities are noted so that manpower appropriations for future work can be better gauged. In the section of records dealing with spare parts procurement, the entries should be so arranged that extraordinary parts requirements that become repetitious are highlighted. By intelligent use of a comprehensive record, forecasts and projections of reasonable accuracy and credibility can be made to the benefit of the owner, plant, and consumer.

Planning and forecasting go hand-in-hand, and both rely heavily on the plant's recorded data. The quality of the forecast can be only as good as the recorded engine history. Superior forecasting, planning, and scheduling of overhaul work yield many important advantages. If the forecasting is good, the parts requirements will be met without any delays being inflicted on the work schedule. Again, if the parts forecast and procurement are good, there will be an absolute minimum of surplus parts on hand by the time the overhaul is complete.

No attempt can be made to build a work plan that has the activity of another, separately controlled plan as one of its components.

32.5 BASIC ELEMENTS OF A PLANNED AND WRITTEN PROCEDURE FOR EQUIPMENT CARE

For an engine care system to work, based upon the previously advocated principles, numerous tasks and duties must be performed by the personnel charged with maintenance and overhaul planning. Although maintenance and overhaul work have rather different objectives, the planning and scheduling of both activities must be correlated and so arranged that service to the power consumer is not impaired. The frequency, scope, and effectiveness of maintenance work can have a marked effect on the timing and extent of the overhaul work on the same machine. The work done in preventive maintenance or an overhaul must be totally necessary; If it is not, the scheduling and planning can only be deemed inefficient and uneconomic. Several basic elements are essential to the conduct of a workable planned maintenance scheme.

1. *Scheduled maintenance and overhaul* of the machinery and equipment in the most cost-efficient manner possible with minimum downtime and without associated stoppage incidents. Any engine, or any other machine for that matter, can be successfully and economically kept in good serviceable order for long and predictable time spans provided that it is well maintained and periodically overhauled. The cycles of maintenance and overhaul applied to any particular engine will be extremely variable and dependent upon the nature of engine service, the quality of the fuel used, and the stated requirements of the owners.

The nature of overhaul is quite different from that of preventive maintenance work: Whereas preventive maintenance work routines serve to keep the engine in good running order on a day-to-day basis, periodic overhaul is the time when wearing parts that have reached the end of safe working life are renewed, reworked, or replaced. Partial or total dismantling of the engine may be necessary, and it will be dependent on the running hours accumulated and the extent of the work planned. Internal inspections are carried out; wear readings are recorded; and, following assembly, the engine is fit for a further period of reliable service. After a major overhaul, the engine will be virtually as good as new. During its life, an engine may go through several overhaul cycles.

Major overhaul work may be carried out by the OEM representative. In that case, quality control is maintained and the owner's warranty position is solidified. If the owner's personnel also participate in the work, a measure of on-the-job training is gained by the owner's employees.

2. *Coordinated planning* is carried out on behalf of one or more plants by a central planning bureau in conjunction with OEMs, dealers, contractors, company management, and plant superintendents following standard practices as laid down by the owner.

The scope of the duties devolving upon the engine upkeep planner will differ greatly from one organization to another. The person charged with the maintenance and overhaul planning may only look after the machinery and equipment of one plant. In that case, his task will be relatively simple and straightforward; it is probable that he will have several other duties. In an organization with several generating plants, it is not uncommon to find one or more people who devote their energies exclusively and directly to planning the upkeep of plant machinery.

Although all OEMs provide maintenance manuals for use in conjunction with their equipment, many owners prefer to compile an overhaul program that fits better into their corporate needs. Engine utilization factors, budgetary requirements, climatic limitations, and other factors will pressure the development of overhaul programs that are perfectly adequate but are peculiar to an individual company. In the case of owners with a large array of equipment, it is quite possible to identify the wear and failure rates of specific components in some detail. That enables schedules to be developed for timely component changeout with considerable economic advantage. In the smaller companies the formation of history relating to the machinery will not accumulate so quickly, but consultation with owners of similar equipment will provide valuable support data.

Whatever the size of the operating company, very workable overhaul schedules and routines may be worked out from a mixture of past experience, OEM recommendations, and individual engine histories. The schedules may be revised from time to time as gained experience indicates, operational conditions change, and fuel quality alters. The OEM recommendations do tend to be weighted in favor of the parts supplier rather than the owner's natural frugality.

3. *Overall procedural guidelines* in which the detailed work requirements, quality controls, acceptance levels, and reporting procedures are stated. The procedures must be so arranged that they will not inhibit future expansion in the scope of work. Many owners and their planners rely on OEM recommendations for the timing, content, and frequency of equipment overhaul routines, and these recommendations are usually found in the manual supplied with the machine. Quite frequently, the manufacturer will issue supplementary instructions or bulletins calling for modifications of, additions to, deletions from, or other changes in the original instructions. It is important that the planner give these notices due

regard in the preparation of his procedures, and that the same notices are circulated to all the plant personnel concerned.

32.6 PREVENTIVE MAINTENANCE: PURPOSE AND SCOPE

The maintenance routines on large and small generating sets fulfill the same purposes, and the range of the work is not greatly different. Nevertheless, there are considerable differences in the amounts of work that must be done. In general terms, it must be said that the larger units demand a much higher and broader degree of skill on the part of the operator, maintainer, overhauler, and supervisor.

It will be necessary for a planner to develop, refine, and issue written procedures for the governing of equipment operations, maintenance, overhaul, and other specific works in accordance with the owner's policies and directives. The procedures devised, together with any subsequent revisions thereto, must be communicated to all the parties concerned in the equipment upkeep program.

The planner should make every effort to prepare the guiding procedures so they have a common application throughout the owner's organization. They should be written in a format that makes it easy for any employee or contractor to become familiar with the procedural requirements and follow them as he moves from one job or workplace to another within the ogranization. Such people need to know with accuracy and certainty what has already been done and what has to be done. The terms of the procedures should be so written that they are not open to interpretation or speculation.

Preventive maintenance is part of the overall planned maintenance scheme; it serves to keep the engine, together with its attached or auxiliary machinery, in an efficient and operable condition so that it can give continuous and reliable service between each overhaul. All machines must be subject to maintenance, and the scope of this activity is usually outlined in the OEM manuals. The actual maintenance work done is at the discretion of the maintenance planner in conjunction with the station superintendent. The work finally done may be more than suggested or it may be less, but anything that is done must be timely, necessary, and correct.

Regular maintenance provides the opportunity to carry out the eyeball inspection that is of so much signficance in maintaining the reliability and integrity of the machine and its protection system. Properly scheduled and reported inspection activities contribute greatly to keeping the incidence of equipment failures in check. An analysis of engine failure rates that compares the frequency of difficulties experienced in one plant with that experienced in another will usually show an inverse performance in inspection, activity, and reportage.

The basic tasks of maintenance on diesel generator sets include sampling lubricating oil for analysis, monitoring the unit's vibration, examining filter elements following routine servicing, cleaning the cooling system radiators, verifying the protective devices, and adjusting V-belt tensions. Much of this maintenance work can be performed by the operating personnel if they have been properly trained. Before anyone is directed to carry out such maintenance work, he must be instructed in how to work safely, how to recognize and report the unusual, and why the work he does is so important. It is recommended that all scheduled maintenance activity be written out in detail and be posted where all participants can see the requirements.

Typical Maintenance Schedule for a 300-kW Unit

An example of a typical maintenance routine for a 300-kW generator set in continuous service is outlined below. It is expected that most of the work will be done by a plant operator.

Daily or Every 8 h, Whichever Occurs First

1. Inspect the lubricating oil dipstick and take notice of the appearance of the oil.
2. Inspect the air filter element if it is visible. Note the position of the indicator.
3. Inspect the fuel tank level, check for leaks, drain the water, and read the fuel meter.
4. Inspect the governor oil level; add makeup oil as needed.
5. Record the kilowatt-hour meter reading and, in conjunction with the fuel meter reading, establish the kilowatt per gallon, or per liter, performance for the preceding period.
6. Wipe down the entire machine while having full and due regard for safety. Note any defects.
7. Drain any water from the fuel-water separator.
8. Clean the drip trays.
9. Check the oil level in the pedestal bearing.
10. Check the coolant level and color when possible.
11. Peform a walk-around inspection of the entire plant including the fuel system, cooling system inside and outside, waste-heat recovery system, ventilation system, exhaust system, switchboards, transformer yard, and outbuildings. Record all noticed defects.
12. Perform standard housekeeping duties.
13. Write up the log sheet. Note any incidents.

Weekly or Every 100 h, Whichever Occurs First

1. Check the electrolyte level in batteries.
2. Drain water and sediment from fuel filter housings.
3. Take and record vibration readings from set points.
4. Exercise all standing units.

Biweekly or Every 250 h, Whichever Occurs First

1. Change the lubricating oil, change the lubricating oil filters, and clean out the filter housing. (The frequency of these activities may be longer or shorter, but it must be in accordance with the OEM manual.) Examine the lubricating oil filter element fabric for metal presence.

2. Examine the air filter element.
3. Clean the crankcase breather element.
4. Clean the top and terminals of the batteries.
5. Check the specific gravity of the batteries.
6. Check cleanliness of the radiator.
7. Check the tension and condition of fan belts. (For this and item 6, the engine must be stopped).

Monthly or Every 500 h, Whichever Occurs First

1. Check the pH of the coolant. Adjust as necessary.
2. Grease the fan hub bearings, alternator bearings, and tachometer drive.
3. Inspect the generator and exciter brushes. Record sparking and condition.

Bimonthly or Every 1000 h, Whichever Occurs First. At this point, it will be seen that the maintenance duties begin to be shared between the operation staff and the journeymen.

1. Take a sample of lubricating oil for analysis.
2. Lubricate the governor speeder motor.
3. Test all unit protection devices. (J,O)[1]
4. Renew worn generator brushes and dress the commutator.(J)
5. Clean the slip rings.(J)

Biannually or Every 2500 h, Whichever Occurs First

1. Adjust valve lash.(J)
2. Check valve rotation when applicable.(J)
3. Clean out the bottom of the sump.

Every 5000 h

1. Carry out the scheduled overhaul of engine.(J)
2. Carry out scheduled overhaul of alternator.(J)
3. Carry out scheduled overhaul of auxiliary equipment.(J)
4. Check alignment.(J)
5. Test all equipment and prove its ability to perform in accordance with specifications.(J,O)

Typical Maintenance Schedule for a 1.0-mW Unit

For a larger generation unit of about 1.0 MW or greater, the maintenance work will not be very much different from that of a smaller unit. It is probable, how-

---

[1]Work requiring the presence of both a journeyman and the operator.

ever, that an operator will be continuously present when the unit is running and the volume of work will be greater.

Every Hour

Inspect the lubricating oil dipstick. Take notice of the appearance of the oil. Check the lubricating oil levels on the governor, turbocharger(s), camshaft, or any other component with its own independent lubricating oil system. Make at least one full turn on each self-cleaning filter. Write up the log sheet. Inspect the coolant sight glass for evidence of contamination.

Every Shift or Every 8 h, Whichever Occurs First

1. Inspect the air filter element if it is visible. Note the position of the differential pressure indicator.
2. Inspect the daily service fuel tank. Look for leaks, spillage, or water.
3. Take vibration readings from set points and record them in the log.
4. Drain the water from the fuel-water separators.
5. Drain any water from the compressor water separators.
6. Wipe down the assigned equipment while working with due regard for safety. Log any defects noted and perform all other designated housekeeping duties.
7. Clean all drip trays.
8. Check the coolant level when possible.
9. Perform a walk-around inspection of the entire plant including the bulk fuel system, the primary and secondary cooling systems, and the waste-heat recovery equipment, transformer yard, and storage facilities whether internal or external to the main plant buildings.
10. Write up the log sheet.

Weekly or Every 100 h, Whichever Occurs First

1. Check the electrolyte level in the batteries. Clean the batteries as required.
2. Check the cleanliness of the radiator and the tension of fan belts of all standing engines.
3. Drain the water and sediment from fuel filter housings on standing engines.
4. Exercise all standing engines.
5. Lubricate governor linkages.
6. Clean magnetic plugs when fitted. Enter comments in the logbook.

Monthly or Every 500 h, Whichever Occurs First

1. Check the pH of the coolant and adjust it as necessary.
2. Grease fan hubs, motor bearings, linkages, and so on, as described in posted instructions.
3. Change the oil in the oil bath air filters. (Frequency will vary with the local environment conditions.)
4. Change and wash oil-wetted air filter elements.

5. Inspect and service brushes, slip rings, and commutators of standing machinery.(J)
6. Adjust tappet clearances of standing engines.(J)
7. Change lubricating oil in turbochargers.(J)
8. Check condition and adjust compressor drive belts as necessary.

Bimonthly or Every 1000 h, Whichever Occurs First

1. Take a sample of lubricating oil for analysis.
2. Test all engine protection devices.(J,O)
3. Take a set of crankshaft deflections.(J)
4. Change lubricating oil and lubricating oil filter elements, clean filter housings including self-clean filters. Clean out sump. (Oil change frequency should be in accordance with OEM recommendation.)
5. Clean engine and turbocharger breather.
6. Perform planned segments of overhaul cycle.(J)

32.7 OVERHAUL PURPOSE

The objective of machinery overhaul is to enable a piece of equipment to operate for a definite period with optimum reliability, economy, and efficiency. All of the objectives of engine overhaul can be reached if the owner, superintendent, maintainers, and operators act in concert with a uniform desire.

The effectiveness of any overhaul can be judged by comparing the cost over a complete cycle, the machine's reliability and efficiency performance during the cycle, and the required downtime in relation to similar information from other machines. The effectiveness will be measured in terms of the overall cost per kilowatt-hour of capacity operated. The cost per kilowatt-hour produced will not be a valid comparison figure because engine wear is not directly proportional to the loads carried. However, the production cost per kilowatt-hour is the figure that would be monitored for overall plant performance.

The nature of overhaul work is quite different from that of maintenance work. Whereas maintenance routines serve to keep the engine in day-to-day running order, the periodic overhaul is the time when wearing parts that have reached the end of their safe working life are renewed. Partial or total dismantling of the engine may be necessary depending on the hours accumulated and the extent of the work planned. Internal inspections are carried out; wear readings are recorded; and, following reassembly, the engine is fit for a further period of reliable service. During its life, an engine will probably go through several complete overhaul cycles.

The parts changeout schedules may be devised for any engine model, but they must be based on past operating experience, OEM recommendations, and individual engine histories. These changeout schedules may be revised from time to time as the gained experience dictates, operational conditions change and fuel quantity alters. Individual owners may find the experience of other engine owners useful in determining the safe and economic life of components; OEM recommendations tend to favor the parts supplier rather than the owner's economic desires.

Major overhaul work may be carried out by an OEM representative. In that case, quality control is maintained and the owner's warranty position is enhanced. If the owner's personnel also participate in the work, a measure of on-the-job training is gained by the owner's employees. All overhaul work must be conducted in accordance with the manufacturer specifications as laid down in the service manuals and bulletins. The well-planned overhaul will permit the personnel and parts to be on site at just the right time, and the period that the engine to be overhauled will be out of service and without inconveniencing the dispatch schedules and needs will be minimized.

In many diesel engine plants, it is the practice to keep comprehensive records of the work done along with measurements taken from the wear surfaces. By plotting the wear against the hours run by the respective engines, it is feasible to develop quite accurate forecasts of future parts requirements and budgetary needs. Similarly, good record keeping will also provide the information necessary to forestall failure of components. It can be said that the budgetary and practical efficiency of an overhaul cannot be any better than the effectiveness of the record-keeping system. It is also true that an effective, disciplined, and economic overhaul program can virtually eliminate any engine failure caused by normal component wear.

The overhaul cycle of a machine usually includes several overhaul exercises which may vary quite widely in content but will be done at least once or repeated at some point in each successive cycle. The cycle for a typical 1200-rpm 4-cycle 500-kW generator set will be described in detail as an example, but the same format can be applied to any other generator set of between, say, 30 and 1500 kW. Of course, there would be detailed adjustments to cater to for the different marques and models and there would be a shorter cycle for higher-speed engines. But in spite of the many different makes, models, and sizes of diesel generator sets available worldwide, there should not be any great difference between the overhaul needs of one engine over another, and their costs in terms of kilowatts per hour of capacity operated should match quite closely in the final analysis.

The geographic location, type of service, grades of fuel, and standards of attention have profound effects on the engine's wear rates; therefore, the overhaul plan devised for an engine working in polar areas may not be suited to an engine in a tropical area.

By drawing upon previously acquired experience together with the experience of other users in similar environments and taking heed of the OEM recommendations, a basic starting plan for an engine's future care can be written. If good comprehensive records are kept, wear measurements are collected whenever they become obtainable, and the machine performance is reviewed almost continuously, a good understanding of the overhaul requirements for the future may be evolved. The overhaul plan that remains always unaltered is probably not as economic as it could be, whereas the plan that suffers radical, frequent, or abrupt changes is apt to bring trouble in its wake. The trouble may be of either a budgetary or a reliability nature.

32.8 OVERHAUL CLASSIFICATIONS

The scope written into the overhaul schedules for 1200-rpm and 1800-rpm engines is not greatly different, but whereas 1200-rpm units have a 30,000-h cycle, the

1800-rpm unit has a 20,000-h cycle. Starting with a new, unused 12,000-rpm engine, overhauls are carried out at intervals of 5000 operating hours over a cycle of 30,000 operating hours. The content and cost of the overhaul work done at the end of each interval are somewhat variable, but they always culminate in a major overhaul of considerable scope and value. The overhauls done at 5000, 10,000, 20,000 and 25,000 h are usually referred to as minor overhauls; at 15,000 h the work is dubbed an intermediate overhaul; and 30,000-h work is said to be a major overhaul. Similarly, for 1800-rpm units, a 5000-h overhaul is followed by a 10,000-h overhaul (intermediate), a 15,000-h overhaul, and a 20,000-h (major) overhaul.

32.9 OVERHAUL SCHEDULES

The overhaul cycles described are typical of those that have been shown to give a generator set the best practical blend of economy and reliability. Three classes of generator sets are cited as examples, and they are typical of a large number of units in similar service within one particular organization. All the engines perform without any discernible pattern of failure, and from the performance, considerable data and experience have been derived. Each of the units operates for approximately 5000 h per year on a standard low-temperature distillate fuel. The cetane number of the fuel is above 40, and the sulfur content is below 0.2 percent. The machines are generally handled by less than fully skilled labor except when major overhauls are performed.

The work suggested in the following overhaul schedule examples 1 and 2 can be carried out successfully only as capsule jobs. There is no merit to or benefit in attempting to do any of the work on an interrupted or piecemeal basis. However, the engines featured in the third example are generally designed and so arranged that the overhaul work can be conducted on a semicontinuous basis.

The example schedules merely outline the basic requirements of each stage, and they are intended only as a guide. Individual owners and their planners will wish to devise schedules that best fit their specific equipages and utilization demands.

Example 1: 1800-rpm 2-cycle 40- to 200-kW Generator Sets

5000-h Intervals

1. Check and adjust the fuel and valve timing.
2. Check and adjust the valve clearance.
3. Change out the thermostats.
4. Change out the fan belts.
5. Change out the front and rear crankshaft seals.
6. Change out the coolant pump (exchange).
7. Change out the fuel pump (exchange).
8. Change out all flexible and nonmetallic hoses.
9. Drain, flush, and refill the governor.

10. Clean the radiator inside and outside and do the pressure test.
11. Clean the sump.
12. Change out all air, coolant, lubricating oil, and fuel filter elements.
13. Check and adjust the engine-generator alignment.

10,000-h Intervals

1. Change out all injectors (exchange).
2. Change out the fan hub bearings.
3. Service the cylinder heads. Use either exchange or existing cylinder heads depending on the on-site facilities available.
4. Change out the governor (exchange).
5. Replace the governor link ball ends.
6. Change out the blower and turbocharger (exchange).
7. Change out the alternator bearing.
8. Clean the alternator windings.
9. Perform the 5000-h schedule routine.

15,000-h Intervals

1. Perform the 5000-h schedule routine.

20,000-h Intervals (Major Overhaul)

1. Change out the pistons and piston ring sets.
2. Change out the cylinder liners.
3. Change out the connecting rods (exchange).
4. Change out all bearings.
5. Change out all pressure and temperature gauges.
6. Change out all pressure, temperature, and speed switches.
7. Change out the torsional damper.
8. Perform the 5000- and 10,000-h schedule routines.

Example 2: 1200-rpm 4-cycle 300- to 900-kW Generator Sets

5000-h Intervals

1. Check and adjust the fuel and valve timing.
2. Test and adjust injectors.
3. Test all alarm and shut-down devices.
4. Change out all nonmetallic hoses.
5. Change out all air, fuel, coolant, and lubrication filter elements.
6. Change out turboblowers (exchange).
7. Clean the radiator on both sides and pressure–test it.

8. Clean out the engine base, sump, and pan.
9. Remove governor, flush, reinstall, refill, and bleed.
10. Adjust head valve clearances and check rotation by tap test.
11. Check the engine-alternator alignment.

10,000-h Intervals

1. Change out the alternator bearings.
2. Clean the alternator and crack-test the coupling disks.
3. Realign the alternator with the engine.
4. Change out the coolant pump (exchange).
5. Change out the belt tensioner bearings.
6. Perform the 5000-h schedule routines.

15,000-h Intervals

1. Remove and service the cylinder heads (on-site rehabilitation or dealer exchange).
2. Change out the coolant thermostats.
3. Change out the injector nozzles.
4. Change out the fly ball type of overspeed switch.
5. Clean the lubricating oil cooler.
6. Clean the aftercooler.
7. Random-inspect one or more main and bottom end bearings.
8. Change out the governor link ball ends.
9. Perform the 5000-h schedule routine.

20,000-h Intervals

1. Perform the 10,000-h schedule routine.

25,000-h Intervals

1. Perform 5000-h schedule routine.

30,000-h Intervals (Major Overhaul)

1. Change out the cylinder liners.
2. Change out the pistons and piston rings.
3. Change out the lubricating oil pump (exchange).
4. Change out the connecting rods (exchange).
5. Change out all bottom end and main bearings and all rocker shaft and camshaft bearings.
6. Change out all pressure and temperature gauges.
7. Change out all pressure temperatures and speed-protection devices.

8. Change out the torsional damper (exchange if viscous).
9. Crack-test the liner counterbores and cylinder heads.
10. Perform 500-, 10,000-, and 15,000-h schedule routines.

Example 3: 600-rpm 4-cycle 3000- × 5000-kW Generator Sets

The time frames of overhaul schedules are in 3000-h terms or intervals, and the shorter time frames are dictated by the necessity to service the head valves. The complete cycle for engines of this class will be 12,000 h, culminating in a major overhaul at 12,000 h. It will be found most convenient to space the scheduled work throughout the 3000-h time span; for instance, in the example engine there are 24 exhaust valve cages each of which requires servicing at 3000-h intervals. If the first four pairs of valve cages are done over at 1000 h, another four pairs are done at 2000 h, and the final four pairs are done at 3000 h, a good work rotation will be achieved and the period of engine downtime will be minimized.

Much of the other overhaul work can be arranged to take place in a similar manner. In the plant equipped with multiple engines, it may be found possible to take one generator unit out of service for a period of many days or even weeks while the planned work is carried out. In a plant with only two or three units, it will be more difficult to arrange for a single lengthy period of downtime, and the rotating overhaul method can be utilized to better advantage.

3000-h Intervals

1. Service each exhaust valve and cage.
2. Service and qualify each injector.
3. Qualify all protective devices.
4. Clean all aftercoolers on both sides.
5. Read and record the crankshaft deflections.
6. Test-run and qualify the unit against original-acceptance test sheets.
7. Adjust timing chain if fitted.

6000-h Intervals

1. Inspect and service the tappets, rocker levers, and rocker shafts.
2. Inspect and service the camshafts, cam followers, and rollers.
3. Inspect and service the governor and fuel pump linkages.
4. Remove the cylinder heads, decarbonize, and service the air inlet and air start valves.
5. Check all valve springs for height loss.
6. Inspect, service, and time all fuel pumps.
7. Change out turbocharger bearings; clean the unit internally.
8. Inspect, service, and qualify the governor.
9. Clean both sides of the lubricating oil cooler and the jacket watercooler.
10. Inspect and qualify the thermostatic control valves.

11. Clean all areas of the crankcase, timing casing, sump, lubricating oil galleries, filter housings, and sump strainer.
12. Check and adjust the alternator-to-engine alignment; also adjust the alternator air gap if necessary.
13. Inspect the pedestal bearing.
14. Perform 3000-h routine.

9000-h Intervals

1. Perform 3000-h routine.

12,000-h Intervals

1. Perform 3000-h routine.
2. Overhaul all fuel pumps.
3. Remove, decarbonize, and rering the pistons.
4. Inspect all main and bottom end bearings.
5. Check all crankshaft bearing surfaces.
6. Inspect all bottom end bolts for stretch tolerances.
7. Inspect all timing and auxiliary drive gear teeth.
8. Deglaze and then measure the cylinder liners for wear.

15,000-h Intervals

1. Perform 3000-h routine.

18,000-h Intervals

1. Perform 3000-h routine.

21,000-h Intervals

1. Perform 3000-h routine.

24,000-h Intervals

1. Perform 12,000-h routine.
2. Megger-test the alternator and exciter windings.
3. Clean and inspect the alternator and exciter windings.

Some of the value of the aforementioned activities will be lost if the opportunity to record the wear dimensions of components is not taken.

32.10 OVERHAUL SCHEDULE VARIATIONS

The schedules outlined do not necessarily apply to any particular engine make or model; they are intended only as a guide to what is reasonable and practical. The superintendent or planner will therefore need to take into account the particular needs of the engines in each individual plant. He must, however, be mindful of the consequences of curtailing the work scope. At the same time, no part shall be changed on the basis of guesswork.

32.11 DEFERMENT OF CHANGEOUTS

When changeouts of components are recommended, the components may remain in service if their condition permits them to go through to the next cycle stage. At that time, they would again be ready for examination. Thus, for example, if the scheduled change of thermostats which would ordinarily occur at 15,000 h is deferred, the thermostats must be able to remain in the engine for another 15,000 h.

32.12 EXCHANGE COMPONENT REBUILD

When the changing out of a component and the installation of an exchange unit is suggested, there may be a case for the plant personnel to rebuild the worn one. That will be a matter of judgment for the plant superintendent or the planner, who will be able to decide if the necessary skills, man-hours, and shop facilities at the site provide genuine advantages over those provided by dealer exchange methods.

32.13 INJECTOR NOZZLE CHANGEOUT

It will be noted that there is no fixed requirement for a complete injector changeout as part of the planned routines. It is felt that with the help of modern fuel-filtering methods and with regard to the high cost of injector nozzles, there is no economic reason for taking out injector nozzles that are in good working order. Today's injector nozzles will give very extended service—in some cases up to and well beyond 30,000 h. If one injector nozzle fails, it obviously must be changed out, but there is no reason whatsoever to change out complete sets of nozzles at the same time.

32.14 UNPLANNED ACTIVITY

In any machine-operating organization there will always be a residue of unplannable activity. The only unplannable activites are those that are unpredictable, but anything that can be foreseen as possible future trouble can receive the planner's attention before it becomes a failure or a breakdown. It is an unavoidable fact that many of the incidents that result in machine failures, downtime, cost excursions, and other difficulties have their origins in either defective operating standards or work-planning shortcomings.

Broken but yet unreplaced instruments, vibration-monitoring routines that are not enforced, infrequent or skimped inspection tours by any or all personnel from the owner down through the ranks, and log sheets that are neither regularly nor completely reviewed are quite conventional examples of neglect that breed unplannable activity. Any unplanned events should, whenever possible, be turned to advantage by a thorough review of the incident causes, the remedies and the preventive actions developed for the future. Everyone learns a little from his own mistakes; everyone learns faster and more comfortably from the mistakes of others. Many incidents have an excellent teaching value. Therefore, following a breakdown, part of the cleanup should include an analysis of why the incident occurred, what should have been done to avert it, and what must be done to prevent its recurrence. The results of the analysis are set down in a cause-effect-remedy format and then circulated to the operating and maintaining personnel as worthy learning material. The finger-pointing details should be left out.

A portion of the unplannable events will have warranty or insurance claims as their aftermath, and someone, usually the plant superintendent or the planner but perhaps the company's purchasing agent, will be required to follow up the claims. In fairness to the owner and the vendor, accurate and complete records of the incident must be prepared and made available to either party when needed.

32.15 MISCELLANEOUS PLANNED WORK AT SET FREQUENCIES

All work scheduled for intervals of greater than 1000 h on larger units involves the rework or replacement of components having a definite and known life span. Although the plant operating staff may assist in it, this further work is normally handled by specialist journeymen. There are a multitude of engine care activities that must take place frequently but which cannot feasibly be written into a fixed schedule. For example:

- The air filter element should be changed when so indicated by the monitoring instrumentation or when it is visually obvious that the element is plugged. The time frame required for the filter to plug will vary with the load, the seasonal climatic conditions, and changes in local environmental conditions (e.g., adjacent dusty roads becoming muddy roads in the wet season).
- The primary fuel filter will collect more debris immediately after new fuel has been delivered. After the fuel has settled, there will be less sediment movement. The fuel-water separator will also precipitate more water after the bulk supply has been agitated.
- The secondary fuel filter elements will serve for a longer period if the fuel supply is relatively clean, and the plant operating staff must exercise their judgment accordingly.
- Governor bleeding will be necessary from time to time to eliminate any hunting or surging of engine speed. This activity can be carried out only by a journeyman mechanic who is fully conversant with the procedures.
- Battery cleaning will be necessary at intervals. The frequency is dependent largely on the care taken to mop up fluid drops when the hydrometer is withdrawn from individual cells and the batteries are topped up. Geographic location also can have an effect on the frequency. In humid areas and coastal loca-

tions, the moisture in the atmosphere will foster an increase in the deposits of lead oxide and sulfuric acid on the battery terminals.

In many plants, a continuous battle against the effects of corrosion must be fought; in other plants, the equipment does not suffer any corrosion at all but does suffer from sand and dirt invasions.

The planner will on occasion need to consult with the operational supervisors when work that will require the participation of operations personnel for starting, stopping, and test-running of plant and equipment is being planned. It is not recommended that mechanics or electricians take it upon themselves to test-run equipment when operating personnel are present and available to do so.

Although the operator on shift should not normally be included in any work plan that would involve a prolonged hands-on upkeep activity, his availability for incidental equipment running duties is desirable. In some cases it may be expedient for an additional operator to be placed on shift. Whatever happens, the standard shift operator remains responsible for all operation activity in the plant at all times.

The record-keeping system will, if it is comprehensively structured and supports easy retrievability, provide a large amount of data pertaining to the various tasks and the man-hours absorbed in their completion. All of this history is invaluable in establishing the most economical size of work force and in forecasting the manpower needed to handle an extraordinary event.

The roles of the individual participants in a work program should be well identified and described before the work commences. For example, when the maintenance personnel complete an overhaul on a machine, a certain amount of cosmetic after-work will be appropriate. The work plan must state who does the cleanup and garbage removal, the painting, and similar job finishing. All too often, this kind of work is given to one of the operators simply because he seems to be around most of the time. If the operator is to assist with upkeep work, paint, get involved in extensive cleaning efforts, or do anything additional to the standard operational duties, he should be encouraged to do so. Never, however, should that be at the expense of his normal operating routines as established by the station superintendent. Any variance must be only when the shift is under complete control.

To ensure that the skills of the in-house work force are adequate for the tasks that will be planned and assigned in the future, the planner may need to participate to some degree in the preparation of apprenticeship training schemes and upgrading courses for personnel currently employed. He may also usefully review and advise on the selection of replacement crafts people.

32.16 COMPLETENESS OF OVERHAUL

Many owners rely on OEM recommendations for the timing, content, and frequency of engine overhaul routines. Those recommendations are normally found in the manuals supplied with the engine. Quite frequently the manufacturer will issue supplementary bulletins calling for modifications of, additions to, deletions from, or other changes in the original recommendations.

Some owners, however, prefer to compile their own overhaul programs that suit their corporate needs better. Engine utilization demands, budgetary requirements, and other pressing factors can lead to the development of overhaul programs that are perfectly adequate but peculiar to one company. In the case of owners with a large array of equipment it is quite possible to identify in detail the wear and failure rates of many components and plan for their timely changeout along very economic lines much to the owner's advantage. Whatever the plan intended by him, the owner must be able to rely on the personnel assigned to the upkeep work to carry out all of the required work without omission of any item or requirement of the program. Quality of the work is essential if the work done is to be effective.

Subsequent to the completion of any work, the workbook must be written up to show the implementation of any required modification, the total work done, and, in particular, the work rescheduled and not yet done and the reason for its delay. If the details of completed work are recorded, the next person assigned to do work on the same unit at a later date will have a much better idea of what else must be done to keep the unit in first-class running order.

By continuous review of the work done (or not done) the planner is in a position to adjust the overhaul program content to the benefit of the company, the machinery, and the consumer.

32.17 REPETITIVE WORK COMPARISON

When sectors of maintenance or overhaul work are repetitive, the planner will seize the opportunity to compare previous estimates with actual costs, analyze where and why the differences occurred, and make modifications of or alterations to the plans for future work of the same kind. This will result in better costing, improved labor utilization, shorter downtime periods, timely parts supply and other associated gains. The value of the workbook becomes very apparent when the comparison attempts are made.

32.18 UNPLANNED REPETITIOUS WORK

One of the prime purposes of the OEM shop manual is to provide written descriptions of the many tasks that comprise an overhaul or repair activity. Each step of a particular job is described in some detail, together with the correct tools being named, the torque values, loadings, tensions, clearances, and tolerances. Today it is virtually impossible to perform a quality-controlled task on any equipment without multiple references to the OEM shop manuals. The superintendent and maintenance planner will regard work done without the benefit of the manual with considerable suspicion of its ultimate quality.

The mechanics and electricians chosen for overhaul work must be proved able to perform diagnostic work, deduce from worn parts and other evidence where the roots of present or future failures lie, identify defects in areas beyond the immediate planned work scope (when they become evident), and be capable of both recommending and implementing appropriate remedial action. Whatever is asked or expected of the overhauling journeyman, his work must be of such quality that the engine or machine can be totally relied upon to run for a definite planned period of time following completion of the overhaul.

Although the necessary repeated work plans can be groomed for their maximum effectiveness, a review of the tasks that recur intermittently and outside a planned routine may lead to the true identification of a problem and its elimination. A one-of-a-kind incident may carry no meaning, but if the same unplanned incident recurs, it must be investigated and the cause eliminated. It is worthwhile to review the hour intervals between such incidents, the loads being carried, and any other coincidental features, particularly if they occur on similar model engines.

32.19 QUALITY CONTROL

Each of the available labor sources has its advantages and disadvantages which will bear close examination before the overhaul work plans are finalized. Obviously, if the owner is satisfied that his permanent staff is perfectly capable of performing all the overhaul work, there is only the problem of matching the man-hours available against the forecast machine downtime to be considered.

The owner and his planner will do well to consider in some depth the merits of insurance coverage for equipment overhaul and repair. The difficulties of making a claim following a postoverhaul failure is a major encouragement to the owner to use OEM representatives or other contractors for any work involving great expense. An owner may be prepared to accept the apparently higher contracted labor charges knowing that he will gain a superior warranty coverage. These are the sort of tradeoffs that must be considered by the planner and the owner.

The real worth of a warranty is debatable and difficult to estimate, and there is no doubt that any warranty is profitless to both the vendor and the client. Claims under warranty must be properly justified by the owner and fully honored by the OEM. However, repeated warranty claims will be regarded with some suspicion by the OEM, and hence they must be properly administered by the owner.

32.20 PREPARED PARTS LISTS

In a single organization that operates a large number of generating sets, the maintenance planner can save much effort if he develops parts lists for repetitive work such as minor and intermediate scheduled overhauls. The items in the lists will exactly match the parts needs of immediately intended and future overhauls on specific engine models. In practice and after some detailed adjustments, it will be found that there are no parts left over upon completion of the work nor will the job have been delayed while minor or missing parts were delivered. In addition, it will not be necessary to incur freight charges to return unused parts to the supplier nor will restocking charges be incurred. Unused parts relegated to the owner's storeroom represent unemployable money.

The prepared parts list also serves as a prework commencement checklist that the journeyman will wish to see before he actually starts the overhaul work.

32.21 PARTS CLASSIFICATION

The parts and materials used in a maintenance program fall into four distinct classes: consumable parts, exchange parts, rebuilt parts, and replacement parts.

Consumable Parts. Those parts and materials used in the preventive maintenance routines are the consumable parts. They include paint, greases, cleaning solvents, wiping rags, fuses, light bulbs, and the like, all of which are used on an as-needed basis. Fan and other V-belts have variable life spans which are often determined by the type of service in which they work. It is not unreasonable to limit the service span of a V-belt to a maximum of 5000 h by which time the effects of heat, tension, and material fatigue will have aged the belt to a point of questionable reliability. The replacement cost is small, particularly when compared with the potential expense of a breakdown originating in a V-belt failure. Because such belts are subject to renewal as indicated by routine inspection, they are generally classed as consumables.

In some organizations, fabric or nonmetallic hoses also are classified as consumables, but in other companies they are treated as replacement parts. In any case, all nonmetallic hoses and other connections have no further utility once they have been taken off the engine and should not be retained for any other purposes. Many of the hoses used around an engine have reusable couplings or fittings, however, and all that is then needed is lengths of replacement hose of appropriate composition. Replacement hoses made up from the old connection fittings and new hoses are usually a great deal less expensive than all-new assemblies.

Exchange Parts. The parts that are removed from the engine during its overhaul and are recovered for factory rebuild are the exchange parts. It is common practice for the OEM or the dealer to take back worn parts or equipment, rehabilitate the parts to original specification, and offer them to the customer at prices considerably lower than those of new items. The warranty coverage of OEM exchange components is usually less extensive than that applied to new components, but the final cost is advantageous in most cases. Also, exchange parts are offered by jobbing suppliers other than OEMs. These rebuilds may have a parts and labor input that is below OEM specifications, but they are offered at quite low prices. They may be perfectly satisfactory, but they usually carry a minimal warranty. Furthermore, the OEM dealers contracted to do overhaul work will sometimes refuse to use them. For all normal and practical purposes, OEM exchange parts exclusively should be used for overhaul.

Rebuilt Parts. The spare components that have previously been rehabilitated and are ready for use in the rotating overhaul process are the rebuilt parts. Those removed from the engine are serviced by the resident mechanics in readiness for the next phase of the overhaul. One would expect to find perhaps one turbocharger, two cylinder heads, four exhaust valve cages, three or four fuel pumps, and a similar number of injector bodies typically available for a 2.5 to 7.5 MW 600-rpm engine, but if two similar units were installed, it would not be necessary to increase the numbers of rebuilt components available. It is anticipated that these components would be serviced by the station mechanics only.

The removed components are renovated immediately and placed in the ready-to-use rack awaiting the next use cycle. The manpower needed to work a continuous component renovation system must be taken into account when developing the work programs for the maintenance and overhaul work force. It is of the greatest importance that the components of this class be processed without delay. All too often, the delay degenerates into neglect, and neglect is well recognized as a fertilizer of tomorrow's problem.

No components for engines of the above class should be consigned to the scrap heap until their recoverability has been closely reviewed. At the same time, no rehabilitable part should be returned to the stockroom until it has been properly rebuilt or qualified as reusable by the OEM or dealer.

Replacement Parts. The parts that are installed during an overhaul and that replace all of the worn-out items which will have no further use are the replacement parts. Valve seats, gaskets, bearings, and valve guides and so on, are examples.

32.22 SPARE AND OVERHAUL PARTS

In an ideal parts procurement and material control setup no parts will be left over at the end of a given task nor will there have been any shortages during the task execution. It is always possible to come to some arrangement with the supplier for the return of unused items for credit, but good planning and scheduling will keep the return of such excess parts to a minimum.

Before any parts are purchased, their source should be investigated, because there are often cost advantages in going to the original supplier rather than to the engine manufacturer or dealer. That is particularly the case for items like turbochargers, fuel injection equipment, and other components not normally made by the engine manufacturer. The buyer must be cautious when this route is followed, for there may be certain warranty impediments attached to parts so obtained.

All of the renewal work done in an overhaul should be totally necessary and the maximum amount of service should be extracted from the parts before their replacement occurs, but it may be found expedient to change out certain parts a little earlier if, by doing so, there is an improvement in the efficiency of the total overhaul plan. An occasional review of the repair parts demand will provide useful information. Some engines will use more parts than others, and there will also be variations in the extraordinary parts requirement of individual plants. Again, patterns will be discernible and perhaps trigger suitable action. Defective rebuilds, repeatedly fitting incorrect replacement parts, improper operating practices, and poor quality control of work are commonplace causes of breakdowns. The overhaul planner is usually in the best position to recognize this sort of defect trend.

Another important area of activity in spare parts planning control is the marking up of the engine parts book together with the insertion and deletion of OEM manual service bulletins.

32.23 CONSIGNMENT STOCK PARTS

The availability of parts at the OEM equipment dealer and the lead time needed for parts acquisition must be taken into account by the planner. In most instances, parts are immediately obtainable from the dealer's stock, but for machines that are older and that have been produced in limited quantities, there may be delays of some degree in the delivery of ordered parts. Similarly, there may be some waiting period when the parts must be imported.

All diesel power plants have some spare parts on hand. Some of the parts cannot really be regarded as spares because they are more of a consumable nature.

Air, fuel, and lubricating oil filter elements, fan belts, standard fasteners, and the like fall into the consumable class and their rate of use over a given period of time is usually very predictable. Exhaustion of a stock of this class is inexcusable.

Now there are some parts, such as turbochargers, fuel pumps, cylinder heads, exhaust valve cages, governors, and similar items in the larger diesel generating facilities, which are rotated on and off the engines on an almost continuous basis and rebuilt in the plant in accordance with a well-regulated overhaul work plan.

For new installations, the OEM will often offer to provide a stock of parts on consignment. If the parts are accepted, the user must be cautious with them and ensure that each is paid for as and when it is used. Consignment stocks are not free of charge, and sooner or later the supplier will present an invoice for the entire stock remaining that he supplied. The terms under which consignment stocks are supplied usually call for the parts to be completely paid for after a certain period of time whether they have been used or not, unless they have earlier been returned as unneeded. The planner responsible for monitoring any consigned stock must be aware of the payment deadlines and return the unwanted parts if he does not wish to pay for an unwanted and unjustifiable stock. Generally speaking, consignment stocks are not recommended. They represent an unwanted workload for the plant storekeeper, and they offer no economic benefit to the plant owner.

32.24 MANPOWER REQUIREMENTS FOR OVERHAULS

Although no hard figures are available, it is thought that approximately one fully skilled mechanic man-hour is required for mechanical overhaul purposes for each 50 mWh of capacity operated. This requirement may be entirely met from the plant work force, made up of partly in-house labor supplemented by contractor's labor, or filled wholly by contractor's labor. Each arrangement has its advantages and disadvantages, and although the contracted labor route may appear to be the more expensive, the warranty coverage will be superior. These man-hours will not vary greatly from one engine to another in the case of four-cycle units. The mechanic time needed for two-cycle units will be in the vicinity of 1 man-hour for every 120 mWh of capacity operated, and the electrician man-hour input will remain at an estimated 1 man-hour for every 50 mWh of capacity operated.

The owner-planner will do well to review in some detail the advantages of engine insurance and the difficulties of making claims relative to failed overhaul work before deciding whether to use company or contractor labor for planned overhauls and repair work. The total costs of an engine overhaul cycle will vary with the method chosen by the owner and his planner.

Before any particular overhaul method is chosen, the planner must review the human resources available within the organization. The resources will be in terms of man-hours available and the degree of skill present at both supervisory and practical levels.

At the commencement of his studies, the engine care planner must determine the limitations of the work forces available to him. It is fundamental to the success of any operational, maintenance, overhaul, or repair activity that the work force utilized have the necessary strength and ability. The planner must therefore consider the skills present in the in-house work force, the standards by which replacement personnel are recruited, and the level of quality control that may be applied by the on-site job supervisor.

The considerations given to those aspects of the work planning will largely determine how the work will be allocated. At no time should any work plan or procedure that is evolved be dependent for its execution on the skill, presence, or ability of one particular individual.

The standards of workmanship must be set by the planner, but the on-site supervisor must see to it that the standards are observed. Depending on the nature of the organization, it could happen that the planner and site supervisor are the same person. Whatever the timetable or complexity of the overhaul plan that is evolved, the owner must be able to rely upon the personnel assigned to the job to be able to carry out all of the tasks without the omission of any item or requirement of the program. Quality control of the work is essential in all respects.

Undoubtedly, the cheapest method is for the owner's personnel to do all of the work called for. That practice can be effective, however, if the overhaul personnel have a high degree of finesse and can be wholly spared from any other routine or emergency duties for extended periods during the span of the overhaul. The completed work will be without warranty.

The second alternative is to use the owner's employees for the minor overhaul duties and OEM representatives or contractors for the intermediate and major overhauls. This method is quite economic insofar as the minor overhauls can usually be handled by the owner's personnel without the personnel being unduly diverted from routine maintenance duties. Furthermore, none of the minor overhaul routines require the overhauling personnel to disturb any bearings or other highly sensitive engine internal components.

The third course that may be followed is to have contractors or OEM representatives do all of the overhaul work. That will entail a greater immediate expense, but the cost may be offset by a much shorter machine downtime, greatly improved quality control, and a better warranty position. The standards of workmanship required must still be set by the planner, and the supervisor must also see to it that the standards are upheld.

To ensure that the skills of the work force are adequate for the tasks that will be planned and assigned in the future, the engine upkeep planner may need to participate in some degree in the preparation of apprenticeship training schemes and of upgrading courses for people presently employed. He may also advise on the selection of replacement personnel.

It will be necessary for the planner to continuously monitor the immediate and future manpower availability that will be required to fulfill the intended work. It may be expedient to have casual labour, contractors or other short-term workers employed temporarily to meet specific or extraordinary objectives.

The record-keeping system, if it is comprehensively structured, will provide a considerable amount of data relating to various jobs and the man-hours absorbed in their completion. All of this history may be used to forecast the manpower demands of future work.

CHAPTER 33
SPARE PARTS AND MATERIALS

33.1 SPARE PARTS STOCKS

Spare parts stocks at plants are a subject of continuing debate. Some plants get to hold more spare parts than others, and some will want to hold "emergency spares." Such a desire is somewhat illogical. If a list of emergency parts is compiled, the parts will permit the emergency to be handled very swiftly, but if the necessary parts can be forecast, so can the type of emergency. Obviously, a "predictable emergency" will be avoided by proper operational maintenance and overhaul practices.

In an organization in which there is a great diversity of equipment models, the problem of equitable parts allocation is virtually unresolvable. As the years advance, the need to consider spare parts stocks recedes. Communications are continually improving, and today it is perfectly reasonable to expect 2- to 3-day deliveries on practically any part for current series North American–produced engines. The parts can be moved almost overnight from the OEM or dealer to even the remotest location.

It has been demonstrated over and over again that a plant with a stock of miscellaneous spare parts will tend to consume the parts over a period of time (and outside the overhaul program). All too often, the spare parts get used on the basis of "let's see if this will do any good." Successful or not, the effort is then usually followed up by an order for a replacement for the part used. A similar plant with practically no spares stock will have a very small requirement for parts over a comparable period. As stated earlier, the question of spare stock is a vexing one with no overall solution. It can be said that the rate of use of anomalous spares can possibly be taken as an index of the care given to an individual plant.

33.2 PART NUMBERS

If the part numbers do not agree with what was ordered, check with the supplier before returning the parts. There is always a good chance that the parts are correct but the original part numbers have been superseded by improved parts or new numbers. The part number is invariably on the package containing the part, and it is probably unnecessary to remove the part from the packing to check it.

The part should not be stored without its package, and the package is disposed of only when the part within is used.

Parts and part numbers must be checked for completeness and correctness upon arrival at the site. If there are any omissions or errors, they must be found before the actual repair or overhaul work for which they are intended commences. Part numbers often change with advancing serial numbers. They can also change when the OEM makes product improvements. The supplier must always be contacted in cases of part number discrepancies or reconciliation before returning apparently incorrect parts.

If items of similar appearance but for different engines are to be held in stock, it is recommended that they be identified with suitable marks made with a felt-tip pen. (For example, Caterpillar D353 and Caterpillar D399 turbochargers should be so marked.) Be aware that the part number may change when an item is rebuilt. The part number for a rebuilt part is not normally found in the parts book. The color of the paint on a spare part cannot be used to identify OEM origin.

33.3 USED SPARE PARTS

In the normal course of events, all the parts removed from an engine during an overhaul are handled by one of three basic methods:

1. Items like piston rings, gaskets, locks, split pins, bearings, and bushings are to be sent away from the workplace as garbage or scrap metal.
2. Cylinder heads, water pumps, and lubricating oil pumps will either be returned as cores to the original supplier or be rebuilt on-site for use again at a future date. (The rebuild of the parts should be charged as part of the overhaul cost and time.)
3. Fuel injection equipment, governors and, in many cases, turbochargers must be returned to the supplier for credit because it is simply not possible to rework equipment of this nature in any but the most sophisticated plant. There are certainly no economic or quality control benefits for the owner in trying to do otherwise.

For the individual small plant, the guiding rule could well be: "If it was worth changing out the part, the changed out part is not worth keeping".

33.4 VARIANT PARTS

As engines age, various anomalous parts get fitted, and it is important that they be easily recognized or identifiable. Whenever work that is done on an engine results in nonstandard parts being fitted, a metal plate describing the component and its date of fitting should be made and attached to the engine in the vicinity of the engine identity plate. For example: "Number 6 cylinder liner chromed, standard size, 81.10.06 61274 hours" or "Number 4 bottom end bearing 0.025 in, O.S., 80.05.08 31125 hours." Such legends can easily be stamped on thin sheet metal and either screwed or glued to the engine.

The provision of such information permits the personnel currently working around the engine to plan ahead, order, and have on site the correct parts for the job. Any variant components known to be fitted but not tagged should be marked retroactively.

33.5 COMPONENT INTERCHANGEABILITY

1. In a diesel power-generating complex or organization, it may be found convenient from time to time to take components off one engine and install them on another.
2. Close attention must be given to the details carried on engine equipment, like fuel pumps, injectors, governors, and turbochargers, which are supplied by specialist manufacturers.
3. The manufacturers of such specialty equipment generally use common basic components in the production of the equipment speciality. They alter internal details to suit the final requirements of the engine manufacturer.

For example, just because two governors look very much alike does not necessarily mean they will perform with the same characteristics. Somewhere on one governor there will be a speed range which may differ from that on the other. Similarly, the specification numbers of two look-alike turbochargers may indicate that there are differences in the compressor size or the nozzle configuration. It is these differences that will have a limiting effect on the interchangeability of the items. If there is any doubt about the suitability of a piece of equipment, the manufacturer of the engine or his representative should be contacted. If equipment of incorrect specification is blindly used, there is considerable risk of damaging the host engine.

33.6 O-RING SELECTION

Certain manufacturers supply O-rings in different colors. The colors denote the types of service for which the rings are intended, and they also indicate that different materials are used in ring manufacture. The user must be careful in his selection of colored O-rings. The colors used by one manufacturer are not necessarily those used by another.

When any O-ring is to be changed for a new one, the related parts book must be consulted to see that the correct ring is used. It must never be assumed that because an O-ring of a particular size and color was taken off it must be replaced by one of similar size and color.

33.7 GASKET STORAGE

Whenever possible, all paper and fiber gaskets should be stored in a horizontal position. The hanging up of gaskets on nails or hooks cannot be recommended. No gasket should ever be folded in order to save storage space. When more than one pattern of gasket is to be placed on a shelf, cardboard separator sheets should be used so that the different patterns are kept apart and repeated search handling, which is so damaging to stored gaskets, is minimized. Cardboard or thin plywood (say, ⅛ in) cost a lot less than gaskets of a similar size, and such storage materials should be acquired and put to use.

33.8 CONSUMABLE MATERIALS

Certain parts and materials used in a diesel generating facility can be classified as consumable spares. In this classification are fuel oil filter elements, lubricating oil filter elements, air filter elements, coolant filter elements, coolant conditioner elements, coolant conditioner chemicals, V-belts for pumps, radiator fans, and so on, cleaning solvents and rags, valve cover gaskets, filter housing gaskets, battery acid, distilled water, fuses, light bulbs and tubes, lubricating oil sample containers, litmus paper, hand cleaners, paper towels, chalk, pencils, notebooks, first-aid supplies, eyewash, gasket dressing, gasket materials, governor oil, grease, thread, dressing, carbon brushes, and mops.

The materials listed above range from direct engine maintenance materials to housekeeping goods, but all are consumed at a discretionary rate by the operator. Their continuous use at an even rate usually indicates a well-conducted and reliable plant. If the supply should become exhausted or interrupted, it can be assumed that the all-important maintenance sector of the plant upkeep work is being impaired. For that reason, it is recommended that a regular check be made on the stocks and the status of their replacement.

The range of material needed in any one plant is not extensive, and it will be found useful if a permanent checklist is posted in the consumable material storage area. The checklist should include all of the particular plant-related consumables together with part numbers where applicable (e.g., filter elements). The minimum and maximum holdings can also be stated as guides to the quantities needed. There is no reason at all why any plant should ever run out of consumable spares and material, all of which should be part of the annual bulk supply.

There are other commodities, like fire extinguisher refills, peat moss, granular oil absorbent material, and cotton waste, which should not be regarded as consumables. Instead, they are supplies for use in emergencies, and no usage for everyday purposes is permitted.

33.9 PARTS CONTROL (STOCKTAKING)

Stocktaking of the parts warehouse is best accomplished at the end of the year's major work exercises. This timing will permit the reordering of parts requiring a long lead time and the recycling of recoverable parts. Also, it allows the planner to maximize the use of existing stock. In comparison with previous stocks, any unusual use trends will be identified and investigated without interference to the next stage of major work. Surplus stocks can be reduced, and obsolete stocks can be disposed of in the most economic manner.

CHAPTER 34
GENERATOR OPERATION

34.1 GENERATOR VOLTAGE REQUIREMENTS

The generator output voltage must be kept as close as possible to the rated voltage as shown on the generator data plate. High voltage, low voltage, or voltage fluctuation (hunting) can cause damage to the generator and its control equipment. Also, of course, abnormal voltage outputs can cause damage to the power consumer's equipment.

34.2 GENERATOR SPEED REQUIREMENTS

The generator speed must be maintained at that shown on the data plate. The frequency of the generator output depends on rotational speed, and a frequency variation will affect control equipment. If the generator rotates too slowly, its voltage will drop. The automatic regulating equipment will then try to maintain the voltage by forcing the field, thereby thermally overloading itself.

34.3 COMMON GENERATOR PROBLEMS

Each of six common generator problems is followed by a list of possible causes:

The Generator Will Not Develop Voltage

- Open circuit in exciter shunt field.
- Loss of residual magnetism in exciter field poles.
- Failed insulation in stator windings.
- Open circuit in manual voltage control circuit.
- Short circuit in generator output leads.
- Malfunctioning voltage regulator.
- Insufficient excitation.
- Partially shorted field.

- Blown fuses.
- Tripped protection devices.

Fluctuating Voltage

- Improper speed regulation of the generator prime mover.
- Unstable load.
- Loose terminals in control circuits.
- Excitation voltage unstable.
- Short circuit in field coil.
- Defective bearing.
- Blown or grounded resistors or diodes.
- Open circuit in exciter field.
- Open circuit in alternator field.
- Grounded or blown surge suppressors.
- Failed automatic voltage regulator.

The Alternator Will Not Produce Full Voltage

- Defective voltage regulator.
- Excessive load.
- Low speed.
- Overload.

Abnormally High Voltage

- Excessive speed.
- Automatic voltage regulator defect.

Generator Overheating

- The generator is overloaded.
- The generator is insufficiently ventilated.
- The surrounding air temperature in the plant is high.
- The phase loading is unbalanced.
- Generator is misaligned.
- The stator windings are shorted.
- The field current is excessive.

Noisy Generator

- The bearing(s) may be failing.
- The coupling is loose or out of line.
- The rotor touches the stator.
- The spring isolators are springbound.

34.4 GENERATOR DEFECT RECORD

When a generator trips off the board, there is always a reason. That reason must be identified; the causative fault must be corrected; and the incident must be logged before the unit can be returned to service. If this procedure is not followed, there is a considerable risk of creating further damage in the electrical system or endangering personal safety in the vicinity of the distribution system. All protection device flags must be inspected as a first step in each incident, and all fallen flags must be noted in the log. If the operator has any doubts whatsoever as to the correctness of his actions in restoring service, he should discuss the problem with either the station superintendent or the superintendent's delegate before reclosing any tripped breaker.

34.5 AUTOMATIC VOLTAGE REGULATORS (AVR)

Automatic voltage regulators are provided so that a diesel generator set can operate continuously without it being necessary to manually adjust the output voltage at frequent intervals or whenever the load changes. A typical system is illustrated in Fig. 34.1. The regulation obtainable on a permanent utility type of generator set will be within approximately 0.25 percent of rated voltage, but on a temporary construction or industrial type of generator set approximately 0.5 percent may be acceptable. The regulator continuously monitors the output voltage of the alternator. Through its sensing capabilities, the regulator varies the exciter field current and thereby controls the ac generator output voltage.

34.6 AUTOMATIC VOLTAGE REGULATOR SWITCHING

If the operator knows nothing else about automatic voltage regulators, he must know that the regulator must always be switched off if the generator set is run below synchronous speed. The automatic voltage regulator senses the ac generator output voltage when it is switched on. At less than synchronous speed, the ac generator voltage will be depressed. Sensing that, the automatic voltage regulator will attempt to correct it by increasing the exciter field current. It will thereby overload itself and often the exciter field as well, with initially destructive effects.

34.7 AUTOMATIC AND MANUAL REGULATION SWITCH

Almost all diesel generator units have an automatic voltage regulation feature, and all automatic voltage regulators can be switched to a manual mode if necessary. When the voltage regulator is in good working order, the governor is oper-

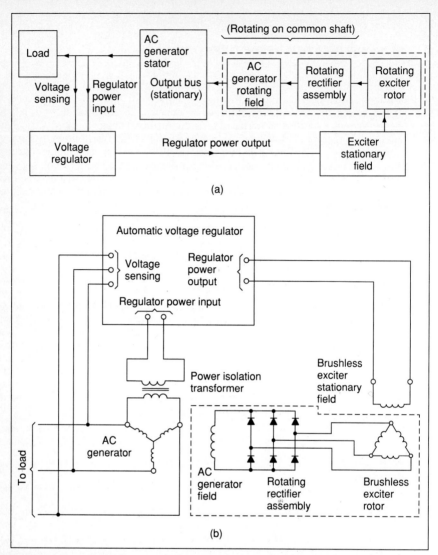

FIGURE 34.1 Typical brushless generator-voltage regulatory system. (*a*) block diagram; (*b*) schematic diagram.

ating correctly and the load is reasonably steady, the supplied voltage will not vary more than approximately 0.5 percent.

34.8 AUTOMATIC VOLTAGE REGULATOR TROUBLE IDENTIFICATION

Although there are many possible contributions to automatic voltage regulator malfunctions, there are numerous steps that the operator can take without consultation with or assistance from a skilled electrician. In the event of a suspected automatic voltage regulator difficulty, the operator should review the elementary symptoms, possible causes, and suggested remedies before calling for help. If the problem remains, the electrician must consider the situation, and his actions will very much follow the recommendations and guidelines of the pertinent OEM manual.

The automatic voltage regulator is self-adjusting, and it requires no manual alteration of settings. If any alterations do appear to be necessary, they can be performed only by a suitably trained electrician.

Troubleshooting Guide

Voltage Does Not Rise to Rated Level. Voltage shutdown switch open: close switch. Prime mover not at rated speed: bring unit up to synchronous speed.

Voltage Is High and Uncontrollable by Voltage-Adjusting Rheostat. Transfer switch is in automatic position: place in manual.

Voltage Is Low and Controllable with Voltage-Adjust Rheostat. Prime mover unit is not running at synchronous speed: increase speed.

Poor Voltage Stability. Frequency is unstable: review governor performance, review engine performance, review load fluctuation by reference to the kilowatt meter.

Voltage Recovery Is Sluggish after Load Change. Poor governor response: review governor performance.

The above suggestions are merely elementary steps toward problem resolution. Section 18.8, which deals with governor operation and difficulties should also be consulted (Fig. 34.1).

34.9 LOAD DIVISION

When generators are operating in parallel, the kilowatt load should be divided between the running units proportionally to the units' ratings. On ac generators, the load can be moved from one unit to another only by speed control, not by manipulation of the voltage regulator rheostat. Alteration of the voltage regulator rheostat will alter only the power factor. Thus, the current output of the generators results in undesirable crosscurrents.

When running in parallel, the load carried by a unit running at less than full load can be increased by transferring from one unit to another merely by operating the governor RAISE control. The frequency will rise also, and it can be normalized by operating the governor LOWER control on the unit that is to surrender load.

After the kilowatt load has been allocated to the units in good proportion, the reactive (VARS) or wattless load must also be proportioned. If the generators are of the same capacity, their output amperages should be similar. If there is a difference in the output amperage of one machine and another for the same load, the presence of crosscurrents is probable. The difference should be eliminated by adjusting the voltage regulator rheostats on the respective units. Only minute movements of the adjusting knob should be made, and enough time between adjustments should be allowed for the meters to settle. If the generator sets are of different sizes, the loads will be divided proportionately and the amperages will differ, but the reactive load (VARS) must be balanced if circulating currents are to be avoided.

When the load distribution between the running units has been established, very little subsequent adjustment will be needed if the load increases or decreases. Thereafter, the load apportionment will be handled by the regulating governor's speed and the reactive load will be handled by the crosscurrent compensation feature of the voltage regulator.

34.10 PHASE AMPERAGE

As the load grows, the operator should switch the ammeter selector switch to the highest phase. In that way, the possibility of overload on one phase is reduced.

34.11 CROSSCURRENT COMPENSATION

When two generators are operating in parallel, their field excitations should be balanced. If the excitation current in one becomes excessive, a current will flow between the generators. This current, known as a circulating current, will appear as a lagging power factor (inductive) to the generator with the excessive excitation current and as a leading power factor (capacitive) load to the other generator. A parallel compensation circuit in the automatic voltage regulator will influence the field excitation of both generators, minimize the circulating currents, and tend to balance the power factors.

34.12 QUALITY CONTROL OF PRODUCED POWER

The ordinary power consumer does not really care about the method used to generate the power that he uses. The power can be generated through the use of diesel engines, steam turbines, gas turbines, water turbines, windmills, or any other means of energy conversion. All that is of interest to the average user is that the cost is reasonable, the supply is continuous, and the quality is first-class.

In comparison with the cost of steam- or water-generated power, that of diesel-generated power appears to be high. However, for remote locations or for standby installations, diesel-powered plants offer great economies in capital costs. The diesel power plant owner will naturally wish to offset the higher cost of diesel-generated electricity by selling a product that is highly reliable and is of first-class quality. The contented user will not complain if he receives a quality product at a reasonable price.

TABLE 34.1 Generator Phase Amperage Guide, (0.80 Power Factor)

KVA Rating	kW Rating	Voltage, V		
		600	2400	4160
6.3	5.0	6.1		
9.5	7.5	9.1		
12.5	10.0	12.0		
18.7	15.0	18.0		
25.0	20.0	24.0	6.0	3.5
31.3	25.0	30.0	7.5	4.4
37.5	30.0	36.0	9.1	5.2
50.0	40.0	48.0	12.1	7.0
62.5	50.0	61.0	15.1	8.7
75.0	60.0	72.0	18.1	10.5
93.8	75.0	90.0	22.6	13.0
100.0	80.0	96.0	24.1	13.9
125.0	100.0	120.0	30.0	17.5
156.0	125.0	150.0	38.0	22.0
187.0	150.0	180.0	45.0	26.0
219.0	175.0	211.0	53.0	31.0
250.0	200.0	241.0	60.0	35.0
312.0	250.0	300.0	75.0	43.0
375.0	300.0	361.0	90.0	52.0
438.0	350.0	422.0	105.0	61.0
500.0	400.0	481.0	120.0	69.0
625.0	500.0	602.0	150.0	87.0
750.0	600.0	721.0	180.0	104.0
875.0	700.0	842.0	210.0	121.0
1000.0	800.0	962.0	241.0	139.0
1125.0	900.0	1082.0	271.0	156.0
1250.0	1000.0	1202.0	301.0	174.0
1563.0	1250.0	1503.0	376.0	218.0
1875.0	1500.0	1805.0	452.0	261.0
2188.0	1750.0	2106.0	528.0	304.0
2500.0	2000.0	2406.0	602.0	348.0
2812.0	2250.0	2710.0	678.0	392.0
3125.0	2500.0	3005.0	752.0	435.0
3750.0	3000.0	3610.0	940.0	522.0
4375.0	3500.0	4220.0	1055.0	610.0
5000.0	4000.0	4810.0	1204.0	695.0

34.13 GROUND FAULT INDICATION

Three ground fault indicator lamps are set up in the power plant: one for each phase. If a ground fault occurs on one phase, the problem should be rectified by a qualified electrician without delay, but there will be no direct danger to personnel. However, if more than one ground fault indicator lamp lights up, an immediate effort must be made to isolate the fault location by a process of elimination: switching out individual breakers at the MCCs and similar switch points. The unattended double ground fault presents considerable danger to personnel. It cannot be permitted, and it must be rectified by a competent electrician.

34.14 POWER QUALITIES

A basic function of the diesel power plant operator is to ensure that the quality of the power produced is the best possible at all times. Electric power can be qualified by its continuity, voltage, frequency, cost, and adequacy.

Continuity

The continuity of an electrical supply system is unnoticed until the supply is interrupted. Many of the reasons for supply failures, such as lightning strikes, trees falling across distribution lines, traffic-demolished distribution poles, and vandalism, are beyond the operator's control, but the operator with good ability will be able to minimize the duration of such an interruption when it occurs.

The able operator will be a valuable participant in any planned power outage that may be necessary for construction or development reasons. Being able to shut down and restart the power plant with alacrity, he will minimize the length of inconvenience to the customer. He may also be required to play a large part in the subsequent testing and evaluating of the new or revised equipment. In all planned outages, the consuming community must have been fully advised well in advance of the event, probably by the station superintendent or possibly by his delegate.

All planned outages are intended to improve service to the consumer. It is the unplanned or overfrequent interruption of power supply that irritates the consumer, and it is fair to say that in many situations repetitious interruptions are a result of inadequate or faulty training of the plant operators. That defect may not necessarily be the operator's own fault; it may be due to deficient or nonexistent training programs and methods.

The other major contributors to unscheduled power stoppages are poor equipment care practices, which again may be beyond the operator's control. However, the operator must be able to recognize trouble in the making and take all possible actions to preserve continuity of supply. The frequency of interruption incidents in any community is a direct reflection of the individual plant's operational effectiveness and, to a certain degree, the awareness of the owner.

Voltage

Almost all power-generating facilities have provisions for the automatic regulation of voltage. All power-consuming equipment is dependent upon a given voltage as selected by its designer. Minor variations in the supply voltage will impair the performance of the consumer's equipment, and major variations in the supplied voltage will be damaging. Under normal circumstances, the consumer can expect the supply voltage to be maintained within a standard that permits his equipment to work without any difficulty.

Voltage variations at the generator will be accentuated at the point of consumption. Therefore, it is necessary for the operator to pay strict attention to the generated voltage. Automatic voltage regulators generally have the capability of controlling within ±0.5 percent but this standard could certainly not be achieved with manual regulation. Any failure in the automatic voltage regulation system should be rectified without delay if for no other reason than the maintenance of good customer relations.

Frequency

Insofar as the frequency of the power generated is totally dependent upon the rotational speed of the generating equipment, the quality of frequency is very much within the operator's control. Frequency fluctuations will certainly have effects on the performance of the power user's equipment and appliances, but inconvenience rather than damage will be the end result.

To maintain an optimum frequency, the operator must take care of the prime mover's speed and governing system; he must adjust the governor controls in accordance with the OEM manuals and the notes on governing to be found in this handbook.

Cost

The cost of power can be regarded as a major quality. Many factors combine to affect the final cost. Many of the factors, such as the price of fuel and parts, capital costs and amortization charges, are totally beyond the control of the plant staff, but the day-to-day operating practices can have a considerable bearing on the cost-effectiveness of the plant, which in turn is reflected in the rates charged. The bulk of the notes in this handbook are devoted to the details of good operating practice which result in maximum economy.

Adequacy

The adequacy of power supplies permits any consumer served by the system to have his power requirement met upon demand, assuming all of the installed generating equipment is in order and has sufficient capacity to meet the demand.

The adequacy of power supplies is beyond the control of the operator, but the day-to-day accumulation of information by the operator relating to observed loads provides the data upon which the planners will base their load forecasts. Such forecasts are then used to determine the future equipment additions needed in individual plants.

CHAPTER 35
GENERATOR OVERHAUL AND UPKEEP

35.1 ELECTRIC MACHINE HEAT AND TEMPERATURE

Any generator or other electrical machine will heat up naturally when it is working. On the generator data plate there is often a reference to *temperature rise*. This is the designed temperature rise in the winding when the machine is working at full capacity. The heat generated in the windings is transmitted in part to the stator and rotor laminations and in part to the stream of ventilating and cooling air (Fig. 35.1).

The stated temperature rise does not mean that the stator housing or the rest of the machine will become so heated. However, the operator and the electrician should become familiar with the normal operating temperature characteristics of the electric machines under their care. Overheating motors and generators often advertise their condition by very noticeable odors, and they will require immediate removal from service before they are burned out. The remedial work necessary to prevent further overheating is usually quite simple. The most usual causes of machine overheating are dirty windings, running under overload conditions, and operating in an insufficiently ventilated environment.

35.2 GENERATOR AND MOTOR WINDING DAMPNESS

If any moisture is present in the atmosphere surrounding an electrical machine, the winding insulation may be affected. Constant exposure of a machine's windings to a damp atmosphere will contribute to a reduction in the insulation resistance. The condition is accentuated in a saline atmosphere.

Because of the higher voltages prevailing in it, the stator of a generator is susceptible to moisture-induced damage. The rotor, with its generally lower operating voltages, is less likely to suffer such damage. Every reasonable effort must be made to prevent rain, fog, and spray from being inducted into the machine's cooling airstream if the risk of damp or dirt damage is to be minimized.

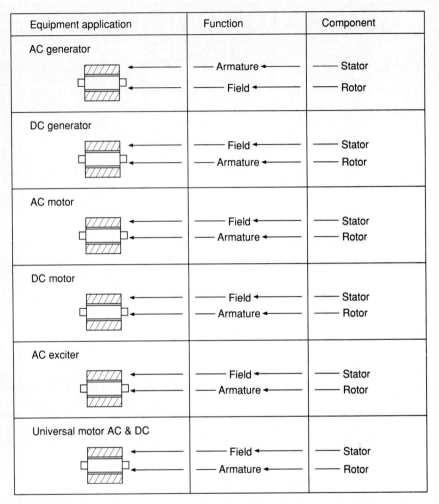

FIGURE 35.1 Rotating electric machine nomenclature. Note that all output power, either mechanical in the case of a motor or electrical in the case of a generator, is derived from the armature.

35.3 GENERATOR WINDING INSPECTION

When the generator windings are inspected as part of the routine maintenance check, the bindings, straps, and other tying material should be examined for looseness or separation. If anything unusual is found, repairs must be made before the machine is restarted. Any cracking or apparent peeling of winding wrapping tapes must be fully investigated and repaired and the insulation thereafter qualified.

35.4 GENERATING WINDING TERMINAL FAILURE

The tails of generator coil windings sometimes fracture. The most usual reason for such failure is vibration. Whenever such an incident occurs, the vibration characteristics of the generation unit should be compared with the original vibration signature after the repair is complete. Both the linear and the torsional aspects should be checked out.

Generator windings also will suffer metal fatigue and subsequent breakage whenever it is possible for a vibratory movement of any part or section of the coil(s) to take place. Although the movement of the coil copper may be quite minute, continuous working will harden the copper and lead to metal fatigue with eventual fracture. The most likely breakage locations are at the coil ends of either the field pole interconnectors or at the armature collector ring terminals.

35.5 GENERATOR AND MOTOR MAINTENANCE

The insulation values of generator and motor windings must be assessed annually. The amount of dirt, fumes, and other foreign matter induced into a rotating electrical machine depends a great deal upon the nature of the local environment, the general condition of the machinery in the plant, the effectiveness of the housekeeping work, and the mode of plant ventilation.

The values of the winding insulation will be affected in part by the nature of the contaminants that come to rest on the winding insulation. The insulating materials can be affected by the cleaning agents used, and that is why great care must be given to the selection of electrical cleaning fluids. It must be remembered that all cleaning materials have negative effects on paint, insulating material, and varnishes. The frequency of generator cleaning is a matter requiring some discretion, and it should be regulated in conjunction with hi-pot or polarity index testing (Fig. 35.2.)

35.6 TESTING GENERATOR AND MOTOR WINDING INSULATION

When a generator is to be cleaned, it is preferable that the windings be meggered both before and after the cleaning process has taken place. "Good" insulation may be taken as that which provides an acceptable level of resistance to current flow. If the level of resistance in a machine's insulation is known, any deterioration can be gauged by comparative megger testing. Such testing should be a normal part of the overhaul and maintenance routines of all electric motors and generators.

The insulation in electrical machines may deteriorate under various influences and circumstances. As part of the preventive maintenance routines, the plant electrician should make regular visual inspections of the machine windings. If these inspections are neglected, there is a high probability of the machine getting

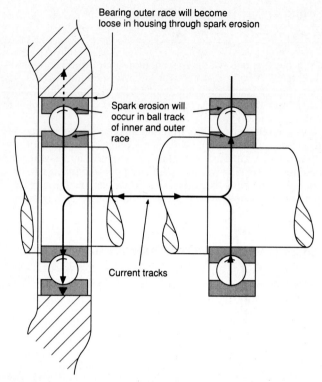

FIGURE 35.2 Alternator rotor: half sections showing coil windings and pole.

into difficulties at a later date. Dirt, heat, corrosion, vibration, as well as certain cleaning methods and materials, will have some degree of deterioration effect on the machine insulating materials.

35.7 GENERATOR AND MOTOR CLEANING

Any generator, exciter, or other rotating electrical machine must be kept as clean as possible at all times. Internal cleaning must be periodically carried out under the supervision of a qualified journeyman. Cleaning is best effected when it is integrated with the prime mover's planned overhaul. Whenever convenient, suction methods should be employed, but the use of low-pressure compressed air in conjunction with solvents will provide satisfactory results. The solvent must be used sparingly, because all electrical cleaning agents have some ill effect on the winding insulation. Metal tools must not be used to gouge or poke dirt from the windings, laminations, or connections.

The only solvents recommended for electrical machine cleaning are of petroleum origin, and they will have been approved by the official regulating body controlling safety matters in the locality of the given plant. Whatever cleaning product is contemplated, the plant superintendent must read all of the product literature and usage instructions very carefully indeed. Many products offered for cleaning purposes are hazardous to health and property if handled carelessly. Cleaning agents containing trichlorethylene ($CHCl:CCl_2$) or carbotetrachloride (CCl_4), which is also known as tetrachloromethane, were widely used for electrical cleaning purposes at one time. Now products containing those substances must be avoided because medical evidence has shown them to be extremely harmful to health.

No attempt must be made to clean a machine in motion, nor must any machine be cleaned until it has been immobilized and a clearance ticket has been issued. The air pressure used to blow dirt out of the machine must not exceed 35 psi. (Air pressures above 35 psi can lift insulating tapes off the protected areas.) The insulating resistance of the windings must be checked both before and after the cleaning operation.

Dirty generators are prone to overheating and early failure. Generators get dirty only when worked in dirty environments. It is necessary to pay attention to the sources of generator dirt if the cleaning process is to be fully effective.

Most of the dirt inducted into an electrical machine is airborne and is derived from fuel oil, lubricating oil vapors, welding fumes, cleaning solvent, paint spray, road, mining, and milling dusts, atmospheric humidity, rain, and exhaust soot. Part of the routine maintenance that goes on continuously in a plant must be aimed at limiting the ingress of dirt, and well-conducted housekeeping activities will be the primary factors in this effort.

Formations of dirt on the surfaces over which the cooling air passes will impair the transfer of heat from the windings. Some of the dirt, particularly that of fuel oil or lubricating oil origin, may have a decaying effect on insulating varnishes and coatings.

35.8 ALIGNMENT IN CONJUNCTION WITH ANTIFRICTION BEARINGS

A correct alignment, if necessary, is required to be sure that the drive and driven shafts of engine-generator sets rotate about exactly the same axis. Any misalignment will result in fretting of the coupling disks, fretting of the stub shaft spigot, improper crankshaft deflections, or overload of the generator bearing(s).

The air gap between the stator and the rotor must be approximately equal all around, but any effort made to adjust the air gap clearance may affect the alignment. It follows that the alignment process and the air gap setting are part and parcel of the same exercise.

35.9 AIR GAP ON GENERATORS AND MOTORS

The air gap between the rotor and the stator of any rotating machine should not vary more than 10 percent with the measurements taken 90° apart. Whenever possible, the air gap measurements should be made from both ends of the machine. A significant change in the air gap will require an inspection of the generator bearing(s).

35.10 EFFECT OF ELECTRIC ARCING ON ANTIFRICTION BEARINGS

When an electric current passing through an antifriction bearing is interrupted at the contact surfaces between the races and the ball, arcing results. This phenomenon produces very high localized temperatures. Spark erosion takes place on both the race and the ball. Eventually a condition known as *fluting* occurs and, as it develops, the bearing gives off increased noise and vibration.

The causes of the current may be static electricity derived from belt drives and similar attachments or from electrical leakage past faulty insulation. The current may be of very low value and difficult to detect. If the current is of higher amperage, such as that which may be expected from a partial short circuit, the ball track of the bearing will display a rough granular appearance when it is dismantled (Fig. 35.3, 35.4).

Heavy currents passing through the bearing may produce a welding effect with migration of metal from the race to the ball. The noise and vibration of the bearing will be very noticeable. There is some thought that similar spark erosion may

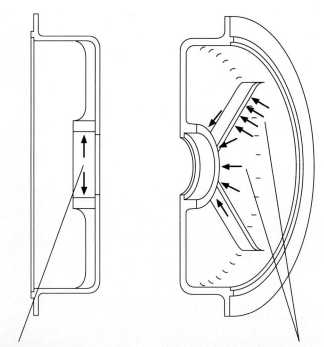

Fretting in the bearing housing   Cracks in the webs around the bearing housing

FIGURE 35.3 Antifraction bearing damaged by spark erosion.

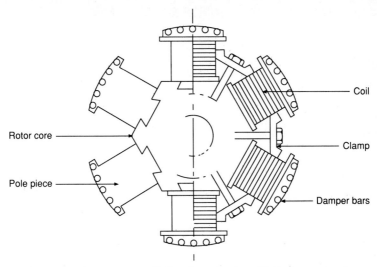

FIGURE 35.4 Alternator rotor: Half sections showing coil windings and pole.

be responsible for the outer race of the bearing becoming loose in the bearing housing. It is suggested that whenever there has been trouble in the motor or generator windings or a failure of the exciter, an examination of the antifriction bearings is appropriate.

35.11 LUBRICATION OF ANTIFRICTION BEARINGS

Antifriction bearings must be served with the correct type of grease as specified by the OEM. Grease types and grades may not be mixed. If the type of grease is to be changed for any reason, the bearing must be thoroughly cleaned before the new grease is applied. Antifriction bearings that have been cleaned must not be spun by hand until they have been relubricated.

If when the bearing is dismantled the grease has a stiff or caked appearance or has changed in color, it is an indication of bearing failure. The grease will have a smell of burned petroleum, and it will have lost its lubricating qualities. Occasionally the grease will have a varnish-like brittleness.

Usually the first indication of antifriction bearing lubrication failure will be a rapid rise in the bearing temperature. At the same time, the bearing will begin to give off a whirring or whistling noise. If the bearing is allowed to continue running, the temperature will continue to rise, the bearing hardness will be lost, and total failure will ensue. A brown or bluish discoloration of some or all of the bearing parts will be indicative of excessively high temperature operation. Such a bearing must be discarded.

35.12 GENERATOR AND MOTOR BEARING LIFE

For rotating electrical machine applications, the life of an antifriction bearing may be taken to be approximately 30,000 h under good clean conditions, but it will generally be found convenient to change out engine-driven alternator bearings on the same frequency as the prime mover main bearings. However, if the cleaning and inspection routines of the machine are scheduled to take place at lesser intervals, it is quite reasonable to fit new bearings at that time. Early changed bearings will have a greater reliability than late changed bearings.

35.13 ELECTRIC MOTOR AND GENERATOR INSPECTION AFTER MAJOR REPAIR OR RELOCATION

Inspect the newly delivered unit for warning tags, labels, and other paperwork. Remove all wedges, packing, chocks, and blocking. Review the nameplate voltage, frequency, and direction of rotation and be sure that it corresponds with the established generator requirements.

After delivery and before reconnection commences, each lead must be identified and tagged so that a correct match-up of the generator and system leads is obtained. The resistance of the windings will require checking with an insulation resistance meter. The value obtained should not be less than 1000Ω (ohms) per volt of rated voltage; for example, a 600-V generator will show at least 0.6 MΩ (60,000 Ω), and for 2300V, 2.3 MΩ should be obtained.

If the insulation resistance is below the required value, the windings must be dried out by the application of heat. The temperature of the windings should not be raised above 85°C (185°F), for if water is present in the windings, it must not be permitted to boil. Before, during, and after the drying process, the insulation resistance must be measured so that the drying trend can be gauged. If drying is necessary, it is recommended that the manufacturer's advice and comments be sought and followed.

35.14 START-UP OF A ROTATING ELECTRICAL MACHINE FOLLOWING REPAIR, RELOCATION, OR CLEANING

The initial test run of a generator must be of an unenergized mechanical nature only with all generator and exciter terminals disconnected. At idle speed the unit is inspected for any unusual or unexpected noises. If all is satisfactory, the unit may be run up to synchronous speed and the vibration level may be checked with a hand-held vibration monitor. Record the readings and compare with the last reading taken from the machine before the repair or cleaning commences.

Having obtained a set of vibration readings that are within its toleration range, the machine may be stopped. The alignment of the engine and generator is again checked. If it is satisfactory, connection of the terminals may commence. The connections of a three-phase motor or generator can be made only by a qualified electrician using either phasing sticks or a phase-sequence meter.

35.15 GROUNDING STRAPS

Grounding straps provide a path of low impedance for fault currents that flow when insulation breaks down. Protection devices thus have to be in place to minimize the risk of injury to personnel or further damage to equipment. When grounding straps are improperly connected, stray currents may pass through engine components with resultant damage. If current passes between the crankshaft and the main bearings, some pitting of the bearing surfaces will occur; the roughness thereby induced will contribute to greatly increased wear rates. Ultimately the bearings will be destroyed. Electrical discharge damage can be identified in its early stages as random streaks on the crankshaft journals, and the bearing shells will be completely worn. In ball bearings there may be evidence of burning along the line of ball contact in the cage. In roller bearings there may be signs of axial lines on both cage and roller. Electrical discharge damage can be prevented by maintaining the grounding straps in good order.

CHAPTER 36
SWITCHGEAR

36.1 SWITCHGEAR DEVICE FUNCTION NUMBERS

The purpose of switchgear device function numbers is to provide a means of quickly identifying the main purpose of any device used as a part of the switchgear equipment when displayed in a blueprint or diagram. This numbering system, first developed in connection with automatic switchgear, is now applied to all types of switchgear equipment. It is based on the allocation of a standard number to each of the several fundamental functions performed by the component elements of a complete switchgear array.

These device functions may refer to the actual function the device performs in equipment or they may refer to the electrical or other medium to which the device is responsive. Hence, there may in some instances be a choice of the function number used for a given device. The preferable choice in all cases is the one which is recognized to have the narrowest interpretation so that it most specifically identifies the device in the minds of all individuals concerned with the design and operation of the equipment.

The device function numbers, with appropriate suffix letter or letters when necessary, are used on electrical diagrams, in instruction books, and in specifications.

36.2 STANDARD DEVICE NUMBERS LIST

Device numbers, names, functions, and descriptions are listed in Table 36.1. The numbering system does not necessarily apply to any particular utility company or organization, and some caution must therefore be exercised. However, it is recommended that the numbers be used to form the basis of a standard plant numbering system. For the basic electrical symbols commonly used in diesel electrical power system diagrams, see Fig. 36.1.

TABLE 36.1 Standard Devices Used in Diesel Electric Power Plant Systems

Number	Name	Function and description
1.	Master element	The initiating device, e.g., a control switch voltage relay or float switch, which serves either directly or through a permissive device such as a protective or time-delay relay to place equipment in or out of operation.
2.	Time-delay starting or closing relay	A device that functions to provide a desired amount of time delay before or after any point of operation in a switching sequence or protective relay system except as specifically provided by devices 48, 62, and 79.
3.	Checking or interlocking relay	A relay that operates in response to the position of a number of other devices (or to a number of predetermined conditions) in equipment, to allow an operating sequence to proceed or to stop, or to provide a check of the position of the devices or of the conditions for any purpose.
4.	Master contactor	A device generally controlled by device 1 or the equivalent and the required permissive and protective devices. It serves to make and break the necessary control circuits to place a piece of equipment in operation under the desired conditions and to take the equipment out of operation under other or abnormal conditions.
5.	Stopping device	A control device used primarily to shut down equipment and hold it out of operation. It may be manually or electrically actuated, but it excludes the function of electrical lockout (see device 86) in abnormal conditions.
6.	Starting circuit breaker	A device whose principal function is to connect a machine to its source of starting voltage.
7.	Anode circuit breaker	A device used in the anode circuit of a power rectifier for the primary purpose of interrupting the rectifier circuit if an arc-back should occur.
8.	Control power—disconnecting device	A device such as a knife switch, circuit breaker, or pull-out fuse block used to respectively connect and disconnect the source of control power to and from the control bus or equipment.
9.	Reversing device	A device used to reverse a machine field or perform any other reversing function.
10.	Unit sequence switch	A switch used to change the sequence in which units can be placed in and out of service in multiple unit equipments.
11.	Reserved for future application	
12.	Overspeed device	A direct-connected speed switch which functions on machine overspeed.

13.	Synchronous-speed device	A centrifugal-speed switch, a slip-frequency relay, a voltage relay, an undercurrent relay, or any type of device that operates at approximately the synchronous speed of a machine.
14.	Underspeed device	A device that functions when the speed of a machine falls below a predetermined value.
15.	Speed- or frequence-matching device	A device that functions to match and hold the speed or the frequency of a machine or system equal to or approximately equal to that of another machine, source, or system.
16.	Reserved for future application.	
17.	Shunting or discharge switch	A switch that serves to open or close a shunting circuit around any piece of apparatus (except a resistor), e.g., a machine field, machine armature, capacitor, or reactor.
18.	Accelerating or decelerating device	A device used to close or to cause the closing of circuits by means of which the speed of a machine is increased or decreased.
19.	Starting-to-running transition contactor	A device that initiates or causes the automatic transfer of a machine from the starting to the running power connection.
20.	Electrically operated valve	A valve used in a vacuum, air, gas, oil, or similar line. It is electrically operated or has electrical accessories such as auxiliary switches.
21.	Distance relay	A relay that functions when the circuit admittance, impedance, or reactance increases or decreases beyond a predetermined limit.
22.	Equalizer circuit breaker	A breaker that serves to control or to make and break the equalizer or the current-balancing connection for a machine field or for regulating equipment in a multiple-unit installation.
23.	Temperature control device	A device used to raise or lower the temperature of a machine or other apparatus or any medium when it falls below or rises above a predetermined value.
24.	Reserved for future application	
25.	Synchronizing or synchronous check device.	A device that operates when two ac circuits are within the desired limits of frequency, phase angle, or voltage to permit or to cause the paralleling of the two circuits.
26.	Apparatus thermal device	A device that functions when the temperature of the shunt field or the amortisseur winding of a machine, a load-limiting or load-shifting resistor, a liquid or other medium exceeds a predetermined value or when the temperature of the protected apparatus, such as a power rectifier or any device, decreases below a predetermined value.

36.3

TABLE 36.1 Standard Devices Used in Diesel Electric Power Plant Systems (*Continued*)

Number	Name	Function and description
27.	Undervoltage relay	A relay that functions on a given value of undervoltage.
28.	Reserved for future application	
29.	Isolating contactor	A devices used expressly to disconnect one circuit from another for the purpose of emergency operation, maintenance, or test.
30.	Annunciator relay	A nonautomatically reset device that gives a number of separate visual indications upon the functioning of protective device and which may also be arranged to perform a lockout function.
31.	Separate excitation device	A device that connects a circuit such as the shunt field of a synchronous converter to a source of separate excitation during the starting sequence or that energizes the excitation and ignition circuits of a power rectifier.
32.	Directional power relay	A device that functions on a desired value of power flow in a given direction or on reverse power resulting from arc-back in the anode or cathode circuit of a power rectifier.
33.	Position switch	A switch that makes or breaks contact when the main device or piece of apparatus which has no device number reaches a given position.
34.	Motor-operated sequence switch	A device, such as a motor-operated multicontact switch or the equivalent or a programming device such as a computer, that establishes or determines the operating sequence of the major devices in a piece of equipment during starting and stopping or other sequential switching operation.
35.	Brush-operating or slip-ring short-circulating device	A fitting for raising, lowering, or shifting the brushes of a machine for short-circuiting a machine's slip rings, or for engaging or disengaging the contacts of a mechanical rectifier.
36.	Polarity device	A device that operates or permits the operation of another device on a predetermined polarity only or verifies the presence of a polarizing voltage in an equipment.
37.	Undercurrent or underpower relay	A relay that functions when the current or power flow decreases below a predetermined value.
38.	Bearing protective device	A device that functions on excessive bearing temperature or on other abnormal mechanical conditions associated with the bearing such as undue wear which may eventually result in excessive bearing temperature or failure.

39.	Reserved for future application	
40.	Field relay	A device that functions upon the occurrence of an abnormal mechanical condition (except that associated with bearings as covered under device 38) such as excessive vibration, eccentricity, expansion, shock, tilting, or seal failure. A relay that functions on a given or abnormally low value or failure of machine field current or on an excessive value of the reactive component of armature current in an ac machine indicating abnormally low field excitation.
41.	Field circuit breaker	A device that applies or removes the field excitation of a machine.
42.	Running circuit breaker	A device whose principal function is to connect a machine to its source of running or operating voltage. This device may also be used as a contactor in series with a circuit breaker or other fault-protecting means primarily for frequent opening and closing of the circuit.
43.	Manual transfer or selector device	A manually operated device that transfers the control circuits in order to modify the plan of operation of the switching equipment or of some of the devices.
44.	Unit sequence starting relay	A relay that starts the next available unit in a multiple-unit equipment upon the failure or nonavailability of the normally preceding unit.
45.	Reserved for future application	A device that functions upon the occurrence of an abnormal atmospheric condition such as damaging fumes, explosive mixtures, smoke, or fire.
46.	Reverse-phase or phase-balance current relay	A relay that functions when the polyphase currents are of reverse-phase sequence, are unbalanced, or contain negative phase-sequence components above a given amount.
47.	Phase-sequence voltage relay	A relay that functions upon a predetermined value of polyphase voltage in the desired phase sequence.
48.	Incomplete sequence relay	A relay that generally returns the equipment to the normal or OFF position and locks it out if the normal starting, operating, or stopping sequence is not properly completed within a predetermined time. If the device is used for alarm purposes only, it is designated as 48a (alarm).
49.	Machine or transformer thermal relay	A relay that functions when the temperature of a machine armature or other load-carrying winding or element of a machine or a power rectifier or power transformer (including a power rectifier transformer) exceeds a predetermined value.
50.	Instantaneous overcurrent or rate-of-rise relay	A relay that functions instantaneously on an excessive value of current or on an excessive rate of current rise, thus indicating a fault in the apparatus or circuit being protected.

TABLE 36.1 Standard Devices Used in Diesel Electric Power Plant Systems (*Continued*)

Number	Name	Function and description
51.	AC time overcurrent relay	A relay with either a definite or an inverse time characteristic that functions when the current in an ac circuit exceeds a predetermined value.
52.	AC circuit breaker	A device that is used to interrupt or close an ac power circuit under normal conditions or to interrupt the circuit under fault or emergency conditions.
53.	Exciter or dc generator relay	A relay that forces the dc machine field excitation to build up during starting or which functions when the machine voltage has been built up to a given value.
54.	Reserved for future application	
55.	Power factor relay	A relay that operates when the power factor in an ac circuit rises above or falls below a predetermined value.
56.	Field application relay	A relay that automatically controls the application of the field excitation to an ac motor at some predetermined point in the slip cycle.
57.	Short-circuiting or grounding device	A primary circuit-switching device that functions to short-circuit or ground a circuit in response to an automatic or a manual command.
58.	Power rectifier misfire relay	A device that functions if one or more anodes of a power rectifier fails to fire or to detect an arcback or on failure of a diode to conduct or block properly.
59.	Overvoltage relay	A relay that functions on a given value of overvoltage.
60.	Voltage or lance relay	A relay that operates on a given difference in voltage or current input or output of two circuits.
61.	Reserved for future application	
62.	Time-delay stopping or opening relay	A relay that serves in conjunction with the device that initiates the shutdown, stopping, or opening operation in an automatic sequence or protective relay system.
63.	Liquid or gas pressure, level, or flow relay	A relay that operates on given values of liquid or gas pressure or on given rates of change of these values.
64.	Ground protective relay	A relay that functions on failure of the insulation of a machine, transformer, or other apparatus to ground or on flashover of a dc machine to ground.

65.	Governor	The assembly of fluid, electrical or mechanical control equipment used for regulating the flow of water, steam, or other medium to the prime mover for such purposes as starting, holding speed or load, or stopping.
66.	Notching or jogging device	A device that functions to allow only a specified number of operations of a given device or equipment or a specified number of successive operations within a given time. Also, a device that functions to energize a circuit periodically or for fractions of specified time intervals or that is used to permit intermittent acceleration or jogging of a machine at low speeds for mechanical positioning.
67.	AC directional overcurrent relay	A relay that functions on a desired value of ac overcurrent flowing in a predetermined direction.
68.	Blocking relay	A relay that initiates a pilot signal for the blocking of tripping on external faults in a transmission line or in other apparatus under predetermined conditions or cooperates with other devices to block tripping or to block reclosing on an out-of-step condition or on power savings.
69.	Permissive control device	Generally, a two-position manually operated switch that in one position permits the closing of a circuit breaker or the placing of equipment in operation and in the other position prevents the circuit breaker or the equipment from being operated.
70.	Electrically operated rheostat	A variable resistance device used in an electric circuit. It is electrically operated or has other electrical accessories such as auxiliary, position, or limit switches.
71.	Reserved for future application	A relay that operates on given values of liquid or gas level or on given rates of change of those values.
72.	DC circuit breaker	A circuit breaker that is used to interrupt or close a dc power circuit under normal conditions or to interrupt the circuit under fault or emergency conditions.
73.	Load resistor contactor	A contactor that is used to shunt or insert a step of load-limiting, shifting, or indicating resistance in a power circuit or to switch a space heater in a circuit or to switch a light or regenerative load resistor of a power rectifier or other machine in or out of circuit.
74.	Alarm relay	A relay other than an annunciator (as covered under 30) that is used to operate in connection with a visual or audible alarm.
75.	Position-changing mechanism	A mechanism that is used for moving a main device from one position to another in equipment. An example is shifting a removable circuit breaker unit to and from the connected, disconnected, or test position.
76.	DC overcurrent relay	A relay that functions when the current in a dc circuit exceeds a given value.

TABLE 36.1 Standard Devices Used in Diesel Electric Power Plant Systems (*Continued*)

Number	Name	Function and description
77.	Pulse transmitter	A device used to generate and transmit pulses over a telemetering or pilot-wire circuit to the remote indicating or receiving device.
78.	Phase angle measuring or out-of-step protective relay	A relay that functions at a predetermined phase angle between two voltages or between two currents or between voltage and current.
79.	AC reclosing relay	A relay that controls the automatic reclosing and locking out of an ac circuit interrupter.
80.	Reserved for future application	A relay that operates on given values of liquid or gas flow or on given rates of change of those values.
81.	Frequency relay	A relay that functions on a predetermined value of frequency (either under or over or on normal system frequency) or rate of change of frequency.
82.	DC reclosing relay	A relay that controls the automatic closing and reclosing of a dc circuit interrupter, generally in response to load circuit conditions.
83.	Automatic selective control or transfer relay	A relay that operates to select automatically between certain sources or conditions in an equipment or performs a transfer operation automatically.
84.	Operating mechanism	The complete electrical mechanism or servomechanism, including the operating motor, solenoids, position switches, and so on, for a tap changer, induction regulator, or any similar piece of apparatus which otherwise has no device number.
85.	Carrier or pilot-wire receiver relay	A relay that is operated or restrained by a signal used in connection with carrier current or dc pilot-wire fault directional relaying.
86.	Lockout relay	An electrically operated hand or electrically reset relay or device that functions to shut down or hold an equipment out of service or both upon the occurrence of abnormal conditions.
87.	Differential protective relay	A protective relay that functions on a percentage or phase angle or other quantitative difference of two currents or of some other electrical quantities.
88.	Auxiliary motor or motor generator	A device used for operating auxiliary equipment such as pumps, blowers, exciters, and rotating magnetic amplifiers.

89.	Line switch	A switch used as a disconnecting, load-interrupter, or isolating switch in an ac or dc power circuit when that device is electrically operated or has electrical accessories such as an auxiliary switch or magnetic lock.
90.	Regulating device	A device that functions to regulate a quantity or quantities such as voltage, current, power, speed, frequency, temperature, and load at a certain value or between certain (generally close) limits for machines, tie lines, or other apparatus.
91.	Voltage directional relay	A relay that operates when the voltage across an open circuit breaker or contactor exceeds a given value in a given direction.
92.	Voltage and power directional relay	A relay that permits or causes the connection of two circuits when the voltage difference between them exceeds a given value in a predetermined direction and causes those two circuits to be disconnected from each other when the power flowing between them exceeds a given value in the opposite direction.
93.	Field-changing contactor	A contactor that functions to increase or decrease in one step the value of field excitation on a machine.
94.	Tripping or trip-free relay	A relay that functions to trip a circuit breaker, contactor, or equipment or to permit immediate tripping by other devices or to prevent immediate reclosure of a circuit interrupter if it should open automatically even though its closing circuit is maintained closed.
95 to 99.	Used only for specific applications on individual installations for which none of the assigned numbered functions from 1 to 94 is suitable	

36.9

Item	Symbol	Description
Ground	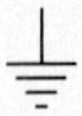	Ground
Lamp	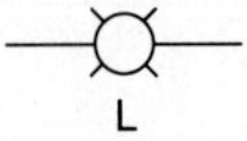 L	Indicating light
Meters, instruments	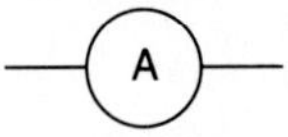 A	
Potential (voltage) transformer	3 PT 2400/120 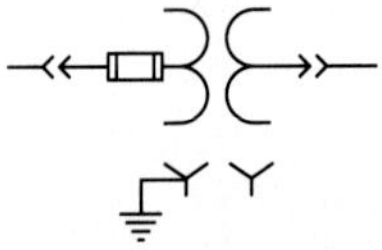	With drawout feature
Rectifier half wave (dry-type)	Rect. 	
Rectifier full wave (dry-type)	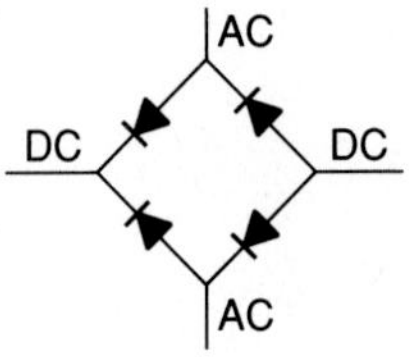 AC / DC / DC / AC	
Relays, operating coil	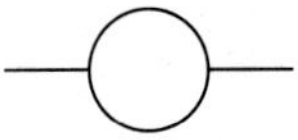	The relay - device or function number should be placed within the circle.
Rheostat	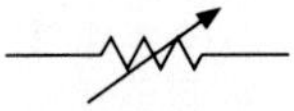	Rheostat (high power)

FIGURE 36.1 Basic electrical symbols (*Extracted from Y-32.2.3-1949, rev. 1988, Graphical Symbols for Pipe Fitting, Valves and Piping, with the permission of the publisher, the American Society of Mechanical Engineers, 145 East 47 Street, New York, N. Y. 10017*).

Item	Symbol	Description
Battery	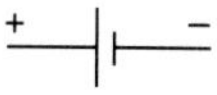 + ─┤├─ −	Single cell (multicell repeated)
Cable termination	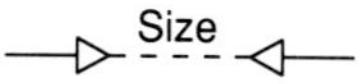 ─▷ Size ◁─	Cable underground or in conduit
Capacitor	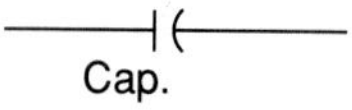 ─┤(─ Cap.	
Circuit breaker (air)	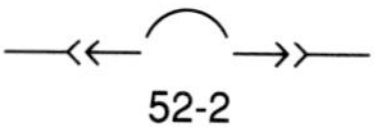 ─«←⌢→»─ 52-2	ACB: generator 2
Circuit breaker (power)	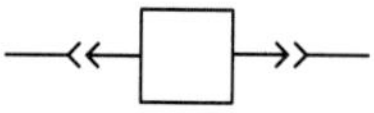 ─«←☐→»─	Breaker with drawout feature
Current transformer	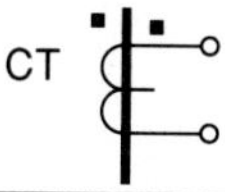 CT	Instantaneous polarity markings
Diode	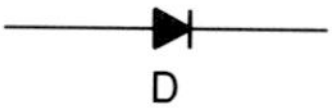 ─▶├─ D	
Fuses	 ─▱─	Fuse disconnecting switch
Generator	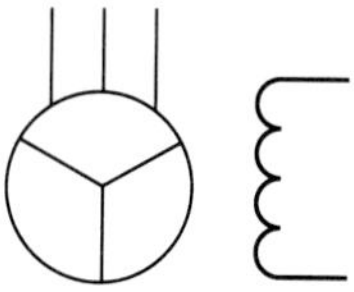	3-phase synchronous generator

FIGURE 36.1 (*Continued*) Basic electrical symbols.

Item	Symbol	Description
Switches		Flow speed actuated
	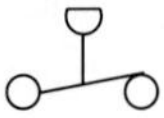	Pressure actuated
		Temperature actuated
	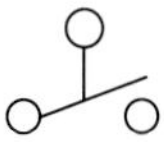	Liquid level actuated
	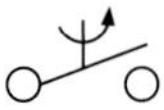	Speed level actuated
Tripping devices		Thermal overload tripping device
Transformer winding connections	△	3 - phase 3 - wire delta
		3 - phase solidly Y - grounded neutral

FIGURE 36-1 (*Continued*) Basic electrical symbols.

36.3 STANDARD ABBREVIATIONS USED IN ELECTRICAL DIAGRAMS

AC	Alternating current
ACB	Air circuit breaker
AIL	Amber indicating light
AM	Ammeter
AMP	Ampere
ANN	Annunciator
ARM	Armature

AS	Ammeter switch
AUTO	Automatic
AUTO-TR	Autotransformer
AUX	Auxiliary
BAT	Battery
BAT CGG	Battery charger
BC	Back-connected
BCT	Bushing current transformer
BIL	Blue indicator light
BKR	Breaker
BPD	Bushing potential device
BV	Back view
CAP	Capacitor; capacity; capacitance
CAT	Catalog
CC	Closing coil
CKT	Circuit
CONN	Connect
CONT	Control
CS	Control switch
CT	Current transformer
D-C	Direct current
DIAG	Diagram
DIFF	Differential
DISC	Disconnect
DM	Demand meter
DPDT	Double-pole double-throw
DPST	Double-pole single-throw
DS	Disconnecting switch
DWG	Drawing
ELEM	Elementary
EQUIP	Equipment
EXC	Exciter
F	Farad
FC	Front connection
FDR	Feeder
FLD	Field
FM	Frequency meter
FU	Fuse
FV	Front view
GEN	Generator
GIL	Green indicating light

GOV	Governor
GRD	Ground
HC	Holding coil
HP	Horsepower
HTR	Heater
HV	High voltage
IMP	Impedance
IND	Indicate
INST	Instantaneous
INV	Inverse
KV	Kilovolt
KVA	Kilovolt-ampere
KVAH	Kilovolt-ampere-hour
KVAR	Kilovar
KW	Kilowatt
KWH	Kilowatt-hour
LA	Lightning arrester
LC	Latch checking switch
LS	Limit switch
LT	Light
LTG	Lighting
MAM	Milliammeter
MAN	Manual
MAX	Maximum
MCC	Motor control center
MECH	Mechanism
MFD	Microfarad
MG	Motor generator
MIN	Minimum
MISC	Miscellaneous
MOT	Motor
MRBCT	Multiratio bushing current transformer
MV	Millivolt
NC	Normally closed
NEG	Negative
NEUT	Neutral
NO	Normally open; number
NP	Nameplate
OC	Overcurrent
OCB	Oil circuit breaker
OPER	Operate

PAR	Parallel
PB	Push button
PB SW	Push-button switch
PF	Power factor
PFM	Power factor meter
PH	Phase
PM	Polarity mark
PNEU	Pneumatic
PNL	Panel
POS	Positive
POT	Potential
PRI	Primary
PT	Potential transformer
PWR	Power
RCD	Reverse-current device
RE	Receptacle
REA	Reactor
REC	Recording
RECL	Reclosing
RECT	Rectifier
REG	Regulator
RES	Resistor; resistance
REV	Reverse
RHEO	Rheostat
RIL	Red indicating light
SEC	Secondary
SECT	Section
SEQ	Sequence
SH	Shunt
SOL	Solenoid
SPDT	Single-pole double-throw
SPST	Single-pole single-throw
STA	Station
STD	Standard
STR	Structure
SUB	Substation
SUM	Summary
SW	Switch
SWBD	Switchboard
SWGR	Switchgear
SYM	Symbol

SYN	Synchronism; synchronizing
SYN SW	Synchronizing switch
TB	Terminal board
TC	Trip coil
TD	Testing device; time delay
TDC	Time delay closing
TDO	Time delay opening
TELE	Telemetering
TEMP	Temperature
TERM	Terminal
TPST	Triple-pole single-throw
TRANS	Transformer
UV	Undervoltage
UVS	Undervoltage device
V	Volt
VA	Volt-ampere
VARM	Varmeter
VM	Voltmeter
VS	Voltmeter switch
W	Watt
WHM	Watthour meter
WIL	White indicating light
WM	Wattmeter
YIL	Yellow indicating light

36.4 FUSES

No fuse may be replaced by one of greater capacity without OEM approval. The temporary use of pipe fittings, short lengths of copper tube, or any other makeshift device or substitute for a proper fuse of known capacity is extremely dangerous and must not take place.

All fuse breakdowns must be noted in the incident record part of the daily log. Their cause and remedy must be noted, and any repetition must be thoroughly investigated and rectified.

CHAPTER 37
BATTERIES

37.1 BATTERY

A key to good performance of an electric starting system is the battery bank. The batteries must be well maintained and replaced at suitable intervals if full starting power is to be delivered. Ordinary lead-acid storage batteries are most frequently used in engine starting systems, although other types such as nickel-cadmium (nicad) batteries with longer life characteristics are becoming more popular in spite of their higher initial cost.

Because electric starting motors draw very heavy currents of the order of several hundred amperes for short periods of time, the 20-h ratings are useful primarily as a guide to capacity. A 100-Ah battery could not deliver 400 A for 15 min or 1200 A for 5 min, as might be expected from the rating.

A battery is composed of a number of individual cells connected in series to provide the rated output voltage. For lead-acid batteries the terminal voltage of each cell is 2.14 V; for nicad types it is 1.34 V. Individual batteries also can be connected in series to produce the voltage required. Terminal voltages will drop as the cells are discharged.

Battery energy storage capability is rated in ampere-hours (Ah). The current drawn from a fully charged battery multiplied by the number of hours until the battery is discharged gives the ampere-hour rating. However, because of internal losses, a battery discharged at heavy currents will not deliver as much total energy output as it would at a lower current draw. The rating is therefore expressed in terms of a given discharge time, usually 20 h. For example, a battery rated at 100 Ah could deliver a 5-A output current for a period of 20 h.

Battery efficiency is adversely affected by cold. For a graph of the effects of low temperatures on new and used batteries, see Fig. 37.1.

37.2 BATTERY TYPES

Of the two basic types of battery in common use, lead-acid batteries are encountered most frequently. They are very similar to the batteries used in automotive applications. Nickel-cadmium batteries are used in many power plants; they have a performance superior to that of lead-acid batteries. However, they require much better frost protection because they freeze at −10°C (13°F) and are much more expensive to replace.

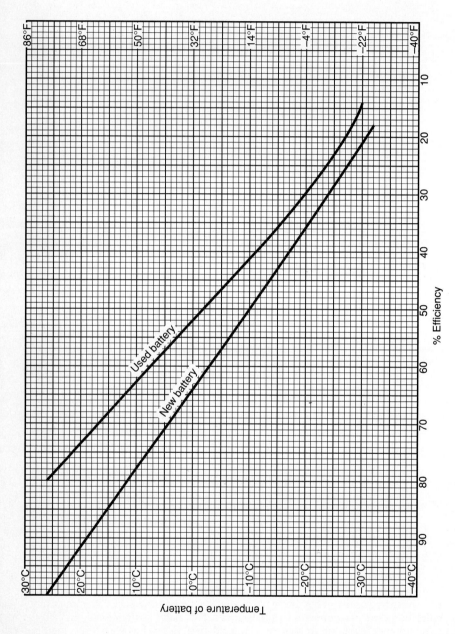

FIGURE 37.1 Effect of temperature on battery efficiency.

37.3 BATTERY ELECTROLYTE DIFFERENCES

The fluid in any battery is known as the *electrolyte*. In a lead-acid battery, the electrolyte is sulfuric acid and water; in a nickel-cadmium battery, it is a solution of potassium hydroxide. Mixing or incorrect application of electrolytes will shorten the life of a lead-acid battery and destroy a nickel-cadmium battery. Both battery electrolytes are corrosive, and care must be exercised in handling them so none is spilled on clothing, skin, or sensitive equipment.

37.4 DISTILLED WATER FOR BATTERY SERVICING

The container used to insert makeup distilled water into lead-acid batteries *must not* be used to service nickel cadmium batteries. Similarly, the container used to insert distilled water into nickel-cadmium batteries must not be used to service lead-acid batteries. Any container used to handle distilled water for battery service must be reserved for distilled water only and must be kept immaculately clean both inside and out. Unfortunately, many of the distilled water feeder containers are made from a nonconductive polyethylene material which is very susceptible to surface scratching and thus becomes very difficult to keep clean. The effort must be made to do so, however, if contamination of all battery electrolytes is to be avoided.

37.5 CARE AND CHARACTERISTICS OF LEAD-ACID BATTERIES AND DC SYSTEMS

The four most common faults that contribute to shortening life of lead-acid batteries are:

1. *Contamination:* Dirt must not be allowed to get into the battery. Small amounts of iron rust will ruin the cell. Clean the caps before removing them, and be sure contaminants do not enter the battery while the caps are off. Be sure that testing and filling equipment is clean. *Do not use equipment which has been used around nickel-cadmium batteries.*

2. *Electrolyte level:* The electrolyte level must cover the plates. The specific gravity of the electrolyte must be checked prior to filling, because a false reading will be obtained until the water and electrolyte are thoroughly mixed.

3. *Overcharging:* Cell temperatures should not be allowed to rise above 43°C (110°F). If overcharging continues for too long, gas bubbles may form in the spongy plate material and cause some breakaway from plate structure. Particles of the plate material will then precipitate to the bottom of the battery casing and form a sludge. Buildup of the sludge will cause short-circuiting of the plates, and plate contact through bulking also will occur. Overcharging can usually be identified by an increased need for makeup water.

4. *Standing while discharged:* Allowing the battery to be left discharged for a

long period of time may result in the formation of lead sulfate on the plates, which will reduce the overall capacity of the battery.

37.6 HYDROMETERS AND OTHER BATTERY TOOLS

Where engine cooling systems or storage batteries are in service, there will be hydrometers. It is important that each hydrometer be clearly identified and reserved for its specific purpose. The hydrometer used for checking the frost protection quality of coolant antifreeze must be used for *only* that purpose. The hydrometer used for checking the gravity of a lead-acid battery can be used *only* for checking lead acid batteries. A hydrometer is not necessary for checking the gravity of nickel-cadmium batteries.

To use any hydrometer for other than its particular purpose will result in some degree of damage to the cooling system or the respective battery.

Use tools with insulated handles when working around batteries. Never use pliers or grips of another type to tighten the battery cell caps. Finger tightness is perfectly adequate. Do not place any tool, wire, part, or other metal object on top of the battery at any time. Do not wear rings, bracelets, or wristwatches when working on batteries, because severe burns can occur if contact is made with the battery posts.

Do not smoke when servicing batteries.

37.7 BATTERY TERMINAL PULLERS

In all cases when it is necessary to remove a battery cable terminal from the tapered type of battery post, terminal pullers must be used. If prying or twisting methods are used to loosen the grip of the terminal on the battery post, there is a grave risk of damage to the battery plates or the battery casing. Inevitably, there will be some internal distortion. The use of regular terminal pullers contributes greatly to extended battery life.

37.8 EXTERNAL PROTECTION OF BATTERIES

The batteries used for starting purposes are frequently set on the floor adjacent to the engines they serve. In that position they provide handy steps for persons to climb up on when working on or inspecting the units, but stepping on any battery for any reason at all is bad practice and cannot be permitted.

All batteries should be set in boxes so that they are fully protected from damage by handcarts, trolleys, rollaway tool boxes, barrel movements, or other passing objects. They must also be protected by a wooden lid so that their tops are protected against dropped tools, tails of chain falls, boots and shoes, and so on. Not only can the tops collapse and damage the plates within; there is a good chance of a total discharge of the battery if a metal object bridges the terminal posts.

37.9 BATTERY WIRING

A considerable portion of the available starting energy in a battery can be lost in cable and terminal resistance. The heavy copper cables used to connect the battery to the starting motor must be examined occasionally for evidence of insulation breakdown, kinking, corrosion, strand severance, or looseness of connection. If a replacement cable is fitted, the length of the new cable must be kept to a minimum, and the correct size cable must be used.

If a starter disability is noticed, one of the first troubleshooting steps must be to check the battery terminals for continuity and then for tightness. Tight terminals do not always mean that there is good contact between the battery post and the cable terminal.

37.10 BATTERY CHARGING

Batteries can be recharged from engine-mounted alternators or, more usually, from external charging equipment. Line-operated chargers can be connected to the generator set output or, in the case of a standby system, be switched between ac power mains and the generator set output, whichever is supplying power at any particular time. The use of some form of external charger is recommended for standby systems to maintain a full charge in the batteries over long periods during which the generator set is not operated.

Trickle chargers are the most common form of line-operated charging devices. For lead-acid batteries, they supply 2.33 V per battery cell. The full rated charger output current is delivered when the battery is completely discharged. As the battery acquires charge, the current tapers off to about 0.25 A.

When long battery life is an important consideration, a float type of charger can be installed. The float types include line and load regulation and current-limiting devices. In order to impart maximum battery life, they usually supply voltage to the batteries at 2.17 V per cell (for lead-acid batteries), but periodically switch to 2.33 V per cell for short times to equalize the cells.

37.11 BATTERY MAINTENANCE

If batteries are to provide long and reliable service, provision should be made for their maintenance. They should be located off the floor and be kept as dry and clean as possible. The electrolyte level should be checked, and only distilled or deionized water should be added if needed. Measurement of specific gravity with a battery hydrometer should be made periodically to determine battery condition and state of charge.

37.12 THE USE OF HYDROMETERS

- It is not necessary to have a hydrometer for the care of nickel-cadmium batteries because the specific gravity of the solution has no effect on the charge.

- Because the hydrometer reading of a nickel-cadmium battery does not give it, a voltmeter must be used to get the state of charge.
- The hydrometer readings taken from a lead-acid battery must be temperature-compensated if they are to be fully accurate.
- The hydrometer used to take battery readings must be perfectly clean and dry. Any foreign matter will inhibit the free movement of the float.
- After sufficient electrolyte is drawn into the hydrometer and the float has assumed a stable position, the gravity may be read.
- Empty the hydrometer immediately after taking the reading. Being sure to return the drawn electrolyte to the original cell.

CHAPTER 38
COMPRESSED AIR SYSTEMS

38.1 COMPRESSOR UPKEEP

Air compressors are installed to serve air-starting systems. Most compressors are started and stopped automatically by pressure-sensing switches, and over a period of time these compressors do a great deal of work. Because of their relative simplicity and reliability, their gradual deterioration often goes unnoticed and their overhaul requirements are overlooked. All too frequently, when they do fail, they need extensive rebuilding, repair, or even replacement.

In the compression of air, a great deal of heat is passed from the air to the cylinder, pistons, cylinder head, and valves. The thermal stresses imposed on the valves are not conducive to long life, and therefore the valves require regular servicing at not more than 500-h intervals. An hour meter is a very useful item to have connected to the compressor motor.

All compressors require maintenance and overhaul. The maintenance consists of regular oil changes, adjustment, and occasional changeout of the drive belts. The fins of the cooling coils must be inspected and cleaned if necessary. The air intake filters must be inspected, cleaned, and changed if so warranted. The overhaul work will include servicing the suction and delivery valves, renewing the piston rings, verifying the safety valves, cleaning out the lubrication system including the sump and the suction strainer, and changing out the filter when one is fitted.

The high temperatures developed at the upper end of the cylinder will also contribute to the accumulation of carbon formation on the valves and the piston rings. Regular cleaning is necessary to remove those carbon deposits if the compressor is to continue to work effectively.

Compressors will readily overheat if the air filters are dirty, if the suction, and particularly the delivery valve, is leaking, or if the air cooler is fouled. The operator must include an inspection of the compressors during his tour of the plant at least once per shift and record in the log sheet or workbook notes about deficiencies found on inspection.

38.2 COMPRESSOR MAINTENANCE SCHEDULE

Every Shift, Operator's Duties

- Drain water from trap or separator (if such a device is fitted).

- Drain air vessels of water.
- Check compressor lubricating oil.
- Check engine lubricating oil.

Every Week, Mechanic's Duties

- Tip relief valves to test operability.
- Check belt tension.
- Clean air intake filters.

Every 3 Months, Electrician's and Mechanic's Duties

- Change lubricating oil in compressor (or 250 h).
- Inspect motor or engine thoroughly.
- Inspect for frayed or cracked belts.
- Check function of pressure switches.

Every 6 Months, Mechanic's Duties

- Overhaul cylinder heads; lap LP and HP valves.
- Inspect valve springs.
- Remove crankcase cover; clean lubricating oil strainer.
- Inspect lubricating oil pump and bottom end bearings.
- Inspect intercooler for dirt outside and carbon inside; clean as required.

On multiple-belt-drive systems, do not attempt to renew one belt only. All belts must be changed together.

38.3 COMPRESSOR DIFFICULTIES

Most compressor failures are attributable to lack of inspection and irregular maintenance, and most failures result in the almost complete destruction of the compressor. Ensure by reference to the equipment manual or maker's data plate that lubricating oil of the correct specification is used. In most cases, it is not possible to use engine crankcase oil. If lubricating oil appears to be emulsified or foamy, it should be changed and the cause of the oil deterioration should be determined before restarting the compressor. Probable causes are a mixture of oils, overheating, excessive blowby, water inclusion, wrong oil specification, and overfilling.

Overhot running of the compressor may be traceable to leaking valves (LP delivery, HP suction and delivery in particular), worn piston rings, plugged air filter, or fouled cooler. Excessive carbon on the LP and HP valves is usually caused by poor piston rings (scraper in particular) or air drawn from an oily atmosphere. Slow pressure buildup can be caused by worn piston rings, worn-down bearings, or badly seated valves. Compressors must not be operated without their belt and pulley guards being properly fitted.

38.4 PORTABLE AIR COMPRESSORS

These units must be inspected regularly, and the motor and compressor must receive proper attention. Check for any dents or other damage to the receiver; do not use a unit with a deformed receiver. Wheeled or portable air compressors must not be lent out to any noncompany organization or personnel. All portable air compressors must be fitted with a grounded power cable and all belt guards, and so on, as supplied by OEM.

38.5 ENGINE-DRIVEN COMPRESSORS

In plants equipped with both motor-driven and engine-driven air compressors, the motor-driven compressor should be used for everyday service and the engine-driven compressor reserved for emergency use. It is recommended that the engine-driven unit be exercised on a weekly basis. If a gasoline engine is fitted, a gasoline supply should be kept at some point remote from the engine and preferably external to the power plant. Great caution must be exercised in filling the engine fuel tank. Gasoline supplies are not to be kept in the power plant.

The lubricating oil sump of the engine, be it gasoline or diesel, should be checked by the shift operator on every shift. (Rarely used engines are prone to coolant leaks or fuel drip flooding the sump.) If the compressor engine is liquid-cooled, the cooling system must be inspected for coolant level and frost protection. If it is air-cooled, anything that might impede the free passage of air around the engine must be removed. The drive belts must be in good order, and so must the safety valves, the pressure gauges and the pressure shutdown. The compressor engine should be supplied with the correct type and grade of lubricating oil with reference to its fuel.

38.6 AIR VESSEL UPKEEP

Most air vessels are subject to annual government inspection. None of the original qualification marks stamped on the vessel may be painted over, ground off, or otherwise defaced. The vessels must at all times be maintained in a condition that could pass such an inspection. Pressure relief valves must always be in first-class working order. Fusible plugs must be inspected where fitted. If any deformation is apparent, they must be renewed.

Every year, each vessel must be blown down, the hand-hole doors must be removed, and the interior must be thoroughly cleaned and inspected for corrosion pitting. At this time, the blow-down pipe must be removed and checked visually for blockage and corrosion. Pressure gauges on all vessels must be checked monthly for working condition. (It is illegal to knowingly operate a certified pressure vessel with a defective gauge.) Any replacement gauges must be of best quality, and the discarded gauges must be destroyed.

The importance of regular and proper air vessel maintenance cannot be overstressed. Extremely dangerous conditions can develop in air vessels and the air lines connecting them to the engines if water, oil, or carbon deposits are allowed to accumulate in the system. It is essential that any replacement fitting used in an air pipe system be suitable for the working pressures. Standard pipe fittings are not normally acceptable.

No attempt should be made to carry out adjustment or change any component on a compressed-air system without fully and totally relieving the pressure therein beforehand. Any work on a certified pressure vessel must be carried out by a journeyman mechanic.

38.7 PRESSURE-RATED FITTINGS

Any replacement fittings or additional features installed in the compressed-air system must have a working pressure rating commensurate with the system maximum pressure.

38.8 ISOLATING VALVES

None of the isolating valves in the compressed-air system should need a wrench or other tool to open or close them. It must be possible to easily open or close all valves by hand, and any valves that leak or are too hard to turn should be serviced without delay. Hard-closing a valve will not prevent a leakage past a defective valve or seal.

38.9 AIR VESSEL ISOLATION

For economy's sake, it is suggested that the valves of the air receivers be kept closed whenever possible. If a control air supply is necessary, only the one service receiver should be in use. It should not be necessary for the compressors to run several times daily just to maintain storage pressure.

38.10 SHOP USE COMPRESSED AIR

The pressure of compressed air for shop use or for cleaning purposes must be kept to the absolute minimum. Clinical tests show that medical hazards exist when the skin is exposed to a direct blast of air at greater than 240 kPa (35 psi). *At no time should a compressed air blast be directed toward any person, nor should it be used to blow clothing clean. Horseplay of any description with an air hose must be strictly forbidden.*

CHAPTER 39
STARTING SYSTEMS

39.1 STARTING METHODS

Starting systems rely on energy stored over relatively long periods of time and released quickly when the engine is cranked. Since only a limited amount of energy can be economically stored, the starting time of the engine must be held to a minimum. Once the stored energy is used, it may take anywhere from several minutes to a day to accumulate enough energy for a second start; the actual time will depend on the condition of the batteries, charger, and compressor. In some systems, as in automotive practice, the energy storage device is recharged by equipment driven by the engine itself once the engine has started. Energy must be available from an outside source if the engine fails to start at the first attempt.

Three basic types of starting motors are available for generator set engines: electric, air, and hydraulic. These motors operate on energy stored in auxiliary devices such as batteries and pressure tanks. The energy can be supplied by equipment driven by the generator set engine or separate charging equipment.

Any one of the three systems can be connected to a control system for remote or automatic starting of the generator set. This feature is particularly useful in unmanned standby power systems which start automatically upon the loss of electric power from the prime source.

39.2 AIR-STARTING SYSTEMS

At least once per shift, the operator should check the starting air pipes on the engine at points both before and after the air-starting distributor for signs of overheating. If one or more of the pipes is unusually hot, it is probable that an air-starting valve is not properly closed. If and when a condition of this sort is suspected, it must be logged and the log statement acted upon without delay. Considerable difficulties in subsequent engine starting will be experienced if remedial action is not initiated. Cylinder head air-start valves will leak if foreign matter is passed through the system.

It is important that the operator drain the starting air receiver of any oil, water, or other substance during each shift. Water in the air receiver will cause corrosion. Further, if a slug of water is introduced into the air-start piping, it will be very dangerous. Moving at a considerable velocity, the water may burst the air-start pipe, particulary at a bend or elbow, and any person in the vicinity will be at risk.

The air receivers or vessels must always be closed once the start-up process has been completed. If the discharge valve is left open, there is always a possibility of air leaking to the cylinder head start valve and building up enough pressure to lift the starting air valve off its seat.

Lubricated air-starting motors are fitted with oil collector jars. The jar must be emptied regularly and always before a start. If the jar is allowed to fill, lubricating oil will carry over into the air motor exhaust and an oil mist will be sprayed all over the area, which will then have to be cleaned up by hand (Fig. 39.1).

39.3 AIR-STARTING VALVES

Leakage of the starting air inlet valve in the cylinder head can be detected by checking the temperature of the pipes connecting the air-starting distributor to the cylinder heads. Carbon formations will develop in air-starting valves and their cages when the valves are improperly seating, which will permit combustion products to get past the valves when the engine is running. The valves will also leak when the starting air supply remains open and some pressurization of the air-starting system takes place to a level that negates the function of the air-starting valve springs. A third source of carbon in the air-starting valves and their cages is lubricating oil that is entrained in the starting air-stream. This third cause will occur only if draining of the starting air receivers and the oil separators is neglected. The air-starting valves must be gas-tight if they are to enable successful starts to take place.

39.4 ELECTRIC STARTERS

Starting a generator set involves cranking the engine at a relatively low speed until combustion occurs in the cylinders. Cranking is accomplished with a motor which is gear-coupled to the engine flywheel (Fig. 39.2). After the engine starts, the gear drive is automatically disengaged and the motor is stopped.

The starting system requirements of diesel and gasoline engines are essentially the same, although variations in ambient temperature will have a greater effect on the cranking time required by a diesel engine. The diesel engine relies on the high temperature produced by the compression of air in the cylinders to ignite the fuel. When the engine is cold, much of the heat of compression developed in the initial strokes is lost and the engine must be cranked longer or at higher speeds to reach combustion temperature. Gasoline engines start at about 50 to 60 rpm, whereas a 1200 to 1800-rpm diesel engine starts at approximately 100 rpm.

In diesel engines fuel is delivered to the cylinders almost immediately, especially if the injector system is fully primed. They therefore usually have better start-up response than gasoline or natural gas engines in subzero temperatures. The average cold-starting time for diesels is 5 to 7 s, but the starting time will be increased at low temperatures because the viscosity of the lubricating oil is greater. That makes the engine more difficult to crank.

Electric starting allows fast, convenient, push-button starting with relatively lightweight, compact, engine-mounted parts. The system is ideal for automatic or remote starting. It is not adaptable to low ambient temperatures, however, unless starting aids are provided. At 18°C (0°F) battery output power drops to about 50

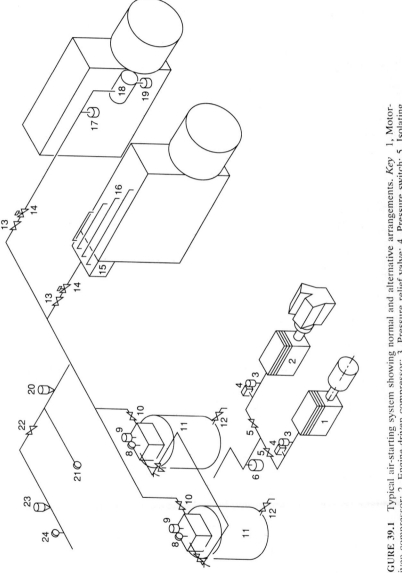

FIGURE 39.1 Typical air-starting system showing normal and alternative arrangements. *Key* 1, Motor-driven compressor; 2, Engine-driven compressor; 3, Pressure relief valve; 4, Pressure switch; 5, Isolating valve; 6, Oil separator; 7, Inlet valve; 8, Pressure gauge; 9, Relief valve; 10, Outlet valve; 11, Air vessel; 12, Drain valve; 13, Isolating valve; 14, Nonreturn valve; 15, Distributor; 16, Cylinder connections; 17, Lubricator; 18, Air-start motor; 19, Exhaust oil trap; 20, Pressure-reducing valve, AUX services; 21, Pressure gauge; 22, Isolating valve; 23, Pressure-reducing valve, shop services; 24, Pressure gauge.

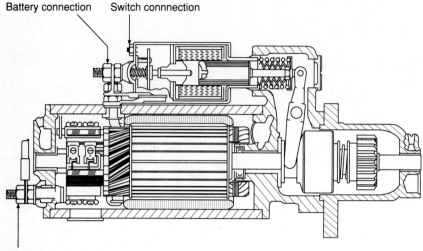

FIGURE 39.2 Starter connections.

percent of the 27°C (80°F) rating. The batteries in low-temperature conditions require frequent maintenance, regular charging, and heating. Electric systems are not suitable for explosive environments (e.g., oil well drilling, oil refineries, and mining). Jacket heating, lubricating oil heating, and battery blankets are standard cold weather–starting aids.

A standard electric starting system includes the starting motor(s), an overrunning clutch to disengage the starting motor when the engine fires, a solenoid switch for control of the motor, and, for a diesel engine, a fuel priming pump. The electric starter motor, clutch, pinion, and solenoid are supplied as a single unit. Starter motor voltage requirements differ with size and type. Common voltages are 12, 24, and 32 V dc.

The electric starters employed in internal-combustion engines are generally similar in concept, but there are a few differences between the design of light vehicular types and that of heavier, industrial engines. Starter motors are very robustly built and are very durable. Because of the heavy current draws, the armature and field coils are wound with heavy-gauge wire and thus have few turns per coil. Similarly, the commutator has only a small number of segments.

A small engine starter is normally of the pinion shift type, and it can be readily identified by its pinion outboard of the bearing. When a start is initiated, the entire armature moves forward within its stator to engage the attached pinion with the flywheel teeth and then rotates. In the inertial gear drive type of starter, a solenoid external to the starter operates a forked lever to engage the pinion with the flywheel teeth while the armature remains in the one axial position when rotating. This type of starter is usually identified by the shroud carrying the rear bearing ahead of the pinion.

A defective starter motor may have enough power to turn an engine over but not fast enough to start it. This condition must not be confused with defective or discharged batteries. If an engine does not rotate freely during the starting sequence, there will be an abnormal draw on the batteries or air receivers. The en-

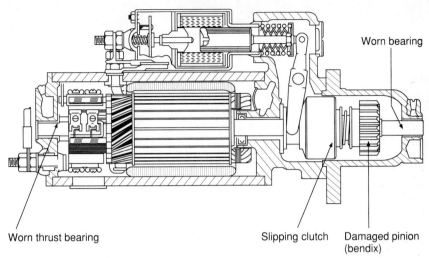

FIGURE 39.3 Common starter mechanical troubles.

ergy normally imparted to the engine to turn it over will be inadequate, and the battery voltage or air pressure will drop quickly. This characteristic must be recognized early in any subsequent fault-identifying routine (Fig. 39.3).

39.5 STARTER DEFECTS

Common defects found in starter motors are rough commutator surfaces, worn or badly seating brushes, misplaced brush springs, shorted windings, bent shafts, damaged clutch springs, and defective solenoid coils or lever mechanisms.

The mica insulation between the commutator segments must be undercut. If it protrudes, it will prevent the brushes from making full contact with the commutator surface. There will be much sparking, undue copper wear, and brush edge chipping. Most heavy-duty starters have a removable plate or plug for the inspection of the brushes. The brush spring loading must be adjusted to keep the brushes in good contact with the commutator (Fig. 39.4).

Poor lubrication will lead to failure of either or both bearings; failure may also be attributed to a sluggish pinion return. The pinion remaining in engagement with the flywheel starter ring will cause the motor to be driven at excessive speed after the engine has started. A starter may fail to disengage from the starter ring. It is quite probable that this defect has its beginnings in one or the other of the following:

- Sticking or dirty start switch contacts
- Sticking or dirty solenoid contacts
- Sticking starter relay where fitted
- Loose or improperly positioned wiring on the solenoid

When an electric starter has remained in engagement after the engine has accelerated beyond cranking speed, damage to the starter may include deformation

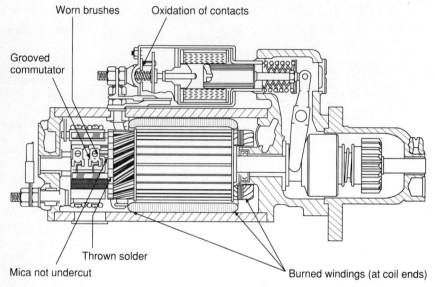

FIGURE 39.4 Common starter electrical troubles.

of the starter pinion, degraded bearings, fractured rear bearing housing and pinion support bracket, armature overspeed defects, and breakdown of coil insulation.

If an electric starter is damaged through disengagement failure, it is not enough to change the damaged starter. The cause of the problem must be cured. Following a starter pinion (Bendix) failure, it is always necessary to examine the entire flywheel starter ring gear for damage. It is not uncommon to find percussion damage to gear teeth in groups at 90° or 120° apart. If the damage is not too severe, the leads on the teeth can be dressed and use of the ring gear can be continued pending the fitting of a replacement. Ordinarily, however, it is advisable to change a damaged ring without delay. Ring gears are cheaper than starters, but they require more time to install. All damaged electric starters can be traded in for exchange rebuilt units.

The primary switch contacts in the starter solenoid must be clean. The inability of these contacts to function is often initially mistaken for a defective start switch or a flat battery. Dirty or burned solenoid contacts may prevent current from being passed to the starter. There can be as many as eight external terminals on the starter, the starter solenoid, and the starter switch, any one of which may be loose.

The appearance of bright solder on the commutator or armature coil ends with or without solder splatter in the starter housing is indicative of gross overheating.

Before a starter which is thought to be defective is turned in for exchange, it should be thoroughly examined. The defect may be quite minor and easily field-fixed.

A jammed starter can sometimes be freed by moving the engine in a reverse direction. Starter jamming is frequently caused by loose fastening screws or by worn leads on the ring gear teeth.

39.6 JUMP STARTING (BOOSTING)

If an engine has a discharged battery, it can be started by using an auxiliary battery, a procedure known as *boosting* or *jump starting*. Jump starting can be successfully carried out if certain precautions to prevent personal injury or property damage are observed:

1. Wear eye protection when connecting a booster battery; in fact, wear eye protection whenever servicing, cleaning, connecting or disconnecting batteries. Never wear rings, bracelets, or wristwatches when working on batteries.
2. When attaching the jumper cables to the battery terminal posts, be certain that the jumper cable clamps do not touch any other metal or each other. The same precautions must be taken when removing the jumper cables.
3. The risk of battery explosion can be minimized by keeping open flames, welding sparks, lighted cigarettes, and other means of ignition away from the battery area. That is particularly important in the case of discharged batteries, when the danger of ignitable gas will be greater.
4. The voltage of the booster battery must match the voltage of the electrical system to which the battery is being connected. If there is any doubt whatever, *do not make the connection*. Before any connection is made, the polarities must be verified. When the connections are made, they must be positive to positive and negative to negative.

39.7 JUMP STARTING (AUXILIARY EQUIPMENT)

Before attempting a start, switch off all possible battery-powered auxiliary equipment. In some installations a battery-powered prelubrication pump may be required to operate as part of the starting sequence. The prelubrication pump should be isolated, the control circuit bypassed, and the lubrication system primed by means of the hand pump. The electrical drain on the auxiliary battery must be reduced to an absolute minimum.

39.8 JUMPER CABLES

The jumper cables must be very firmly attached to solid metal and must be double-checked to see they are clear of any part of the engine that will be in motion after the attempted start. They must be of sufficient capacity to carry the intended starting current, and it is recommended that a set suitable for in-plant service be selected and placed in a permanent position of readiness (i.e., on the shift operator's tool board). If jump starting by using a vehicle with its battery in place is contemplated, the operator must be absolutely certain that the vehicle is not in any metal-to-metal contact with the unit that is about to be started.

39.9 STARTING DIFFICULTIES ON STANDBY ENGINES

When there is difficulty in starting a standby, emergency, older, or other rarely used engine, the blame is often put on an inadequate starting air or battery capacity. The original installation without doubt had sufficient starting capacity, and the operator must therefore look to the starting air pressure, the battery voltage, the fuel system priming, and the status of the protective devices.

Either the time must be taken to fully prime the fuel system by hand prior to the start attempt or the engine must be exercised at intervals. Some engines will lose their prime much more readily than others. The whole fuel injection system must be totally loaded with fuel before the starting sequence is commenced. The protection system must be checked out to see that no devices have been worked upon since the last run. All the fuel supply valves must be checked.

39.10 STARTING AIDS

Occasionally an engine will develop hard-to-start characteristics, which can be classified as either one-time events or habitual difficulties. The one-time events will usually be identified by reference to the symptom table in Sec. 5.3, Possible Faults, but the habitual events will probably develop as the engine ages. They result from premature wear or incomplete or inadequate upkeep practice.

It can be said that manufacturers do not supply engines that need starting aids and that any engine that requires a starting aid once it has been in service or in its subsequent lifetime is deficient. An engine in such a state will need attention.

An unethical but frequently used starting aid is an ether aerosol. Ether can have very detrimental effects on the engine, and it should be used only in the most extraordinary circumstances. It is a powerful solvent, and it will destroy the lubricating oil film on a cylinder liner upon contact. It is also a highly volatile fuel, an intoxicant, and a major safety hazard.

Before attempting to use ether to assist a start, the operator must ask himself: "Is it safe to attempt this start?" If the engine was previously difficult to start, he must follow up with the question: "Has anything been done to improve the starting ability of this engine?" If he then decides to utilize ether in the start-up, he must have a helper standing by with a CO_2 fire extinguisher and suitable contoured blanketing material.

As soon as the engine fires, both the operator and helper must be ready to strangle the engine if it runs up to overspeed by applying the blanketing to the air filter intakes. Engines that have frequent ether-assisted starts have notoriously short lives. Their early demise can be attributed to scored liners and pistons, broken rings, excessive valve guide wear, and burned valve seats.

Ether is a highly volatile substance, and it must not be used in the presence of an open flame. It should be stored away from the power plant and under lock and key. It should not be used by anyone other than a responsible person who is fully trained in its handling and its attendant hazards. Juveniles should not be permitted to have access to it or its used containers at any time. Saturation of the air filter element with ether will send the engine into an uncontrollable runaway condition with calamitous results.

Every starting incident that demands ether assistance must be reported to the superintendent immediately after it has taken place and also be recorded in the workbook. The superintendent will then take steps to normalize the starting methods for the engine involved.

CHAPTER 40
ENGINE TURNING

40.1 ENGINE TURNING SAFETY

No attempt should be made to start an engine when the turning gear is engaged. Severe damage and personal injury could result. When the turning gear on an engine is engaged, it is standard practice to place an advisory board stating "turning gear in" over the engine starting controls returning it upon withdrawal to the turning gear where it is normally stored. The superintendent should arrange for such a notice to be provided and used.

An engine should not be turned by means of the turning gear by any one man working alone. Safety demands that two persons at least be present, one on each side of the engine. Both must pay strict attention to the engine movement and be alert for any possible obstruction. The fuel must be positively shut off from the engine; the governor load limit knob must be in the zero position, and the indicator cocks must be open.

Before commencing the turn, the mechanic in charge of the work should inspect the engine and generator for such obstructions as loose parts, tools, and ropes. He must ensure that the turning gear is fully engaged and properly secured. Although it falls within the mechanic's work to insert and withdraw the turning gear, it is the operator's duty to see that the gear is clear of the engine before commencing a start sequence.

No engine is to be turned by "jolting" it with the starter. If the engine has to be turned anywhere from a few degrees up to a full turn or more, it must be done by barring. If the starter is used, there is little control. Also, there is the danger of the engine starting with severe damage to either man or machine resulting. The engine should be turned only in its normal rotational direction. Any reverse turning must be avoided unless the elementary precaution of consulting the OEM handbooks for comment is taken.

40.2 REVERSE TURNING

Most engines can be barred over manually or by using a turning gear in either direction as supplied by the OEM. However, there are engines such that the emergency shutdown mechanism will be damaged and rendered inoperative if the engine is barred over more than one turn backwards.

It is recommended that each engine that is identified as falling into this category be plainly marked so that all personnel concerned are well advised. In any event, the mechanical shutdown devices that would be subject to damage should be thoroughly checked for integrity at each overhaul. If the work makes it essential to reverse-turn, the protective devices that might suffer must be temporarily removed from the unit. If an engine is to be barred round by hand, it should be done as shown in Fig. 40.1.

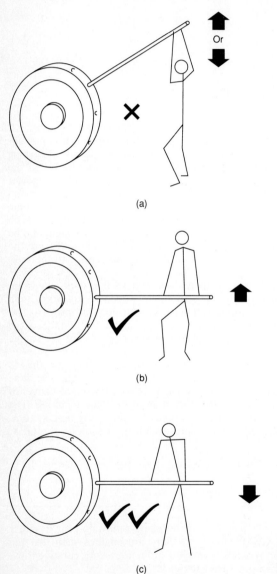

FIGURE 40.1 Engine barring by hand. (*a*) Correct method; (*b*) and (*c*) incorrect methods.

CHAPTER 41
ELEMENTARY POWER PLANT SAFETY

41.1 GENERAL RULES

True *safety* is an awareness on everyone's part of the possibility of an accident at any time. All power plant personnel must have a positive attitude toward safety at all times.

The superintendent is responsible for seeing that the work performed by himself or his personnel is carried out with full regard for the safety of the employees, the public, and the company. To make and keep all company operations safe, both the superintendent and the workers must be aware of the permanent and incidental hazards in the work area. The risk of accident is minimized when the whole job site is regularly inspected and any unsafe conditions are rectified.

When unsafe practices are observed, the superintendent will institute appropriate action. He will not permit contractors or other visiting personnel to work in an unsafe manner, nor will he allow any employee to work when he is impaired by fatigue, illness, or other debility that might affect his attitude toward safety.

Plant safety will be optimal if these practices are followed:

- All personnel must read and understand all warning signs, notices, decals, and plates before operating any of the power plant machinery or doing any work in the power plant.
- All personnel must be sure that their work area is a safe place in which to work and be aware of surfaces that may be slippery. They must ensure that all ladders, walkways, platforms, and so on, are secure.
- All personnel must wear protective eyeglasses and hard-toed, leather-soled footwear in conformity to universal owner's safety attire requirements. They must not wear loose-fitting or torn clothing. To avoid burns, personnel should wear long-sleeved garments with buttoned cuffs.
- All power and hand tools must be in good order. Personnel should reject hammers with loose heads, wrenches that no longer fit the intended fastener properly, sockets that are split, or any other tools that will not work properly.
- The electrician must regularly check that all power tools are properly grounded. Nobody should use any electric tool that may have been wetted by

water or coolant until it has been thoroughly inspected by an electrician. When a grinder is used, care must be taken to see that sparks generated do not go near any crankcase opening, exposed bearing, alternator, or switchgear.

- Assistance must be found when a weight of more than 25 kg (50 lb) has to be lifted. Nobody admires a strong fellow with a damaged back. All lifting equipment such as slings, hooks, and shackles should all be in good order. Do not side-load eyebolts.
- When touch-testing for heat during the walk-around inspection, the habit of using the back of the hand must be developed. If anything is wrong and the hand gets inadvertently burned, repair of the machine will be necessary and the palm side of the hand will be needed to grasp the tools.
- Do not open cooling systems, lubricating systems, or crankcases immediately after the engine has stopped. Allow the engine interior to cool down for at least one hour before removing a crankcase door.
- When removing covers, flanges, or other fittings from the normally pressurized sectors of a system, be sure the component is loose before final removal of the fasteners. Be sure that any residual pressures have been relieved. Similar caution should be given to any spring-loaded components.
- When test-running any machine, all guards, rails, fences and shields must be in place.
- Never disconnect a battery when the battery charger is working. Sparks in the vicinity of batteries can cause the hydrogen-oxygen vapor released from the electrolyte to explode. Handle battery acid or electrolyte with caution; it has harmful effects on flesh, fabric, and most metals. Do not weld near batteries.
- Batteries must be equipped with wooden covers to prevent any metal object that may fall onto them from shorting them out. Batteries shall not be used as steps or work platforms.
- Before commencing work on any machine, be sure that the machine is properly switched off, isolated, immobilized, and flagged.
- When testing fuel injectors, do not subject any part of the body to the atomized fluid spray. Use the correct test fluid to obtain proper injector performance and at the same time reduce the fire hazard.
- Remove all gasoline cans, aerosols, ether spray cans, and other highly volatile media from the power plant operating areas. Place them in cold, remote storage.
- Never test an oxygen, acetylene, propane, or any other gas cylinder by opening its valve in the power plant. Store all gas cylinders chained in an upright position away from the power plant.
- Use tools which are safe and suitable for the job. Keep tools away from energized electrical equipment so they cannot contact live circuits or equipment.
- Keep up to date with procedure modifications for operation of electrical equipment and systems.
- Use the company-approved procedures for switching. All established clearance procedures must be observed without exception.
- No electrical apparatus cabinet may stand with its doors open, covers removed, or otherwise have its internal parts exposed without having a clearance tag showing. All conduit junction boxes, MCCs, busbars, or any other parts of

the electrical system must have their covers or doors in place and securely fastened. They must remain so whenever the equipment is in service.

All work areas must be kept clean of waste materials, rubbish, and other debris. Be sure that such materials are collected and removed from the job site promptly. Oily rags and other flammable materials must be removed from the job site before the end of the standard working day. The work area must be left in an orderly and safe condition at the end of each day.

41.2 ELECTRICAL SAFETY

No matter how experienced or knowledgeable an operator may become, he must always exercise the maximum amount of caution when he gets involved in fault identification and troubleshooting of electrical apparatus or circuits.

Every operator must take the view that any electric circuits are potentially lethal. Until he is absolutely sure of what he is about to do, he must not even open a cubicle panel or junction box. The operator and anyone else with the exception of a qualified electrician must assume that all circuits are live. Before any activity commences, the precautions of opening breakers, pulling fuses, obtaining clearance, and securing assistance must be met.

The following are some of the elementary precautions that must be observed:

- No work may be done on live circuits by anyone other than a certified electrician. Two people must be present, and the work must be authorized by the superintendent in writing.
- Any meter and its attachments used for electrical test work must be clearly suitable for the voltages and currents that may be encountered.
- Entry to live transformer yards and substations must be limited to certified personnel and then only with the written clearance of the plant superintendent in the absence of standard clearance procedures.
- Officially approved nonmetallic hard hats must always be worn when working around and in transformer yards, substations, and transmission structures.
- Wear approved rubber gloves for operating disconnects in the plant or when using a hot stick.
- Hot sticks must be kept clean and dry and used only by authorized personnel. They should be stored inside the power plant when not in use.
- Rubber gloves must be tested regularly, and their use should be restricted to journeyman personnel.
- Use grounded cords for all power tools.
- Keep all electrical equipment and areas clean and dry.
- Keep fuse pullers in an accessible position.
- Keep switchgear doors closed.
- Keep switchgear cubicles externally dry and clean of oil and dirt.
- Keep the provided covers on all electrical apparatus and switchgear.
- Keep covers on motor connection boxes.
- Protect electrical cable insulation from cutting damage or external heat sources.

- Identify the correct locations of buried cables before any digging commences.
- Be sure any equipment is deenergized before it is grounded. Attach ground cable securely. Ground all phases of three-phase work systems. Check newly installed ground cables before use; look for loose connections, frayed wiring, and defective clamps.

41.3 PRECAUTIONS WHEN WORKING AROUND ELECTRICAL CURRENT

All voltages are dangerous. Contact with even a very low voltage can result in a serious accident when the shock causes a workman to fall from a ladder or scaffold. Under certain conditions, human beings have been severely shocked by circuits carrying as little as 50 V.

When switches are opened for repairs, alterations, or examination, they must be locked or blocked open and the fuses removed. After these precautions have been taken, DANGER tags must be attached to all open devices. Tags must remain on open devices until they are removed by the workman handling the work. If he leaves the job without removing a tag, it may be removed by someone else only after thorough investigation.

"Dead circuits" must always be treated like live circuits until they are proved to be dead. That practice will develop a caution that may prevent an accident. When testing circuits to determine the value of any voltage present, always use a voltmeter which is set for its highest range.

Before starting work on any rotating machine that is out of service, make a thorough check of all electrical control and starting devices. When any part of such equipment is remotely controlled, lock open the circuit breakers or switches, provide danger tags, and remove the fuses.

41.4 PURPOSE OF CLEARANCE

Formal clearance procedures are necessary in power-generating facilities to guard employees from injury through the inadvertent activation of equipment when work is being performed. Clearances are also intended to protect machines and apparatus from damage caused by activation when work is being carried out on them. By requiring the use of clearance procedures, the owner is assured that an established degree of safe work practice is being maintained.

All utility companies employ some type of clearance procedures, but the actual drills in each company will vary considerably in detail. All personnel attached to the power plant must be acquainted with the clearance procedures that apply to their own organizations, and any term, casual, or contract labor, must be made aware of the applicable procedures before they are permitted to commence work. It is expected that the individual company's clearance procedures will be regularly discussed at periodic plant safety meetings.

The clearance permit system is so designed that whenever any work is to be carried out on a particular piece of equipment, the equipment is in fact safe to be worked on and will remain in that condition throughout the period of the work process.

A *clearance permit* constitutes permission to do the work. It is issued by an authority, usually the station superintendent, and it states what is to be done to what equipment in concise and exact terms. Any extension of the work beyond the original clearance will require further permission from the superintendent. The superintendent, being fully conversant with the reasons for and the hazards involved in the work, will issue the clearance permit.

The permit is issued only to the person who will do the work, and it must be returned to the issuing authority when the work is complete. The issuing authority is responsible for effecting the necessary safety isolation and for canceling the permit.

To have a clearance permit system operate successfully, it is essential that employees realize that it is primarily for their personal safety and also for equipment protection. A person assigned to carry out work on a particular piece of plant equipment must make application either written or verbal to the issuing authority. The issuing authority is responsible for effecting the necessary safety isolations and for canceling the clearance permit when the work is completed.

41.5 CLEARANCE PROCEDURES

Proper clearances are necessary in power plants to guard against accidental injury to employees through equipment being activated when work is being performed. Some guidelines for safe practices in power plants follow:

Steps to Follow in Issuing a Clearance Permit .

1. The workman requests and obtains a clearance permit from the issuing authority.
2. The operator fills out the clearance permit form and the necessary clearance tags required.
3. The operator deenergizes the equipment and removes the fuses or locks out the breaker as required. Clearance tags are always attached to the locks.
4. The operator closes valves and locks and tags them as required. He depressurizes the equipment if necessary.
5. The operator locks the breaker and attaches the tag to the lock.
6. The workman signs the clearance permit and receives the keys for the clearance locks from the operator.
7. When the work is completed, the workman signs the clearance permit release and hands the keys to the operator.
8. The operator removes the locks and tags and returns the equipment to service.
9. The completed clearance is filed.

41.6 MACHINE GUARDS

To protect personnel from injury, guards are fitted over exposed moving components throughout the plant. The guards, which may be made of sheet metal, wire, expanded metal, fiberglass, or steel tubes, must always be in place when the machinery is operating. At no time should anything be placed upon or leaned against these guards nor must they be used as steps, ladders, or work platforms. No guard is designed to carry any load, and most guards will readily collapse or distort under the lightest stress.

The operator must not permit any person to remove or adjust a guard over any machine that is in operation, nor must he permit a machine to be started up without all of its guards properly and firmly in place.

Guards or other protections will be found on V-belt drives and pulleys, couplings between pump and motors, external viscous dampers, air intake louvers, ventilating fans and radiator fans, exhaust manifolds, uninsulated hot pipes, vertical ladders, gauge glasses, manholes, and cable troughs.

No attempt should ever be made to wipe down the guard over a rotating part of a machine, nor should an attempt be made to wipe down behind the guards of an operating machine.

Report any broken or otherwise defective guards in the daily log, and ensure that the machine involved is tagged accordingly.

41.7 EMERGENCY LIGHTING

In a plant equipped with an emergency lighting system, it is good practice for the operator to test the system once per shift. The plant electrician should check the emergency lighting system batteries once every week. At that time he should also inspect the illuminated exit sign at each doorway.

It is particularly important that the emergency lighting in the remote or less-frequented parts of the plant complex be in good operational order. If the operator experiences an outage, he must be able to proceed to the control area quickly and safely. The operator's flashlight or lantern should be kept in one place only when it is not in actual use so that it can easily be found when needed. Spare dry cells also should be available at short notice. *Candles, propane lanterns, or other flame-lighting methods must not be used at any time within the power plant complex.*

In a plant provided with portable emergency light units it would be proper to station such a unit in any area where work is in progress. If an outage occurs, the workplace with its tripping or falling hazards must be well illuminated to improve the safety factor for the operating personnel.

41.8 FIRE PROTECTION

- The name, address, and telephone number of the local fire department must be permanently displayed adjacent to the power plant telephone that the plant operator normally uses and also externally at the power plant entrance.
- It is essential that all power plant personnel be instructed in the basic use of the plant's firefighting equipment along with the location of each piece of equipment as part of their introduction to the plant immediately following their hiring.
- At regular intervals, say, every 3 months, all of the plant staff without excep-

tion should attend instructional sessions at which firefighting, safety, and protection are discussed along with practice in using one or more items of the plant's firefighting equipment.
- The personnel must be aware of which medium can be applied to fires of specific origins; they must know the established procedures of alarm, communication, and evacuation and understand the factors of personal safety applicable to firefighting.
- If welding or cutting takes place within the power plant, the duty operator or a person so nominated must make himself available as a precautionary fireman and be armed with at least one 10-kg dry powder extinguisher.

It can be said with some justification that all fires in diesel power plants are attributable to some defect in operating or upkeep practices. Most fires have their origins in neglectful attitudes toward cleaning, carelessness in handling inflammable liquid, failure to recognize or limit overloading or overheating, or lack of vigilance and control of events by the operating staff. In some areas, a lightning strike may provide ignition in a potentially hazardous situation, but it is decidedly incumbent on the power plant staff to minimize the opportunities for such a fire to develop.

The refueling of motor vehicles must not take place within power plant buildings. If refueling must take place on the power plant property, it must be done in the open at least 20 m (50 ft) away from any building.

Before the bulk storage tanks are replenished, the operator in charge must ensure that all of the standard protection devices in the fuel area are in a fully operative condition. The community safety officer should also be advised that a major fuel movement is to occur.

41.9 FIRE EXTINGUISHERS

Purely for safety purposes, all diesel power plants will have a certain number of fire extinguishers available for service at all times. These extinguishers should be located in positions of some prominence so they will readily come to hand under emergency conditions. The most favored location for any fire extinguisher is immediately beside an access door in a position of high visibility backed by a colored panel. *No object may be placed on or over a fire extinguisher at any time.*

Any fire extinguisher placed in its position and ready for use must be fitted with a seal that indicates its serviceability. In a safe power plant, each fire extinguisher will also have an inspection record tag to signify the date of the last inspection and overturn. A label on the body of a portable fire extinguisher identifies the class of fire for which the extinguisher can be used. All personnel shall be familiar with such labels.

All power plant personnel must be familiar with the location and intended duty of every fire extinguisher sited around the power plant complex. This knowledge must be acquired by every new employee so that a fire can be attacked without hesitation or fumbling. No matter who fights the fire, he must always be aware that an exit is available to him if the fire gets beyond control.

Danger is just as great following the extinction of a fire as it was immediately before, and therefore all extinguishers must be available at all times. It is recommended that the recharging supplies (i.e., powder and pressure capsules) be divided into two lots and stored in separate locations so that if one lot is lost the other lot will probably survive.

In the event of a fire breaking out within the power plant, the first duty of the operator in charge must be to see that all personnel who are not directly attending the fire leave the building immediately. In any event, the operator in charge is responsible for the evacuation of all people from the building unless relieved of that responsibility by the officer in charge of a regular firefighting agency.

During routine inspections of the plant's firefighting equipment, each dry powder extinguisher should be turned over and shaken to loosen up the settled powder. If that is not done, the extinguisher may not work properly when needed because the released gas will be unable to insufflate the packed powder.

The best fire protection system in any diesel power plant is no better than the care that should prevent the conditions necessary for a fire from developing. The amount of attention given to housekeeping is reflected in the safety of the plant buildings and the equipment housed within.

The following points have to do with the care and use of firefighting equipment:

- All firefighting equipment must be in good working order at all times. If for any reason a fire extinguisher is deficient, it must be immediately recharged, resealed, and returned to its rack.
- When possible, CO_2 extinguishers should be used on electrical fires.
- Avoid directing a CO_2 extinguisher on skin or clothing, because it freezes on contact. (CO_2 will form dry ice.)
- Dry powder chemical extinguishers may be used, but they are not as effective as CO_2 extinguishers. Dry powder use leaves an extensive cleanup job behind.
- When the fire is out, assemble the fire extinguishers for refill. Any extinguisher of either type which has the seal broken should be checked over and refilled as soon as possible.
- Some larger plants have CO_2 units on wheels for use on electrical fires. Keep such extinguishers away from fuel storage areas and switchgear but readily available nearby.
- CO_2 extinguishers must be weighed to be sure they are still usable and have not leaked down and become useless. Weigh each extinguisher regularly and record the weight and the date.

41.10 ANNUAL FIRE EXTINGUISHER MAINTENANCE AND USE

If an extinguisher has to be sent away from the plant for test, refill, or repair and its removal will leave the area unprotected, it must be replaced with a spare of similar capability. If an extinguisher being sent away for hydrostatic testing has a valve at the end of the hose to control the discharge, the hose must be tested at the same time as the extinguisher.

Inspection of CO_2 Hand-held Fire Extinguishers

1. Weigh the extinguisher with hose and nozzle attached. Check with previous stamped and recorded weight.
2. Check the date of refill and hydrostatic test against the plant record. Send the extinguisher away for test if it is out of date.
3. Remove the hose and inspect the hose and nozzle assembly. Inspect internally to see that neither the extinguisher valve nor the cylinder shows signs of corrosion.
4. Ensure that the nameplate operating instructions are legible.
5. Reinstall the hose and nozzle and reseal.
6. Finally, the person doing the fire extinguisher work will record the date and extinguisher weight on the tag attached to the extinguisher, along with his signature.
7. Return the extinguisher to its location. Be sure that the operating instructions are visible and the location is free from obstruction.

Dry Chemical Pressurized

1. Check that the pressure gauge reads normal, the hose and nozzle assembly are in good condition, the container and valve show no sign of corrosion, the nameplate operating instructions are visible, and the seal is not broken.
2. Check the date of the last filling and hydrostatic test against the plant record.
3. Record the date and extinguisher weight; sign the tag; and attach the tag to the extinguisher.
4. Return the extinguisher to its location. Be sure the operating instructions are visible and the location is not obstructed.

Dry Chemical (Rechargeable Type)

1. Check that the extinguisher date of refill and hydrostatic test correspond with the plant record. If necessary, send the extinguisher away for retest.
2. Release any entrapped pressure that might be in the extinguisher and remove the cartridge, hose, and extinguisher cap. Weigh the cartridge and install a new one if the old one has leaked down.
3. Check that the hose and nozzle are in good condition and the valve and container are not corroded.
4. Check that nameplate operating instructions are visible.
5. Check the level of the powder in the extinguisher and agitate the powder so that it will not be caked.
6. Check the expellant tube for damage or clogging.
7. Check all seal areas, reassemble the extinguisher, and weigh the extinguisher

after assembly. Record the date, and the extinguisher weight on the tag. Sign the tag and attach it to the extinguisher.
8. Return the extinguisher to its location. Be sure that the operating instructions are visible and the area is unobstructed.

In-Case-of-Fire Guidelines

1. Call the fire department and supervisor as soon as possible. Emergency phone numbers must be posted by the operator's telephone.
2. If possible, shut off the source of the combustible material.
3. Proceed as far as possible to extinguish the fire while keeping in mind the safety of all personnel. No person should allow the fire to come between himself and the nearest exit.
4. When helpers arrive, instruct them about the nature of the fire and the dangers involved.
5. After the fire is extinguished, assess the situation as to damage with particular reference to what power production can be restored without a recurrence of the fire.
6. Make every effort to remain calm. Discourage panic in others if possible.

Procedure to Extinguish Fires

All fires are divided into three distinct classes. Each must be extinguished with the proper agent as follows:

Class A. Ordinary combustible such as wood, coal, and paper: any material that will leave an ash. This class of fire can be extinguished with water or dry chemical.

Class B. Flammable liquids such as oil, gas, and paint. This class of fire can be extinguished with dry chemical or CO_2 extinguishers.

Class C. Electrical fires; care must be exercised to prevent electrical shock. Class C fires are preferably extinguished with dry chemical or CO_2, although a fog nozzle on a fire hose may also be used. CAUTION: *Do not use a solid-stream nozzle in a Class C fire.*

41.11 FIRE EXTINGUISHER POWDERS

Purple. Potassium bicarbonate is a white powder also known as potassium acid carbonate. When used as a firefighting medium, a purple dye is often included.

Yellow. Ammonium phosphate, a white crystal soluble in water, is a salt of ammonia phosphate and phosphoric acid. It is frequently used as a fire retardant.

White. Sodium carbonate is a white water-soluble crystal also known as baking soda or sodium acid carbonate. It loses carbon dioxide when heated to 270°C (518°F). It is commonly used for fire extinguisher training.

41.12 LIQUID FIRE EXTINGUISHER HAZARD

Certain fire extinguishers utilize a fluid medium that in the past has been used as a cleaning agent. The fluids trichlorethylene ($CHCl:CCl_2$) when heated becomes phosgene gas—the dreaded mustard gas of World War I. Another fluid is carbon tetrachloride (CCl_4), which is also an effective cleaner for electrical equipment. The use of this material is quite hazardous because the fumes will attach to the human liver and the damage is not repairable by either natural or medical means. Therefore, no plant will use liquid-type fire extinguishers employing either of these chemicals.

41.13 SAFE PRACTICES IN FIGHTING ELECTRICAL FIRES

Isolate the Area

Shut off the power supply to the fire-affected area. Should the fire be in a breaker cabinet, open the breaker if you can do so safely. If you are in doubt, it may be necessary to isolate a section of the switchgear.

Generator Fires

When there is a fire in a generator, the excitation must be switched out immediately and the unit stopped at once. If the generator is running alone, there will be no time to start up another unit and transfer the load to it. It is imperative that the fire be controlled before it burns any of the laminations of the rotor or stater.

Should the generator be running in parallel with another, the excitation switch of the burning unit must be tripped followed by the feeder breakers. This will leave the good generator set carrying just the station service load. The burning unit is stopped and the fire is brought under control preferably by the use of CO_2 extinguishers.

Cable Tray Fires

The circuits affected in a cable tray should be disconnected and the fire isolated as well as possible. Be aware that any fire may have contributing sources. A fire in a covered cable tray under a floor may be also fed by fuel, natural gas, or oil. A fire of this nature must be treated as an electrical fire until all affected circuits have been disconnected.

41.14 BURIED CABLES

Many diesel power plants have equipment placed outside the main building and served by underground power cables. The cables may run in conduits or simply be buried at various depths. As part of the regular inspection tour, the plant operator and the electrician must inspect the ground in the vicinity of all buried cables to see that no hazard is developing.

Conditions leading to buried cable damage may be unauthorized parking or passage of overweight vehicles over buried cable runs, skidding heavy equipment over the cables, digging without proper site plan consultation, and water or fuel flooding of the area where buried cables are located. All subterranean cable runs should be clearly identified by surface markers and hazard warnings, and the cable must be protected from vehicular traffic damage wherever it emerges from the ground.

41.15 ELECTRICAL EQUIPMENT PROTECTION

There are numerous devices which protect either individual generating units from distribution system faults or the distribution system from generator faults. The first line of electrical equipment protection in any power plant is the operator's own good sense and knowledge. Jiggling voltmeters, swinging frequency meters, and unbalanced phase amperages indicate something amiss. Odors of burning materials, scorching cloth, and overheated paint are distinctive; they warn of possible electrical trouble in the making.

It is most important that the operator get into a routine of questioning each observed aberration, recognizing its significance, and taking suitable action. He must not fall into the awful habit of relying on alarms and protective devices to do his job for him. He must remember that whereas a person can monitor many hundreds of conditions and situations, only one condition can be monitored by each device.

41.16 PORTABLE ELECTRIC TOOLS

Electric shock is the chief hazard in the use of electrically powered tools. When such tools are used, they are to be grounded at all times. Temporary light or power lines (including extension cords) used for power tools must have at least three wires, one of which must be the ground wire.

Never use defective electrical tools. When tools cannot be repaired immediately, they must be tagged as defective and returned for repair.

41.17 SAFETY INSTRUCTIONS FOR A POWER OUTAGE

The plant superintendent is reminded that he is ultimately responsible for the safety of all persons while they are on the company property. For the benefit of contractors or other visiting work people, it is advocated that prework safety instructions be prepared and permanently posted. Visitors must be aware of what they can or cannot do in the event a power outage or other operational difficulty takes place while they are in the power plant. The instructions will vary considerably from one plant to another and will depend on the size and manning characteristics of the individual plant. Some basic rules that may be applied are:

- Do not occupy the telephone.
- Do not interfere with the operator.
- Attempt to help only if specifically asked to do so.
- Do not start any machine or equipment without permission.
- Keep clear of all machines including those being worked upon.
- Remember that the operator is in charge.
- Do not run.
- Do not shout.
- Stay out of the walkways and control room.

41.18 WELDING AND CUTTING

- Before starting to burn or weld, sweep floors clean, wet down wooden flooring or cover it with sheet metal or its equivalent.
- When doing outside work, avoid letting sparks fall where they may start a fire or burn other workmen. Be certain that there is no flammable material below the floor or work area.
- Do not permit hot metal slag to fall through cracks in the floor or into machines or machine parts.
- No person may work alone with oxyacetylene equipment; a second person must be on hand as watchman, and he must have a fire extinguisher close at hand.
- During the operation, have fire extinguishers and fire pails ready along with a firewatcher standing by.
- After completing a burning or welding operation and for the next half-hour, check the scene of work frequently for signs of smoke or smoldering fires. Also, inspect adjoining rooms and floors above and below the welding site.
- Do not use cutting or welding equipment near flammable liquids or on closed tanks which have held flammable liquids or other combustibles.

41.19 ACETYLENE AND OXYGEN TANKS

Acetylene and oxygen cylinders are to be handled with extreme care because they contain gases under high pressures. Cylinders must always be kept in an upright position either in a cylinder buggy or chained to a vertical structure (a building column for example).

Valve wrenches must be retained where the cylinders are in use. When in use, the acetylene cylinder valve should not be open more than one full turn. The operating pressure of the acetylene must not be more than 15 psi. All connections between the cylinder and the torch must be checked for leaks, and the gauges and hoses must be inspected for any signs of damage. Nobody shall work or watch an oxyacetylene operation without wearing proper approved protective goggles.

Oxygen bottles must not be vented where there is a naked flame or a running engine. Oxygen may not be used to blow clothes, the person, or equipment clean.

Cylinder caps must always be replaced when the cylinders are not in use and when they are in storage or in transit. The regulators must be removed and stored safely and cleanly. Empty cylinders must be returned to the supplier as soon as possible to reduce rental charges to a minimum. Empty cylinders remaining at the plant represent wasted money.

Propane and butane are heavier than air. Leakage of these gases leads to loose concentrations in pits, stairwells, and cable troughs. These gases have a distinctive cat's nest odor that serves as a warning of a potentially disastrous situation. Acetylene has an odor akin to that of onions.

In the event any gas leak is suspected, all gas cylinder valves must be turned off immediately and all flame or spark sources shut down. The area must then be fan-ventilated until all signs of the gas have been dispersed. Only then may the gas leak be searched for.

When loose acetylene gas is mixed with the right ratio of air and there is some form of ignition (hot exhaust, cigarette, or motor spark) a most spectacular and ruinous bang can take place with most destructive consequences. Be cautious!

41.20 HAND TOOL SAFETY

Hand tools must be maintained in good condition at all times. Do not use a pipe wrench or pliers with worn or broken teeth; do not use a socket, box, open-end, or adjustable wrench that is worn, cracked, or broken.

Split or loose handles in hand hammers or sledge hammers must be replaced with new and properly fitted handles before the tools are used. Defective handles must never be wired or taped and kept in use. Hand files should never be used without a proper handle. Files should be kept apart from each other: The wear on file teeth working against each other is far greater than that of their actual use.

Always keep screwdrivers properly ground and their handles in good condition. Never use a screwdriver as a driftpin or cold chisel. Use hand tools only for the tasks for which they are designed. Cold chisels, center punches, and so on, must be dressed periodically to prevent their becoming mushroomed.

The use of gasoline blow torches is prohibited.

Avoid the use of test lamps. Use a voltmeter.

41.21 LIFTING EQUIPMENT

- The overhead crane, when not in use, must be left in such a position that overhead lights are not obstructed, no work in the plant is hindered, and both the crane hook and the control box are parked clear of walkways or moving machinery.
- Crane loads traversing the power plant should be carried at no greater than floor-clearing height whenever possible. Loads should never be carried over and above operating machinery unless it is unavoidable.
- The use of chains and grab hooks for load lifting should not be permitted within the power plant. Nylon slings of correct load rating (SWL) should be used whenever possible.

- The shift operator must be aware of and monitor lifting procedures if they can affect safe power plant operations in any way.
- The crane operator should walk behind the load—never in front. The load must never pass over a person; either the load must stop or the person must move.

41.22 ATTACHMENT OF LIFTING EQUIPMENT

Lifting equipment in the form of come-alongs, chain hoists, or rope blocks should be attached only to structures that are clearly able to take the intended loads. At no time should any pipe system, cable tray or conduit, ladder, stairway, catwalk, or other auxiliary structure be used for load-lifting purposes. The use of the power plants structural steel work for lifting purposes can be permitted only after the load-bearing capabilities of the structure have been carefully defined.

If there is any doubt about safe lifting practices, the district crane inspector should be consulted or the superintendent should discuss the matter with a mechanical engineer at the owner's headquarters.

41.23 SAFE MANUAL HANDLING OF MATERIALS

When any heavy object is to be lifted and carried to another point, first inspect the area and route over which the object is to be carried. Make sure there is nothing in the way that might cause slipping or tripping.

Inspect any object to be lifted to determine how it should be picked up. Make sure it is free of sharp edges, protruding nail points, splinters, or other hazards that might cause injury to the hand or body. Where the sling passes over a sharp edge, wood fiber or metal packing must be inserted between the edge and the sling.

Incorrect lifting methods often cause strain or other types of injury. When it is necessary to lift any object which is overweight or difficult for one person to handle, ask for help.

Care must be taken when opening crates and boxes to avoid injury. Nails protruding from boxes and crates and metal bands and ends of wires must be bent over or removed. Crates, boxes, and covers must be disposed of promptly and not be piled in or near a passageway. All packing material such as excelsior, shredded paper, and straw must be disposed of at once because such material is a fire hazard. All solid material boxes must be stored and stacked in a safe and secure manner ready for future use.

41.24 PORTABLE LADDERS

Only approved ladders that are equipped with nonslip feet or spurs may be used. Because metal ladders are conductors of electricity, their use is prohibited wherever the ladder might contact a live circuit.

It is recommended that all ladders be of either fiberglass or wooden construction. Whatever type of ladder is used, the ladder must be carefully examined for defects on every occasion that it is used. Any ladder found to be cracked, loose, or otherwise unsafe must be condemned. Ladders that cannot be properly and permanently repaired must be replaced. The damaged ladder must be destroyed.

41.25 PERSONNEL WORKING TOGETHER

Employees must not be permitted to work alone on any potentially hazardous operation or in any isolated location. Hazardous operation is defined as any work assignment which involves actual exposure to unprotected electrical circuits, to flammable, explosive, radioactive, toxic, or corrosive chemicals, or to unguarded machine hazards which can cause serious injury. Potentially hazardous isolated locations are those which include manholes, tanks, vaults, underground tunnels, transformer yards, and roofs.

"Working alone" means that the employee cannot be seen and heard by another worker. When it is necessary for one man of a pair to leave for any reason, all work must be discontinued until both are again in contact. When returning to a job in progress on electrical equipment, recheck connecting switchgear, breakers, and so on (on both input and output circuits) to make sure that no part of the equipment has become energized during the absence. Be sure helpers know the hazards of the type of work to be performed. An explanation before starting the work may help prevent an accident.

41.26 PERSONAL HEALTH AND THE POWER PLANT

Dermatitis is an inflammation of the skin. It may be, but is not always, caused by the use of various solvents, exposure to gasoline or diesel fuel oil, some metallic elements such as chromates, zinc, and cadmium, or other products such as fiberglass insulation.

The lighter-complexioned persons who have the drier skins are usually more sensitive to any exterior influence. Human skin is slightly acid with a pH of 6.8, and the presence of a layer of natural oils serves to retard moisture vaporization from the skin. It also has some shielding properties. Therefore, any exposure of the skin to a material that tends to either eliminate the oils or neutralize the acids will cause skin deterioration and encourage a dermatitic situation to develop.

Dermatitis is not common, nor does it have sinister implications. It is avoidable by practicing a little common sense and hygiene. Wash the body and clothes often. Use a "protective barrier" hand cream. Wash hands and arms before eating and at the end of each shift. Report any evidence of skin irritation, and get prompt attention for cuts and abrasions. Do not wear footwear or garments that have become saturated with liquid materials. Do not use gasoline, cleaning solvents, and similar materials for removing oil and grease from the skin. Do not spray petroleum products in the presence of other personnel. Do not use diesel fuel for testing injection equipment; use regular test oil only.

41.27 FIRST-AID KIT

Any power-generating facility will have a first-aid kit accessible to all personnel. The comprehensiveness of the kit is often left to the discretion of the plant owner; as a rule, the larger the plant the more extensive the kit. This usually results in the large-plant first-aid kit carrying six instead of two 1-in bandages, 50 instead of 10 adhesive plasters, and so on, which merely means that the frequency of replenishment remains the same in large or small plants.

It is suggested that the probability of accident to an individual in a larger plant is not much different than the ratio of the number of employees in the large and the small plant. The type of incident will, however, be more varied in the large plant over a given period of time. Therefore, it is suggested that the range of bandage and plaster sizes, medication and other contents of the first-aid kit be supplemented by some more advanced first-aid items. It may be that the plant safety committee will choose to select the supplies or the local first-aid authority will be able to give advice.

A use-record book is generally provided with each first-aid kit, and each use incident must be written up. This document is then focused on at each plant safety meeting: Each incident is discussed with a view to furthering accident prevention. In many localities there is a legal requirement to record all first-aid box uses. Furthermore, a comprehensive accident report must be passed on to the owner following each incident.

Depending on the owner's directives, all plant visitors should be advised of their insurable coverage and liabilities so the owner is protected.

41.28 EYEWASH STATIONS

As a part of the first-aid equipment of many of the larger power plants, eyewash stations will be found in the vicinity of the battery banks or near boiler water treatment areas. These eyewash stations are very rarely used, but they must always be maintained in first-class working order and also at the highest level of cleanliness. It is recommended that regular inspection and cleaning of the eyewash equipment be part of an assigned inspection and maintenance tour of the first-aid and safety provisions at approximately weekly intervals. Dirty eyewash douches may do more harm than good.

Any person who is impelled to use the eyewash station must enter the incident in the first-aid book.

CHAPTER 42
TERMINOLOGY

42.1 TERMINOLOGY USAGE

Science and its governing laws become workable only after the basic definitions and terminology are understood. Today, such terms as *volt*, the *second*, the *gram*, and the *meter* are well known and employed with complete understanding by most people.

Nevertheless there are many words and terms employed in diesel power maintenance and operating activities which are often misapplied, incorrect, or inappropriate. Furthermore, there are many terms which have the same approximate meaning as others. The adoption of a common terminology within the engine upkeep business would go a long way toward the elimination of confusion and misunderstanding. The reader is encouraged to refer to the terminology section whenever possible.

42.2 COMMON TERMS

ac: Current which varies from zero to a positive maximum to zero to a negative maximum to zero a number of times per second. The number is expressed in hertz or cycles per second.

ac generator: See *alternating-current generator*.

acceleration: Rate of speed increase.

acceptance: The activity of viewing a completed project, witnessing its demonstrated performance in accordance with specifications, and accepting the whole from the contractor as safe, efficient and complete. *The owner accepts; the constructor commissions.* (See *commission*.)

actuator: A mechanism that activates control equipment upon a pneumatic, hydraulic, electrical, or electronic signal.

additive (lubrication): Any material added to a lubricating oil during manufacture to improve the oil's suitability for service. It may improve a property already possessed by the lubricant or give the oil properties not naturally possessed. Typical examples are antioxidants, corrosion inhibitors, and foam inhibitors.

aftercooler: A cooler designed to remove excess heat from the aspirating air as the air leaves the turbocharger.

alignment: The adjustment of components in a system to obtain a proper linear relation about a common axis (e.g. an engine crankshaft or a generator rotor shaft).

alternating-current generator: A machine, usually rotary, which converts mechanical energy into alternating-current electric energy. Also known as *alternator* or *ac generator*.

ambient temperature: The temperature of the surrounding air mass.

ammeter: An instrument for measuring electric current flow in terms of amperes.

amortisseur: A short-circuited winding in the rotor of a synchronous generator. It consists of conductors embedded in the pole faces and connected together at both ends of the poles by end rings. Its function is to damp out oscillations or hunting during load changes, and it is an important feature on diesel-driven generators. Also known as a *damper winding*.

ampere: The unit of electric current flow. One ampere will flow when one volt is applied across a resistance of one ohm.

anhydrous: Devoid of water. With regard to a lubricant, one in which there is no water.

antifreeze: A substance added to a liquid to lower its freezing point. (For example, ethylene glycol added to water will prevent freezing to $-40°C$.)

antioxidant: An additive, usually incorporated in a relatively small proportion, to retard oxidation of lubricants including greases and gear lubricants.

apparent power: A term used to describe the product of current and voltage, expressed in kilovolt-amperes. It is the real power, in kilowatts, divided by the power factor.

arcing: Discharge of electricity across a gap.

armature: The part of an electrical rotating machine that includes the main current-carrying windings in which the electromotive force is produced. (See *stator*.)

aspirating air: The air that is induced into the cylinder of an engine and is necessary for combustion of the fuel.

astatic: Having no tendency to change position. (See *isochronous*.)

atmospheric pressure: The natural pressure of air. At sea level it is 100 kPa (14.7 psi).

biodegradable: Capable of being reduced to or broken down to innocuous material by action of living beings, especially microorganisms.

block heater: A heating element normally inserted in an engine cylinder block. It is frequently used in automotive applications. (See *jacket heater*.)

boost: The air pressure being delivered to an engine. The term *boost* is used only in connection with a pressure-charged engine. Also known as the *air inlet manifold pressure*.

brush: A conducting element which maintains sliding electric contact between a stationary and a moving element.

bus: An electrical conductor, generally taken to be the main conductor in switchgear.

capacitor: A device capable of storing electric energy. It consists of two conducting surfaces separated by an insulating material. It blocks the flow of direct current while allowing alternating current to pass.

cavitation: In a fluid circulation system, the collapse of air bubbles on the pressure side of the pumping arrangement, the fluid having been drawn from the suction side. It may also be caused by boiling in a cooling system. Whatever its origin, cavitation causes severe metal damage.

cetane number: A measure of the ignition quality of a diesel fuel. It indicates the rate of combustion following injection.

check (noun): Minute surface cracks apparent in overheated iron or steel bearing surfaces.

check (verb): Verify correctness of sight, sound, touch, or smell.

Circuit: A path for an electric current.

Circuit breaker: A switching device for opening and closing an electric circuit.

collector rings: The conducting rotating rings of a generator rotor connecting windings through brushes to a source of direct current. Also called *slip rings*.

commission: To prove and demonstrate the combined material and equipment of a project as a working entity, free of defects, ready for service, and suitable for acceptance. *The constructor commissions; the owner accepts.* (See *acceptance*.)

condenser: (See *capacitor*.)

conductor: A wire or cable for carrying current.

contactor: A device for establishing and breaking an electric power circuit.

controlled rectifier: (See *SCR*.)

coolant: A generic term for a liquid medium that carries away excess heat.

corrosion: The gradual decay of a mineral or metal surface through chemical action.

cotter: A form of key round in section with a flat machined on one side and tapering off at approximately 0.025 percent per unit length. (See also *cotter pin*.)

cotter pin: Otherwise known as a split pin. (See also *cotter*.)

cps: Cycles per second. (See *frequency*.)

cranking: Turning an engine by means of the starting system.

cranking speed: The speed at which the engine must be rotated by the starting system in order to commence firing.

crankpin: The part of a crankshaft upon which the bottom end (or large end) bearing runs. (See *journal*.)

cross-current compensation: In parallel operation of generators, a system which permits the generators to share the reactive component of the power in proportion to their ratings.

CT: See *current transformer*.

current tranformer: Any instrument transformer, generally with a 5-A secondary, used in conjunction with ammeters and control circuits.

current: The flow of electric power expressed in amperes.

cycle: One complete reversal of an alternating current or voltage from zero to a positive maximum to zero to a negative maximum back to zero. The number of cycles per second is the frequency. Also, one complete vibration movement.

damper: A device which absorbs and dissipates vibration energy.

dc: Current which is unidirectional.

dc generator: A generator which transforms mechanical energy into unidirectional or dc electric energy.

dead band: The speed range in which no measurable correction is made by a governor. The range is usually within 0.5 percent of rotational speed of the engine.

deceleration: Rate of speed decrease.

deenergize: To isolate from electric power source. (See *energize*.)

deflection: The measurable elastic movement of a component under load from its nominal axis.

delta connection: A three-phase connection in which the start of each phase is connected to the end of the next phase to form a configuration resembling the Greek letter delta (Δ). The load lines are connected to the corners of the delta. See also *Y connection*.

dermatitis: An inflammation of the skin.

detergency: Characteristics imparted to a lubricating oil by suitable additives. See *detergent additives*.

detergent additives: Detergent added to a lubricating oil to help suspend sludges, varnish, and carbon in the oil and prevent their deposition on wearing surfaces. Detergents in lubricating oil do not clean existing deposits of contaminants.

detergent, cleaning: A cleaning agent akin to soap in its ability to emulsify oil and suspend dirt. Detergents contain soluble compounds that reduce the surface tensions of liquids and break the interacting tensions between two liquids or a liquid and a solid. Detergents of varying formulation are used in both common and exotic cleaning materials.

detergent, lubrication: See *detergent additives*.

dielectric: An electrical insulator.

dielectric strength: The maximum voltage that a material can support without rupture, usually stated in volts per millimeter of thickness.

diode: In general, a component having two electrodes. Specifically, a solid-state device which allows current to pass in one direction but not in the other. Since it allows only the positive half-cycle of an alternating current to pass, its output will be unidirectional, for which reason it may be considered to be a rectifying element.

distribution panel: The panel to which the output of a generator is supplied and where the current is divided to supply different loads. Generally, it contains circuit breakers and protective devices. More commonly known as a *feeder panel*.

drift: A gradual change in voltage output sometimes caused by an increase in generator temperature or regulator lag.

droop: The difference, expressed as a percent, between the no-load and full-load rotational speeds of a prime mover.

electrolyte: A nonmetallic conductor of electricity in which current is carried by the movement of ions. An example is battery acid: sulfuric acid and water.

emulsion: A mixture of water and oily material in which either very small drops of water are suspended in oil or small drops of oil are suspended in water the whole being stabilized by a third component (such as soap or glycol) called an *emulsifying agent*. Emulsibility is desirable for some products such as soluble cutting fluids.

energize: To apply electric power to a circuit or machine.

engine: A device which converts energy, which may be potential, into mechanical force. Examples are diesel engines, water turbines, and windmills.

ethylene glycol: A colorless dihydroxy alcohol used as an antifreeze for cooling fluids. (See *propylene glycol*.)

exciter: A device for supplying excitation to generator fields. It may be a rotating exciter, that is, a dc generator or an ac generator with rectifiers, or it may be a static device using tubes or solid-state components.

exciter current: The field current required to produce rated voltage at rated load and frequency.

exciter voltage: The voltage required to cause exciter current to flow through a field winding.

field: The windings of the generator rotor which, when supplied with direct current, will establish the magnetic field.

film strength: The ability of a film of lubricant to resist rupture due to load, speed, or temperature.

filter: A device for trapping and retaining solids down to a designated particle size within a fluid flow system. (See *strainer*.)

flash point: The temperature to which an oil must be heated in order to give off sufficient vapor to form a flammable mixture with air under the conditions of the test. The vapor will ignite but will not support combustion.

flexible connection: In pipework, a section of pipe that permits minor lateral offset alignments to be effected.

foaming: The reaction of a liquid, especially an oil, to violent agitation in the presence of air. It may occur if an oil level is too high. The use of antifoam additives helps to break the air bubbles and prevent this tendency. Contaminants, especially water, aggravate foaming.

fresh water: Uncontaminated water fit for human consumption. (See *raw water*.)

frequency: The number of complete cycles of an alternating voltage or current per unit of time, usually a second and then expressed in cps (cycles per second). Also used to express vibration levels.

fretting: A form of wear that takes place between metal surfaces subjected to oscillating or vibrational forces of limited amplitude. Metal particles are detached from the parent metal body. Also known as *false brinnelling* or *friction oxidation* when it occurs in antifriction bearings.

full-load rated speed: The speed at which a machine will run when it is delivering maximum power.

generator: A rotating machine designed to convert mechanical energy into electric energy.

governor: A mechanically, electrically, or hydraulically powered speed-regulating device. (See *actuator*.)

ground: A connection, either intentional or accidental, between an electric circuit and the earth or some conducting body.

grounded neutral: The center point of a Y-connected, four-wire generator which is intentionally connected to ground.

heat: A form of energy measured in kilojoules or British thermal units. (1 kilojoule (kJ) = 0.947 Btu's; 1 Btu = 1.055 kJ).

heat exchanger: A device that permits the transfer of heat from one medium to another without direct contact. (See *intercooler*.)

heat sink: A device which absorbs and dissipates heat especially from solid-state devices such as diodes and SCRs to prevent damage caused by overheating.

hertz: The unit of electrical frequency: alternations per second (Hz).

hunt: Rhythmic variations in speed.

hydrophilic: Having an affinity for water; capable of uniting with or dissolving in water.

hydrophobic: Having antagonism for water; not capable of uniting or mixing with water.

hydrostatic lubrication: A system in which a lubricant, usually petroleum oil, is pumped to a plain bearing under pressure sufficient to maintain a film between the bearing surfaces. Commonly known as a *priming* or *prelube system*, it is commonly used prior to the start-up of heavy engines fitted with plain bearings.

in-phase: Said of alternating currents and voltages in a three-phase system if they both pass through zero and reach their maximums simultaneously.

inhibited oils: Petroleum oils which are fortified by additives to impart resistance to rust, oxidation, and foaming.

inspect: To compare a device with applicable specifications and standards.

intercooler: A heat exchanger in which heat is passed from one liquid to another without direct contact.

isochronous: Single speed: a term applied to governors that limit an engine to one speed regardless of load.

isochronous governor: A governor that keeps the engine speed constant at all loads. Also known as an *astatic governor*.

isolator: A device which limits the transmission of vibration energy.

jacket water heater: A heating facility external to the engine but connected to the engine's cooling system to maintain the engine at a viable starting temperature.

journal: The part of the crankshaft that turns in the main bearings. (See *crankpin*.)

miscibility: The capacity of liquids to form a uniform blend by dissolving into each other. Liquids that will not mix with each other are *immiscible*.

motor: A machine which converts energy into mechanical power. Commonly a motor is energized by electrical, hydraulic, or pneumatic means.

multigrade: Said of lubricating oil which is manufactured to have the necessary viscosity-temperature relationship to place it in more than one SAE classification. Thus, if an oil is within the viscosity bracket of an SAE 10W oil at −18°C (0°F) and also that of an SAE 30W oil at 99°C (210°F), it can be referred to as multigrade, i.e., 10W-30. This reference is to viscosity only; many other characteristics are necessary for satisfactory service.

neutral: The common point of a Y-connected machine or a conductor connected to that point.

ohm: The unit of electrical resistance. One volt will cause a current of one ampere to flow through a resistance of one ohm.

ohmmeter: A device for measuring electrical resistance.

Ohm's law: A fundamental law expressing the relationship between voltage, current, and resistance in dc circuits. In ac circuits, a value called *impedance* is comparable to dc resistance. The law states that $E = IR$; voltage is equal to current times resistance.

open-circuit voltage: The voltage produced when no load is attached to the voltage source such as a generator.

operator: Any person trained for, capable of, and responsible for the start-up, running, and stopping of power plant equipment.

out of phase: Said of waves of the same frequency which do not pass through their zero point at the same instant.

overhaul: A periodic and planned activity during which the worn parts of a machine are replaced or restored to specfications. (See *maintenance*, *repair*, *tune-up*.)

overload rating: The load in excess of the nominal rating which a device can carry for a specified length of time without being damaged.

overload relay: A device which operates to interrupt excessive currents.

overspeed: A machine speed greater than normal operational speed.

oxidation: The combining of constituents of a lubricating oil with oxygen from the air to form harmful acids. It is accelerated by heat and by copper.

parallel: Said of two or more generation units connected to a common bus and with all positive poles connected to one conductor and all negative poles connected to another.

paralleling: The procedure of matching the synchronous speed of one incoming generator in parallel with another and connecting them to a common load.

parallel operation: Two or more generators of the same voltage and frequency characteristics connected to the same load.

phase: The windings of an ac generator. In a three-phase generator there are three windings with their voltages 120° out of phase, meaning that the instant at which the three voltages pass through zero or reach their maximums are 120° apart if one complete cycle is considered to contain 360°. In single-phase generators, only one winding is present.

pour point: The lowest temperature at which a fluid will pour or flow under specified conditions.

power: Rate of performing work or energy per unit of time. Mechanical power can be measured in horsepower, electrical power in kilowatts.

power factor: The extent to which the voltage zero differs from the current zero. In ac circuits, the inductances and capacitances may cause the point at which the voltage wave passes through zero to differ from the point at which the current wave passes through zero. When the current wave precedes the voltage wave, a *leading power factor* results, as in the case of a capacitive load or an overexcited synchronous motor. When the voltage wave precedes the current wave, a *lagging power factor* results. Considering one full cycle to be 360°, the difference between the zero points can be expressed as an angle. Power factor is calculated as the cosine of the angle between zero points and is expressed as a decimal fraction (0.8) or as a percent (80 percent). It can also be shown to be the ratio of kilowatts divided by the kilovolt-amperes; PF = kW ÷ kVA.

propylene glycol: A colorless liquid used as the basis for an antifreeze coolant when there is a possibility of its intrusion into a potable water supply.

pyrometer: An instrument generally used for measuring elevated temperatures from remote points. Otherwise known as *resistance thermometers*, these intruments rely on a change in the resistance of a conductor brought about by temperature alteration. (See *thermometer*.)

raw water: Any untreated water of unknown properties.

real power: A term used to describe the product of current, voltage, and power factor expressed in kilowatts; 1 kW = 1.34 HP.

rectifier: A device for changing alternating current into direct current or unidirectional current.

rectifier bridge: A group of rectifiers (possibly diodes) connected in such a way that a unidirectional (dc) voltage appears across one diagonal when an ac voltage is applied across the other diagonal.

regulation frequency: The value obtained by dividing the difference between no-load and full-load frequency by the full-load frequency. Expressed in percent.

regulation voltage: The value obtained by dividing the difference between no-load and full-load voltage by the full-load voltage. Expressed in percent.

regulator: A generic term used in conjunction with a noun, e.g., *pressure regulator* or *voltage regulator*. (See *thermostat*.)

regulator voltage: A device for holding the voltage constant, regardless of the load, within prescribed limits.

relay: An electromagnetic control operating remotely and automatically in a switching mode within an electric circuit.

repair: An unplanned event requiring replacement or restoration of parts on a machine. (See *overhaul*, *maintenance*, and *tune-up*.)

residual magnetism: The magnetic induction which remains after the magnetizing force is removed.

respirator: A device placed over the mouth and/or nose to protect the respiratory track. Respirators do not necessarily protect the user from any specific or general toxic fumes, vapors, or gases.

rheostat: A resistor the resistance value of which can be varied by rotating a pickup contact through an arc.

rotor: The rotating component of an electrical machine, particularly an ac generator. Also, the rotating component of shaft impeller and blading of a turbocharger assembly.

SCADA: An acronym for *supervisory control and data acquisition*.

SCR: An acronym for *silicon-controlled rectifier*, which is a three-electrode solid-state device that permits current to flow in one direction only and to do so only when a suitable potential is applied to the third electrode, called the *gate*.

seizure: Partial or total welding of metallic components to each other through gross overheating.

sensitivity: The smallest speed change inherent in a governor and expressed as a percentage that occurs when the load alters.

separator: A device for removing one nonmiscible fluid from another, e.g., water from oil.

short circuit: Generally a nonintentional electrical connection between current-carrying wires or components.

sine wave: A wave which represents the sine of angles on the vertical scale and the corresponding angles on the horizontal scale. An ac voltage wave approximates such a curve.

sinusoidal: Varying in proportion to the sine of an angle; the shape of a sine wave.

skid: A wheelless structure, free-standing and capable of permanently supporting a superimposed load, of such design that it and its load can be moved to a site without damage to or distortion of the whole.

sludge: Insoluble material formed as a result of either deterioration reactions in an oil or contamination of an oil or both.

solvent: A chemical agent, often in fluid form, which, when applied to another specific substance, will convert the second substance to a liquid or mobile state.

soot: Impure black carbon combined with oil compounds resulting from the incomplete combustion of fuels.

specific gravity: The ratio of weight, as of a petroleum oil, to the weight of an equal volume of water at the same temperature.

stability: The property of maintaining or quickly reestablishing steady state values after load changes. Also, the ability of a governor to maintain an engine at a particular speed during load alteration.

star connection: See *Y connection*.

static exciter: Means of furnishing direct current to the generator field; it does not have any rotating or otherwise moving elements.

stator: The portion of an electrical machine which contains the stationary parts of the magnetic circuit and their windings. The term *stator* is commonly applied in connection with ac generators but the term *stator armature* also is used. (See *armature*.)

straight mineral oil: A petroleum-base oil which contains no additives or fatty oils. Its use in diesel engines is limited to governor service. More correctly called a *mineral oil*.

strainer: A device for collecting coarse debris from a fluid flow system. (See *filter*.)

surfactant: A soluble compound that reduces the surface tension of liquids or the interfluid tension between two liquids or a liquid and a solid.

surge: A sudden transient variation in current or voltage.

surge suppressor: A device that is capable of conducting high transient voltages and thereby protecting other devices that could be destroyed by the voltages.

synchronizing: To match one wave to another by adjusting its frequency and phase angle until they coincide.

synchronous: Applied to a type of motor or generator in which the relations between frequency in cycles per second and the speed in rpm is fixed and invariable.

test: To prove a machine's function by operating the machine in accordance with the original specifications.

thermocouple: The temperature-sensing element of a pyrometer.

toxic: Relating to or caused by a poisonous substance with reference to a living being.

transformer: A device which changes the voltage of an ac source from one value to another.

treated water: Any water that has been processed chemically by filtration or to remove contaminants. Treated water is not normally regarded as potable.

tune-up: Adjustment of various settings on a machine to permit optimum performance within the original specifications. (See *overhaul*, *repair*, *maintenance*.)

underbase (underframe): A steel structure upon which equipment is mounted. Supportive only of vertical loads, an underbase cannot normally be used as a vehicle for skidded movement. (See *skid*.)

unity power factor: A load whose power factor is 1.0 or 100 percent. This is the case when no inductive loads (motors, transformers, and so on.) are present and only resistive loads (incandescent light, furnaces, and so on) are connected.

vapor: A substance in a gaseous state as distinguished from the liquid or solid state.

viscosity: Resistance to flow, particularly of a lubricant. It is measured by the time taken for the lubricant to flow through a standard orifice at 38°C (100°F) and at 99°C (210°F). The result is expressed in *Saybolt seconds universal* or *kinematic centistokes*.

volatile: Readily vaporized at atmospheric temperatures.

voltage: The electrical potential or pressure which causes current to flow in a conductor.

voltage droop: The difference in voltage at no-load and full-load expressed as a percent of the full-load value.

voltage regulation: The difference between maximum and minimum steady-state voltage divided by the nominal voltage expressed as a percent of the nominal voltage.

voltage regulator: A device which maintains the voltage output of a generator near its nominal value regardless of load conditions.

watt: The unit of power in the meter-kilogram-second system of units equal to one joule per second. The abbreviation is W.

watthour: A unit of energy used in electrical measurements equal to the energy converted or consumed at a rate of one watt during a period of one hour. The abbreviation is Wh.

wear: The removal of materials from surfaces in relative motion.

wear, abrasive: Removal of materials from surfaces in relative motion by a cutting or abrasive action of a hard particle (usually a contaminant).

wear, adhesive: Removal of materials from surfaces in relative motion as a result of surface contact. Galling and scuffing are extreme cases.

wear, corrosive: Removal of materials by chemical action.

wye connection: See *Y connection*.

y connection: An interconnection of the phases of a three-phase system to form a configuration resembling the letter Y. A fourth or neutral wire can be connected to the centerpoint. Same as *star connection*.

CHAPTER 43
JOURNEYMAN NOTES

43.1 PROPER WORK PRACTICE

Although it is part of the operator's duties to keep the generating machinery and the main body of the plant clean, the mechanics and electricians are responsible for the cleanliness and security of their workshops and work areas. They are also required to leave the operating areas in a safe and workable state whenever they leave the job in hand.
  Any machinery that is not to be operated must be clearly tagged and locked out in accordance with normal established procedures, and each journeyman involved must satisfy himself that it is so. If it is necessary to remove the cable trough covers or manholes or make any such temporary changes, the person doing the work detail must provide and install temporary barriers.
  The covers of cable troughs must always be kept in their proper positions. They should never be laid in a staggered or diagonal mode even temporarily. It must be remembered that the operator expects the floor to be where it was last time he was in that vicinity, and he will not be thinking about the floor or the cable trough plates when he is busy with the next blackout.
  The crane hook must always be returned to a position at least 8 ft above the operating floor and be well clear of any machinery when it is not in use. Similarly, the crane control box must be fastened back into a nonobstructive position both within and without the normal workday. It is not permitted to leave chains or slings hanging from the crane hook either when the crane is being moved or when it is parked.
  Each journeyman is responsible for any company-owned tools issued to himself. The senior mechanic and the senior electrician, i.e., those in a particular plant with the longest service with the company, are responsible for the complete inventory of special tools and their condition and repair or replacement.

43.2 DEFECT RECTIFICATION

The diesel power plant mechanic and electrician's duties include the inspection of the current log sheets for notices of defects as reported by the operators. Frequently, they will also be advised verbally of equipment malfunctions. So that each individual problem can be tackled effectively, a systematic attitude and approach are important, and a study of the notes on Engine Fault Identity is recommended. The journeymen must be sure that they fully understand the reported defect, that all of the relevant details have been ascertained, and that all parties are using the same terminology.

The defect report can be confirmed by visual inspection, test running (if both safe and feasible), and examination of logged data. A correct diagnosis of the problem can thus be arrived at. If the terminology is mixed up or the details are incomplete, it will not be possible to accurately identify the source of the malfunction. When the malfunction is identified and rectified, the cause also must be corrected. It is safe to say that the journeymen's maintenance or overhaul work is very repetitive over a period of time, but defect rectification work has only to be repeated if the original work was deficient in some respect.

It is suggested that when a substantial repair or overhaul job has to be done, the journeymen of other trades assist if other duties permit. For example, if a large generator must be dismantled, the electrician will be in charge of the work, and he will be assisted by the mechanic. Conversely, if a complete unit must be realigned, the mechanic will mastermind the work and be assisted by the electrician. In either case, both journeymen will have an opportunity to improve their knowledge and experience.

43.3 IDENTITY NUMBERS AND MARKS

Many engine components carry various numbers, letters, or symbols which may be either cast in or stamped directly on the parent metal. These figures identify surveyor's classifications, part numbers, assembly numbers, rebuild or repairs, data application codes, sizing changes, timing marks, register marks, or build numbers. None of these identities should ever be tampered with. To do so will, in certain cases, invalidate insurance or warranty.

43.4 CANNIBALIZATION OF EQUIPMENT

Cannibalization is the borrowing of parts from one piece of equipment to use on another. Cannibalization of engines or other machines for parts to be used on other engines is a practice that has but little economic benefit and few if any other advantages.

In an emergency situation, it may be permissible to take parts from one engine that is already out of service in order to keep another engine in service, but in most cases such parts will already have been worn to a certain degree. If it is expedient to cannibalize an engine, either replacement parts should be procured without delay and retrofitted or the cannibalized unit should be written off. It is quite probable that more engines have gone out of service forever through cannibalization than have been damaged beyond economic repair by use.

43.5 USE OF CORRECT FASTENERS

Many of the fasteners employed on dynamic equipment have to be replaced once or several times during equipment life. Engine turbocharger exhaust flange nuts and bolts and bottom end bolts are examples that readily come to mind, but many other fasteners may be subject to renewal.

Nut-and-bolt types of fasteners (a generic term that may also include cap screws, setscrews, studs, locknuts, and castellated nuts) are at their most effective tightness when they have been stressed to a stage at which a permanent metal deformation just begins to take place. OEM-provided torque values provide the essential guidance. When a torque wrench is used in conjunction with the OEM specifications, it is likely that a fastener will be stressed correctly provided, of course, that the fastener material also is in accord with the OEM specification.

When fasteners are tightened other than by means of a torque wrench, there is a probability that those below ⅝ in (16 mm) will be overtightened and those above ⅝ in will be undertightened, assuming that the metal grades are similar.

43.6 REPLACEMENT FASTENERS

Sometimes it is necessary to replace a damaged fastener on some portion of power plant equipment. Proper consideration must be given to the quality of the fastener previously used and its intended replacement if the integrity of the reassembled equipment is to be maintained. There are many variations in the material tensile strength, hardness, finish, and dimensional specifications of fasteners that are similar in appearance.

When replacement fasteners are installed, a problem may arise if the material of the replacement does not conform to the OEM specification, and the fastener may subsequently be greatly overstressed. Torque values for OEM fasteners are generally listed in the OEM manual, and the fasteners frequently carry identification marks. Therefore, the replacement fastener can be ordered from the OEM only by quoting the appropriate part number.

Because of the move toward metrication of North American equipment, it is not unusual to find more than one thread form being employed on an individual engine. In many plants several thread forms may be found in use. All threads must be compared for correctness of pitch, angle, depth, and form when replacements are contemplated. For all practical purposes, any replacement fastener used internally on a machine should be of OEM supply.

Plated fasteners may not be used for any engine application. Cadmium and zinc plating will contaminate the lubricating oil or decay under its influence. Chromium plating can detach when the fastener is torqued. The clearance between the fastener thread and the second component will then be incorrect. The only plated fastener that may be used is one that clearly carries an OEM part number.

43.7 SPLIT PINS

All replacement split pins fitted should be checked for proper fit in the drilled hole. A great many replacement or jobber split pins are not round in section and do not fit the full diameter in the hole. Often because of too much freedom of movement when the machine is run, the split pin is in constant motion; fretting commences and the split pin ceases to fulfill its intended function. Fragments of poor-grade pins are one of the more frequently found contaminants of the pan.

Standard-grade split pins may be used for general construction purposes, but for all engines and auxiliary machines first-quality split pins must be used. These pins are fully round and fit the standard drilled holes without any looseness. Nonferrous split pins may not be used except under specific conditions as designated by the OEM. (Never on a fuel system; sometimes on water pumps.)

43.8 THREAD LUBRICANTS

Many compounds called either antifriction dressings or thread lubricants are available. It is not uncommon to find several different ones in any power plant. Great caution must be exercised in the selection of an antifriction dressing for use on an engine, and whatever is used should be approved by the OEM.

Many of the preparations contain substantial quantities of metal powder. Lead and zinc are commonly used in concentrations up to 50 percent, and copper, red lead, graphite, and molybdenum also are in general use depending upon the intended application. All of these thread lubricants have their place in industry, but there is always a possibility of misapplication which, in an engine, could be quite devastating.

Oil sample analysis will detect any metal molecules present in the lubricating oil. Since one of the main objectives of lubricating oil analysis is the early detection and identification of metal loss sources, any metal will be cited for its most likely normal origin. Thus, copper will be said to have evolved from top end bushes or thrust washers, lead from bearing overlays, and so on. If a metal-bearing compound has been used on threads exposed to the flushing action of the engine's lubricating oil, some of the metal powder will inevitably show up in the lubricating oil analysis.

The consequences will be misleading, and the use of antifriction compounds on the threads of fasteners must therefore be approached with caution. Equipment manufacturers have stated that the load imposed upon a fastener that has been served with certain antifriction compounds by a torque wrench may vary from specification by as much as 40 percent. Before any antifriction or antiseize compound is used on such critical components as main bearing studs, bottom end bolts, and cylinder head studs, specific manufacturer's instructions should be obtained. They may be found in the OEM shop manual.

CHAPTER 44
POWER PLANT CONTROL

44.1 POWER PLANT SUPERINTENDENT

Between the bulk fuel storage tank filling valve and the transmission cable that carries away the converted energy, the modern diesel power plant houses a wide variety of devices, machines, and equipment. The ideal diesel power plant superintendent would be a person fully familiar with the construction, purpose, operation, and upkeep of every item within the complex. Unfortunately, the world population of ideal superintendents is very small and there are not enough to go round. However, there are many superintendents who wish to do the very best possible.

It does not matter what his vocational origins are, the average diesel plant superintendent will always find that he has knowledge shortcomings and continuously needs to know more about the equipment in his charge. In the search for and accumulation of knowledge, he becomes a better superintendent.

44.2 PRACTICAL ROLE OF A PLANT SUPERINTENDENT

It is understood that no one superintendent can be fully familiar with all practical aspects of diesel power plant operation. However, it is not unreasonable to expect him to possess a journeyman level of knowledge in one of the major trades associated with power generation. It is indeed rare to find two diesel generating plants with the same design and layout. It is recommended that the newly appointed superintendent, who may have a vast previous experience or perhaps very little background, acquaint himself with all of the plant features at the earliest possible opportunity. It is highly unlikely that the various systems will be exactly the same as anything experienced before.

The superintendent should acquire this knowledge for several very important reasons. He must be able to:

- Discuss plant problems with his upkeep and operations personnel and then make intelligent and correct decisions
- Identify problems in the making and provide appropriate and timely rectifying or preventive action
- Respond to consumer comments in a manner acceptable to both the consumer and the owner in a confident, well-stated manner

In the event difficulties occur in the plant, the superintendent must be able to lead the work force toward solutions. If he can display an excellent overall knowledge of power plant systems and operations, he will command the respect of the owner, plant staff, and consumer. The superintendent who is very familiar with his plant will be able to present his views on improvements, replacements, and alterations in an understandable format to the headquarters with good effect. Without detailed and widespread familiarity with the plant, its equipment, and its operation, the superintendent will not be able to judge the accuracy of information and data passed on to him by the plant staff. In brief, his knowledge of his plant will be reflected in the success of the plant's operations.

44.3 OVERALL CONTROL OF THE POWER PLANT

The superintendent of a diesel power plant remains in charge of the plant and all of the activities conducted therein that can be performed by the normal power plant personnel. Any work done by other company personnel temporarily drafted for specific or individual duties or by contractors doing overhaul modification or renovation work are subject to his ultimate control and responsibility.

When construction work takes place, it may be found expedient for the work to go ahead without the superintendent's detailed participation, but when any activity is likely to infringe on the normal day-to-day operation of the plant complex, the superintendent must be involved and fully informed.

The superintendent must make it his business to inform himself on a daily basis of the progress of any work activities being undertaken in his areas of responsibility and ensure that equipment availability, safety, security, and economy are not being compromised. By his frequent presence in the areas where work of a practical nature is carried out, the superintendent will acquaint himself with any modifications that are being made and thereby maintain his familiarity with the plant and its equipment.

He must also make a point of discussing details of operating practice and the equipment with the operators and journeymen so that all parties both teach and learn. The superintendent who fails to expose himself to the operational personnel in that manner will not have the best possible ability to handle a problem situation when it arises. The most successful superintendents are those who have an excellent knowledge of the plant's equipment, functions, and idiosyncrasies together with some appreciation of the individual abilities of the work force.

44.4 EXCHANGE OF ENGINE KNOWLEDGE AND EXPERIENCE

The users of large diesel engines become members of a fairly exclusive fraternity. The view may justly be taken that an engine is an engine no matter which factory produced it and nothing odd will happen if the manual is followed. It cannot be denied, however, that an occasional conversation with the user of a similar engine can be invaluable in sorting out a difficulty or as part of the ongoing learning process. All kinds of useful information can be passed backward and forward between plant superintendents with benefit to all. A periodic discussion with the manufacturer's service representative will also be useful, and it is highly unusual to come across either a vendor or a user who will not happily discuss his engines for hours on end. It is the feedback to the engine manufacturer service departments that leads to all the modifications of and improvements to equipment throughout the engine's production life.

Talk between one journeyman and another at different plants will frequently provide the answer to a problem that is not discussed in the OEM manuals. Any communications that broaden the power plant staff's knowledge is to be encouraged whenever possible.

44.5 CAPACITY DISPATCH

The superintendent is responsible for the best possible economy in the operation of his plant, and great attention must be given to equipment utilization. Deciding the dispatch of appropriate engines is a complex problem. Consideration must be given to the loads and the equipment available to meet the demands. At the same time, thought has to be applied to the engine overhaul requirements, the necessary downtime, and the target dates for the overhaul work to be done.

Whenever possible, overhaul downtime should coincide with periods of low demand so the work can be carried out without undue or forced haste. Major overhauls represent large expenditures, and it is prudent to arrange the engine dispatch so that no two engines in the same plant fall due for a major overhaul in the same fiscal year. The efficient dispatch of engines must have a high priority in the superintendent's power plant direction duties. The economic advantages in allocating the most appropriate engine to the load demand is profound in terms of upkeep costs and fuel costs.

The load patterns in diesel generating plants are often cyclic; the daily load may fluctuate quite considerably as may the seasonal average load. In such cases, one engine or another may be called upon to generate power on a prolonged basis. The operating staff must develop some awareness of such engine usage and its effect on future overhaul planning. For example, if a given engine which is overhauled at 5000-h intervals and is operated, say, 600 h per month, its overhauls will become due at intervals of 8½ months. If the previous overhaul was in June, then the next will be due in February of the following year. Because of the differences so often found between summer and winter loading, it may be difficult to release the engine from service for the February overhaul. For that reason, the operating staff and the superintendent must establish a pattern of running that will give the best economic advantage in both fuel and parts. At the same time, the pattern must allow the engine to be stood down for overhaul at the most advantageous time relative to the load demands. In short, always run the smallest engine that will carry the load whenever feasible.

The integration of generating equipment into any particular work or overhaul plan is severely complicated by the need to operate with the best possible fuel economy, but an effort to achieve maximum efficiency must be made. The notes on fuel economy to be found elsewhere in this handbook can usefully be taken into account by both the operator and the superintendent. The problems of appropriate dispatch and economic fuel use are interwoven, and they cannot be ignored without very large cost penalties.

44.6 ECONOMIC UTILIZATION

A 1984 survey of engine upkeep costs by a major North American diesel generating authority shows that the cost of replacement parts and contracted services for major overhauls on the prime movers averaged only 0.525 cents (US) per kilowatt-hour of capacity operated. It is well understood that a 100 percent utilization of the generating equipment is an impossible goal, but it is not unreasonable to have a target utilization of 85 percent given that the individual capacities of the plant's generators provide a suitable flexibility.

A mandatory minimum of 75 percent utilization could well be set by the owner interested in basic economy, but it is not unusual to find individual plants operating with utilization factors even below 50 percent. In an example situation where an average of 1000 kWh is generated, the cost differential in parts and service alone between a 65 and a 75 percent utilization rate applied over one year would be:

$$8760 \text{ h} \times \frac{1000}{0.65} - \frac{1000}{0.75} \times 0.525 \text{ cents} = \$9427 \text{ (US)}$$

The $9427 can be regarded as a saving or as waste depending upon one's point of view.

44.7 SHIFT VARIATIONS

Working of back-to-back or double shifts by operators should be avoided. A fatigued operator cannot be expected to perform adequately, and the owner's interests will not be properly served by such a practice. Similarly, the start and finish times of shifts should not be varied to suit individual conveniences. Sooner or later the practice will bring grief to one or more of the operators or other station personnel, who will then turn to the superintendent to straighten things out. Start and finish times for each shift should be posted and adhered to by all personnel. The superintendent will do well to place the start and finish times for all personnel including office and day workers on the notice board.

44.8 SHIFT CHANGE TIMING

In most power plants the shifts are of 8-h duration, but sometimes 10 or even 12 h is worked. Usually the dayworkers, i.e., mechanics, electricians, work the same number of hours per shift as the operating personnel.

The plant superintendent will find it most beneficial to plant operations in general if the starting and finishing times of the dayworkers do not coincide with those of the shift operators. Typically, the operator's shifts should start at 7:00 a.m. and 11:00 p.m. and the dayworker shifts should commence at 8:00 a.m. By staggering the shift changes in that manner, there is an improved communication between the operating staff and the maintenance people. Also, the productivity of the maintenance people will be enhanced, and the people will be better informed on shift affairs.

44.9 STANDING ORDERS

There are numerous activities of an intermittent nature that must be so assigned that they do not always fall into a routine. There are other activities that may not be recognized by a newly hired operator as being part of his duties. The superintendent will find that a list of standing orders clearly and prominently posted will eliminate many grumbles and complaints from the power plant personnel.

The following list of standing orders does not necessarily apply to any one plant, nor does it reflect any particular company's policy. It merely serves as a guide to the preparation of such orders. Individual superintendents may wish to add to, delete from, and modify the list according to their own or their company's requirements.

> Operating shifts shall change at 7:00 a.m., 3:00 p.m., and 11:00 p.m. Oncoming shift operators shall arrive at the workplace and commence their preshift inspection 15 min before the shift-change hour. No shift shall be passed to an oncoming operator who is apparently impaired by some form of substance abuse. No operator shall leave the plant until relieved by the succeeding operator. All defects observed by the oncoming operator shall be noted in the log sheet and initialed by the departing operator.
>
> All garbage receptacles shall be emptied at the end of each shift by the operator.
>
> Nobody other than power plant personnel shall be permitted in any part of the power plant without permission from the operator in charge. All such visitors are to be recorded in the log. The shift operator is in charge of all power plant activity. No machine or equipment may be taken out of service without the permission of the operator.

Another section of the standing orders should define the miscellaneous greasing and oiling points to be visited by a designated person performing the lubrication tours. Perhaps at weekly or monthly intervals, one of the plant operators will be sent on a fixed round to grease and oil the overhead door mechanism, the ventilation louvers and their linkages, the raw water pump, valve spindles, various motors and pump bearings, and any other points requiring regular attention at extended intervals.

If the frequency is a bit flexible, it does not matter too much; what does matter is that all of the points needing some form of lubrication are identified and listed so that there is no excuse for individual points being overlooked or ignored.

44.10 DAILY REVIEW OF LOG SHEETS

One of the most useful and important duties carried out by anybody in a diesel plant is the daily review of the engine log sheets. For the superintendent, this review is absolutely essential. In a well-organized plant of whatever capacity, it is very useful to have various information items relating to the generator set's performance written down so that present and past data can be compared. In the larger plant, a great deal of information is recorded; in a smaller plant less information is recorded; in either one, the more figures recorded the better the interpretation process and result will be.

The interpretive scan will generally be done by the superintendent, who must be capable of understanding the meanings of the logged figures, identifying trends, and initiating timely action. This examination of the engine data is a fundamental element in the early detection of a trend away from optimum performance. Such a trend, if allowed to develop, could possibly lead to inefficient running or accelerated wear. In extreme cases, the trend could develop to the point at which the engine suffers damage and requires repair.

44.11 OPERATING HOURS RECORD

In the record-keeping system, it will be necessary to record the running hours on all machines on a cumulative basis no matter what overhauls, rebuilds, or repairs have been performed. Returning the machine hour counter to zero hours following a particular work action implies that the machine worked upon has been restored to an as-new condition in every respect. The practice can only have misleading consequences; in some instances, it can lead to dangerous operating conditions. If the recorded hours of all machines are cumulative, there will never by any doubt concerning the authenticity of any one machine's service record.

The hours reported and recorded should be only in whole numbers; there should be no decimal, fractional, or minute parts of whole hours. Handwritten numbers have a bad habit of transmitting decimal points as zeros or ones.

44.12 STATUTORY INSPECTION CERTIFICATES

In most jurisdictions, it is a requirement that the certificate for equipment subject to statutory inspection be prominently displayed in the vicinity of the equipment. The certificate should be placed where it can easily be seen. At the same time, it must be so positioned that it does not hinder normal work practices, nor should it be exposed to wet, dirty, hot, or other deleterious conditions. It must not be defaced in any way at all. The superintendent should arrange for the periodic and statutory inspection of the pressure vessels and see to it that the renewal permit is properly posted.

44.13 SAFE WORK PRACTICE

Certain plants have the requirement that specified safety equipment or apparel be worn by the personnel. In such cases, prominent notices should be erected over and outside of the power plant entrance stating, say, Hard Hat Area or Safety Glasses Must Be Worn. Similar signs of a general safety nature may also be useful, for example, No Smoking, No Tools Lent, or No Vehicle Repairs. It is much easier to enforce a rule when the rule is permanently and prominently visible. Moreover, any such rule must be observed by all personnel regardless of rank.

With or without written rules, the superintendent will find himself primarily responsible for any injury to personnel if the rules have not been obeyed.

44.14　SAFETY MEETINGS AND TRAINING

In order to maintain a standard of safety that complies with both corporate and governmental requirements, many companies demand that their employees attend regularly scheduled safety meetings. The conduct and content of such meetings have a great effect on the overall personnel safety record.

For the safety and benefit of the plant and its equipment, each superintendent must conduct similar meetings. In many instances there are close parallels between personnel safety and health and machinery safety and health. Some quite remarkable improvements in the reliability, safety, and efficiency of the plant's equipment will be achieved if regular and open discussions are permitted. For the less experienced personnel, there is an opportunity to learn valuable new knowledge.

At no time should the owner or superintendent ever assume that either a recently hired or a long-term employee is fully aware of all aspects of industrial safety.

44.15　MOTOR VEHICLES

Most diesel plant mechanics get involved in the upkeep of motor vehicles attached to the plant. It is quite commonplace for the priorities to get out of perspective and for man-hours that should be devoted to diesel engine care to be diverted to vehicle maintenance and repair. This presents no problem most of the time, but it must be remembered that the consuming public will not appreciate seeing plant personnel making emergency trips to cold plants in warm vehicles.

Motor vehicle work on the company vehicles can be done only when there is no work left to be done on the power plant and its equipment. The local community always has someone capable of repairing a vehicle; it has nobody capable of repairing the diesel plant.

At no time should any motor vehicle, either company or private, be permitted to be parked within the power plant building for any purpose whatsoever. Included in the term *motor vehicle* are motorcyles, motor tricycles, snowmobiles, cars, vans, and pickups. The use of power plant space for motor vehicle purposes always has evil consequences such as muddy floors, moonlighting, loss of tools, and the temptation to fiddle with a vehicle during a graveyard or evening shift.

44.16　MOTOR VEHICLE FUEL AND OIL CONSUMPTION

All motor fuels, whether gasoline, diesel, or propane, purchased for or issued to a vehicle must be recorded and dated along with the vehicle odometer reading.

Comparison between one vehicle's fuel consumption and another's should show a fairly even consistency, and no one vehicle should show any great variation during its service periods. Similarly, the oil consumption of each vehicle must be logged and compared, and the oil and filter changes and vehicle servicing must be recorded.

44.17 VEHICLE RECORD AND SAFETY

Every company vehicle requires a logbook so that each day's start mileage can be entered. Any fuel supplied also is written in the logbook. Merely entering the start mileage at the commencement of the workday may be taken to imply that the driver has in fact satisfied himself that the vehicle is safe in all respects for the day's use. If the driver assigned is not happy with the safety of the vehicle, he must not drive the vehicle away. To do so may be taken as an acceptance of responsibility for the believed defect. Any vehicle that is to be driven must have the brakes, lights, and steering in good working order.

44.18 OVERHAUL AND MAINTENANCE

For the ordinary light vehicles used by any utility company, an overhaul plan is not normally in being because it is not really necessary. On the other hand, it is imperative that a regular servicing and maintenance plan be applied to each vehicle. When a regular inspection and servicing routine is based on 5000-m stages, it is reasonable to expect vehicles to give excellent service for 75,000 m (125,000 km) over a period of 5 to 7 years at an overall cost of 20 to 22 cents (US) per kilometer or equivalent. That includes the capital cost of the vehicle, fuel, tires, service, and servicing labor (1990).

The expense of a general-purchase light vehicle that serves a diesel power facility is a significant cost which must be accounted for in the price of the generated kilowatt. Although it must be properly maintained, its operating costs must be kept to a minimum. In the larger plants, there are multiple light trucks, digger trucks, pole tractors, and the like. The maintenance and reliability of these vehicles is fundamental to the execution of planned work schedules.

44.19 SECURITY OF VEHICLES AND THEIR CONTENTS

Lock it or lose it. Those who are unable to steal will not become thieves, so any vehicle operator user or owner has a duty to keep his vehicle locked. That will add to the moral level of those who might be tempted from the path of propriety. Windows closed, doors locked, attractive objects in the vehicle covered or out of sight, spare wheel chained and padlocked, and anything of value removed from the trunk or pickup box when the vehicle is parked overnight in the open has to be standard practice.

It is advisable to brand all batteries and tires with the owner's mark immediately following receipt at the plant site, without waiting until they are installed in a vehicle.

CHAPTER 45
ANNUAL INSPECTIONS OF THE POWER PLANT COMPLEX

45.1 ANNUAL INSPECTION TOUR

It is advisable to have an annual inspection of all the facilities within the power plant complex that are not subject to the daily inspections performed by the operators, journeymen, and superintendent. The annual inspection should cover all of the civil facilities, including all buildings, fences, berms, draining systems, hard standings, and docks.

The scheduling of the inspection should be such that there is adequate time for any needed planning and remedial work to be executed without interfering with the upkeep work on the power production equipment. Of course, activities like roofing repairs will be restricted to fine-weather periods and repainting work should be scheduled to be done when little wind, rain, or frost is expected. But roofing and painting are probably better handled by contractors.

The inspection tour should follow a methodical pattern and be as comprehensive as possible. Each of the main trades should be represented on the tour, that is, operator, mechanic, electrician, safety, and supervision. All personnel should notice, remark on, and point out anything of concern or interest. The complete inspection may take several days, and it need not be wholly continuous. The scope of the inspection will vary immensely from one plant to another, and the superintendent may find it useful to start compiling a list of things to be looked at long before the inspection tour takes place so that nothing gets overlooked.

45.2 ANNUAL INSPECTION: PERMAFROST

In plants located in zones of permafrost an additional inspection routine will be necessary. As soon as the ground becomes visible in the spring after the runoff, there must be a thorough tour of the plant environs to look for sinkage, erosion, subsidence, or washout. Any developing watercourses must be considered for the ground-cutting effects of the runoff water and possible destabilization of hillsides and embankments. Diversionary methods must be implemented if any property is endangered.

Particular attention must be given to the ground conditions and fill at the edges of concrete floors, foundations, or pads. Loss of permafrost stability can lead to all sorts of structure or foundation problems which will be costly to repair. The superintendent who identifies a soil problem in the making that may affect the plant complex must not hesitate to discuss the condition with a civil engineer well versed in matters of permafrost control and the appropriate person at headquarters.

45.3 ANNUAL INSPECTION: FLOORS AND FOUNDATIONS

It happens that many diesel power installations, particularly in Arctic regions or where the water table is close to the surface, eventually suffer from movement of the foundations. If a subsidence problem is suspected, it is very useful if the levels of specific points on the floor or foundation can be rechecked against a previously established bench mark.

It is advocated that level reviews of selected structures be conducted as part of the annual civil inspection. The readings are taken with reference to a fixed bench mark external to the structure. Each of the reading points is established in conjunction with a prepared floor map so that each point can be referred to at a later date for check readings. By comparing the level readings, any movement, whether sinkage or heaving, can be detected. It is recommended that a new set of readings be taken as part of the annual inspection routine.

At each annual inspection the crack map should be redrawn and compared with the previous maps. If any cracks appear to be opening, it may be advisable to notify headquarters. In turn, headquarters will advise the constructors and the consultants so that any warranty positions are preserved. Any cracks that appear to be opening should be measured, and the frequency of inspection should be increased. Cracks in the floor that are less than 0.65 cm wide (0.25 in) are acceptable.

All level readings are recorded in terms of either one-hundredth foot or one millimeter.

45.4 ANNUAL INSPECTION: ROOFS

At the time of the annual inspection, the condition of all of the roofs in the power plant complex should be reviewed for cracks, tears, or breaks in the cladding material. Any evidence of erosion, spalling, corrosion, or rot should be noted and scheduled for repair. Accumulations of carbon soot, oil, or other flammables must be removed, as must natural garbage such as birds' nests and leaves, seeds, and other debris.

The ventilating fans are often accessible from the roof. If they are to be inspected, the power to them must be switched off and locked out before the inspection commences (Chap. 41).

The roof inspections cannot be conducted by one person alone. Simply for reasons of safety, two people must be present. Both should be equipped with nonslip footwear, safety ropes, and a modicum of common sense.

45.5 ANNUAL INSPECTION: VENTILATION

The ventilating louvers require occasional inspection. The inspection may show them to be coated with oil and dust or other atmospheric debris that causes them to stick in a fixed position or become heavy and out of balance. Arrangements must be made for them to be cleaned.

In some situations, air ducts offer ideal catchment areas for condensed engine room fumes, birds' nests, dead leaves, and so on, and thus they become potential fire hazards. Any debris must be cleaned out. Barometric dampers must be in good order and rust-free. Each of the ventilation controls should be exercised by the electrician to demonstrate its workability in the presence of the superintendent.

45.6 ANNUAL INSPECTION: ACCESS

The annual inspection that covers the accesses is important primarily from a personnel safety point of view. Included in the access categories are extension ladders, step ladders, mobile work platforms, catwalks, galleries, stairways, duckboards, and railings. Also there are manholes, cable trench covers, gates, fences, and overhead and side-hinged doors to be considered.

It is probable that any of the defects that may impair personnel safety will receive action on a day-to-day basis as they develop, but the annual inspection is intended more as a preventive and planning overview. As a result of the inspection, planned activities may be set in motion to produce permanent improvements in the workplace for the benefit of both the employer and employee.

45.7 ANNUAL INSPECTION: MUFFLERS AND EXHAUST STACKS

Only a little maintenance or overhaul work has to be carried out on engine exhaust pipes and mufflers in the normal course of events. Once per year, however, the pipes and mufflers should be closely inspected for any signs of corrosion so that steps can be taken for their repair or replacement. Corrosion of the muffler is inevitable, but the rate of decay depends on the nature of the fuel being burned. Heavier fuels generally contain more sulfur than lighter grades, and the resultant corrosion rates will be faster.

The muffler must be examined for any signs of perforation, which will probably be more evident on the lower portions of the outer casing. If perforation has occurred, preparations for early replacement will have to be made. Any attempt to patch corrosion holes will be fruitless because the remaining metal of the muffler's outer shell will already be severely wasted. If mufflers are repainted occasionally with a decent grade of weather- and heat-resistant paint (which does not necessarily have to be aluminum-based!) applied to a rust-free surface, a good signal of a corrosion perforation will be seen when rust and carbon streaking begins to show.

The exhaust pipes and their supporting brackets must be reviewed. The rain caps fitted to the exhaust pipes of standby engines in particular and other engines in general must be seen to be in good order if turbocharger windmilling on stationary engines is to be avoided. The drains of all mufflers and exhaust pipes must be checked and left clear.

45.8 ANNUAL INSPECTION: FUEL STORAGE AREA

The fuel storage area and its facilities must be included in the annual inspection tour. The external condition of the bulk fuel storage tanks must be assessed. Yellowish-brown streaking of the tank sides may be indicative of leakage or possible overflow. If leakage is suspected, it will be necessary to formulate long-range plans for repair. In the case of bolted-type tanks, however, it is highly inadvisable to make any attempt to cure leakage by bolt tightening. That activity is strictly for experts!

All vertical tanks should be viewed from the cardinal points N, S, E, and W for signs of tank side bulging and other deformities. The cylindrical surfaces of horizontal tanks and the tops of vertical tanks should be checked for any signs of collapse due to vacuuming. All vacuum-breaking devices in the fuel storage tanks must be checked out. Each of the external fittings—screwlift valves, nonreturn valves, gate valves, and so on—should be individually examined for any sign of damage, and all of the safety chains and padlocks must be in position. The ladders, stairways, and catwalks within the storage facility must be in good repair, secure, and without obstruction.

The inspection party will tour the berms enclosing each tank and/or the storage area and take note of any erosion, washout, or other breakdown. Any garbage work required should be carried out and recorded. Both the fire and fuel spill protection and alarm arrangements will require inspection, as will the fuel off-loading equipment, the security fences, and the gates and roadways.

45.9 ANNUAL INSPECTION: OUTFALLS

The condition of the outfalls where cooling water is returned to the river or lake must be examined. Erosion of the bank or surrounding backfill leading to loss of support for the discharge pipes is not uncommon. Trouble with the water control authorities can be expected if the erosion of the river bank becomes too pronounced. Any evidence of any oil being discharged must be thoroughly investigated and remedied. The run of the pipes carrying the cooling water to and from the power plant should be reviewed. Any sag or apparent movement in the pipe run or any subsidence of the ground in way of the discharge pipe must be reviewed.

45.10 ANNUAL INSPECTION: COOLING WATER INTAKE

Under certain conditions the cooling water intake strainers of plants being served with river or lake water may become fouled by natural debris. Seasonal conditions and locations will have a considerable influence on the problems that can be expected. In some areas such vegetable matter washed down by heavy rain will be picked up by the intake screens. In other areas, spring runoff can promote heavy bacterial growth on the intake pipe systems. Silt may cause some difficulties, and fish, eels, and other marine life can at times cause almost total blockage of the intake filters.

It is recommended that the possibilities of such difficulties be considered and the countering tactics be planned at an early stage in the life of a new plant. A provisional plan that is not needed is much better than having no plan at all when one is needed.

45.11 ANNUAL INSPECTION: FUEL SUPPLY PIPE SYSTEMS

All of the pipe systems are to be inspected annually in conjunction with the other annual inspections. The positioning of the pipes connecting the storage tanks to the power plant on their supporting brackets, ground springs, or pedestals shall be reviewed so that there is no risk of any pipe falling off its mounts. Long runs of pipe must be inspected to see that the original lie of the run has not been so altered by traffic, ground movement, or other effect to the point that the pipe and its fasteners are being strained.

All expansion pieces and flexible sections require inspection and must be without any impediment to their designed movement. At the same time, these fittings must be examined for any posture that will represent undue stress. All sections of the pipe systems are to be capable of natural expansion and contraction without strain to the system or supporting structure.

All valves in the individual pipe systems are to be examined and proved to be in good working order. Each valve shall be opened and closed by hand without difficulty or the use of leverage. No fuel leakage at the valve glands is permitted. The valves on the fuel system located between the bulk storage tanks and the plant are normally chained and locked. These security fittings must pass the inspection.

45.12 ANNUAL INSPECTION: HARDSTANDING AND PARKING AREAS

The power plant parking lot must be inspected at least annually. Signs stating that vehicles are parked on the lot only at the owner's risk should be in place, and the parking of any vehicle should be confined to designated areas.

No scrap or surplus material should be dumped in the parking lot. In the interests of safety and loss control, no company-owned parts, materials, or property should be stored in the vehicle parking areas. No fuel should be handled in the parking areas.

45.13 ANNUAL INSPECTION: BUILDINGS

The skills of the maintenance personnel are not usually suitable for the repair or rehabilitation of civil works. The manning and work allocations within the plant are generally such that the available man-hours are fully used anyway. Thus, for structural steelwork, masonry work, roofing, painting, and other related work the use of contractors experienced in those particular trades is advocated.

Renovations and repairs to the power plant structures require quantities of specialized equipment and material that would not normally be kept on hand by the power plant, and the capital cost of attempting to do so would be unreasonable. Contractors have all the tools that may be used such as power hammers, shotblasting, and airless paint spray equipment, portable compressors, scissor-lift trucks, concrete saws, bitumen heaters, drain augers, drywall and taping tools, scaffolding, and all of the other equipment for speedy and professional work conduct.

Large-scale rehabilitation or modification work must be very well preplanned, and potential contractors must be given good lead time to prepare acceptable estimates of cost and time requirements. Any program of civil renovations or repair work that is to be given to a contractor should be designed to cater to the foreseen needs of the plant for the future 5 years. Work of a civil nature is most disruptive to the operating routines, and interruptions must be kept to a minimum. Infrequent but comprehensive rehabilitation work is much more cost-effective than many small, piecemeal jobs.

45.14 ANNUAL INSPECTION: FIXED AND PORTABLE ASSETS

The annual inspection tour of the plant complex may be a suitable time to review and check all of the plant's assets whether fixed or inventory. All of the assets at the generating site should be listed on a master register. It is suggested that all fixed items and any portable item with a value of over $200 US (1990) or any item with a serial number be registered. The company may choose to stamp its various properties with other identity numbers, and in high-pilferage areas this practice usually has good economic results.

CHAPTER 46
RECORDS

46.1 PURPOSE OF RECORD KEEPING

The purpose of keeping formal records of all aspects of the power plant activities is to enable accurate assessments to be made of the efficiency, reliability, handling, and effectiveness of the plant. Reference to the recorded data on a comprehensive basis will soon reveal the performance differences between one plant and another in both positive and negative aspects and much to the advantage of the owner. The recorded history will support research into plant behavior and permit predictions of future activity to be made. The accuracy of a forecast will be very much affected by the quantity and quality of the recorded details.

All of the information received must be recorded and set down in a format that will permit its use in a comparative or cumulative mode. It must also be arranged in a retrievable manner so that trends become visible and thus can be analyzed, costs can be budgeted, regulated, or estimated, machine failure causes can be examined, suitable remedies can be instituted, and replacement equipment can be selected. At the same time, standard data relating to specific equipment can be easily and quickly determined by scanning the information for detail variations. There cannot be an economic and effective equipment upkeep scheme without accurate, comprehensive and complete data records (Fig. 46.1).

46.2 DIARY-TYPE RECORD KEEPING

A diesel generating plant cannot manage a progressive operation or a maintenance and upkeep system without comprehensive work and incident records. The usefulness of a hard-backed, durable diary-type notebook for the daily entry of all the things that take place that may have a future consequence of any sort cannot be overstated. The operator will keep the original working copy and make individual entries, and the superintendent will keep a clean copy by using information from the original operator's book and adding all such items as contractors' work, owner's instructions, and similar incidents. Personnel memories are faulty at best and absent when convenient. The written word carries a little more continuity and can be referred to by anyone. The operator, mechanic, electrician, or any other plant staff member of the 1990s is expected to write willingly and explicitly as part of his job. What he writes today is often the clue to tomorrow's situation, and for that he receives pay.

Unit # _____   Plant _____

Inspecting personnel _____   Date _____

Total engine hours _____ Load _____ Time _____ AM/PM

Shutdowns	Mech. overspeed	Elect. overspeed	J.W. temperature	Lube oil pressure
Working				
Defective				
Not fitted				

L.O. sample taken	Yes	No	L.O. changed	Yes	No	
Hrs. on L.O.			L.O. filter changed	Yes	No	
Antifreeze reading			Coolant added	Yes	No	
Rad core	Clean		Dirty		Action	
Surge tank sight glass		Clean		Dirty		Action
Hose condition	(state which is defective)					
Fuel leak location				Action		
Water leak location				Action		
Lube leak location				Action		
Water separator drained		Yes		No		
Day tank water drained		Yes		No		
Main tank water drained		Yes		No		
Battery posts	Clean		Dirty		Tight	Loose
Fan belts	Correct tension			Loose		Worn
Exhaust smoke	Clear			Color		
Engine	Clean		Dirty		Action	

Vibration check

Point	Load	11	KW 13
	1		
H			
V			
A			

If reading exceeds
0.55" s velocity
state action taken

Work done on unit since privious report:

Work to be done on unit:

FIGURE 46.1 Typical unit inspection report.

Now what is written must be read; therefore, the relieving operators, the plant superintendent, and the other maintenance personnel must read and initial the incident or workbook entries as seen. After a while, the record becomes history, and from it lessons are taken. The more often the record is referred to, the clearer are the learning and the understanding of ongoing problems. Numbered or dated pages are suggested to deter the loss of passages which might carry a little embarrassment at some point in the future.

46.3 CIRCUMSTANTIAL INFORMATION

A record of incidental events involving each and every piece of equipment is indispensable if the failure rate, breakdown characteristics, and operating difficulties of the company's equipment are to be recognized in their proper perspective and receive timely and permanent corrective attention. In many diesel power plants, much information of this type is written either in the daily log sheet or in the work diary.

It is expected that the superintendent and/or the planner will refer to the defect record on a daily basis and occasionally review the defects written up over an extended period of time with the aim of detecting incidents that keep repeating themselves. These may often be permanently eliminated by modifying the scheduled overhaul or maintenance plan or retraining or changing the operating practices. Similarly, information of this sort when committed to a computerized data system can be readily programmed to identify itself under specific categories of elapsed hours, frequency of occurrence, prevailing conditions at the time of failure, and so on.

It can fairly be said that what is not written down at the time of the occurrence or when the work was done will be lost forever. Nobody has an exact or perpetual memory, and at a later date when someone claims to remember, that memory will at best be incomplete and certainly worthy of doubt.

46.4 GENERATING UNIT HOURS

The planned upkeep of all generating equipment is based upon the hours that the equipment has operated either since new or between major or total overhauls. A continuous and cumulative record of the machine hours operated during the period, usually entered at the end of each calendar month, is an indispensable component in any study of equipment utilization ratios, lubricating oil consumption, overhaul planning, or other, similar work. Only whole hours should be reported. Never report decimals or fractions of an hour, because future confusion and misreading will be inevitable.

For practical purposes, it will be found that a nonadjustable five-whole-digit hour meter mounted on the generator unit's breaker panel will serve longer and much more accurately than any engine-mounted mechanical hour meter counter. Using the previous 12 months running hours of a generator set will enable a monthly average run to be determined. The monthly average divided into the remaining hours until the next overhaul will give a reasonable forecast of when the next overhaul will be due.

46.5 BULK FUEL RECORD

In plants where fuel is delivered frequently a formal record of deliveries is very important to ensure the owner is receiving what he is paying for. Following each delivery notation there should be a statement by the plant concerning the quantity of water or sludge drained from the bulk storage tank.

In all plants, whatever the method of fuel replenishment, a monthly statement will be required. The statement will give the net amount of fuel in stock, the amount of fuel consumed, the gross amount of fuel delivered, the amount of water and sludge drainage, and the fuel lost for some explainable reason. Such records are needed for quality and quantity control of supply, forecasts of future fuel requirements, and efficient monitoring.

46.6 FUEL USE RECORD

The greatest single operating expense in a power-generating complex is that of fuel. Science and technology have been continuously applied for the past 100 years to develop engines that are more fuel-efficient, and great advances have been achieved. At the same time, the generating equipment owner has become more and more selective in his purchases of new equipment. He invariably seeks an engine that is the most fuel-efficient and pays premium prices for the best machines.

It follows that the installed equipment must be dispatched in a capacity order that permits the greatest possible fuel efficiency to result from the equipment chosen to run and the best possible return on the owner's capital and operational investments.

The amount of fuel consumed by one generator set must be compared with that by another, all in terms of kilowatts per gallon. The higher the ratio of kilowatts to gallons the more efficient the fuel consumption. The kilowatt per gallon average is the most important efficiency figure in the power plant.

46.7 OPERATING HOURS RECORD

In the record-keeping system it will be necessary to record the running hours on all machines, be they main generating equipment or auxiliary machines, on a cumulative basis no matter what overhauls, rebuilds, or repairs have been conducted. Returning the machine hour counter to zero hours following a particular work action implies that the machine worked upon has been restored to an as-new condition in every respect. Therefore, *the practice of returning the meter to zero can only have misleading consequences and may in some instances lead to dangerous operating conditions*. If the recorded hours of all machines are cumulative, there will never be any doubt concerning the authenticity of any one machine's service record. Depreciation and amortization formulas can be devised and applied with some respect for reality and a true appreciation of the long-term serviceability of particular machines and their components can be formed.

46.8 MAINTENANCE SYSTEM RECORDS

If a machine-populated company is to have a workable system for the upkeep of its equipment, certain requirements must be established and observed. The system must be cohesive and comprehensively monitored, and all of the guiding procedures developed must be practical. Many of the procedures will call for a reporting response.

The procedures are created so that all of the owner's plants and personnel are able to work in a similar manner with predictable results and costs. To be successful, the full participation and commitment of all individuals throughout the company must be secured or enforced. The basic and traditional principles of an equipment upkeep scheme have to remain generally unaltered in spite of any administrative structure changes.

A comprehensive set of records is necessary for the conduct of any planned maintenance scheme. The scope of the records will depend upon the complexity of the plant concerned and the requirements of the owners. Some companies will have detail variations in the overall requirements.

To be of any value, each piece of recorded information must be easily retrievable at all user points within the organization. The information can be stored on paper or electronically. In either case, it must be understandable without third-party interpretation or explanation.

There must be a use for any particular record that is required to be kept. If the information provided has no apparent purpose, the information provider will feel he can cease to supply it. Ideally, the written procedure that governs the collection of specific information will include a full explanation of why the information is needed, who uses it, when it is needed, what must be recorded, and how it is to be transmitted.

The provision of adequate and accurate information is the lifeblood in any planned maintenance scheme, whether the maintenance is of machinery, accounts, personnel, or any other of the company's assets that may alter with time. The provided information must be accepted by those to whom it is directed. The acceptance and belief in its message is an important component in the information gathering, processing, and dissemination routines.

Practically all of the information that is required regularly is best transmitted in an easily recognized format so it can be quickly identified, picked out, and sorted. It is for this reason that printed, colored, or otherwise easily recognized reporting forms are so frequently used (Fig. 46.2). An equally important reason for using a printed form is to reduce the amount of handwriting on the part of the reporter, simplify the task of the recorder in selecting the essential data in the report form, and obtain uniform reporting. However, the reporter should not feel inhibited about writing additional and relevant comment on a report form even if no spaces or blocks are provided. Too much information is always better than none at all.

The plant staff or ownership who choose to neglect the ongoing completion of a set of records will find it quite impossible to monitor trends, forecast and plan work, develop budgets for money and manpower with any economic accuracy. Also the reluctance or inability of any plant or person to adhere to a unified and systematized method of inspection, reportage, and communication can only have detrimental consequences.

For each piece of machinery, equipment, or structure nominated as an upkeep entity there must be a continuous record of its use and the work done on it depending upon the company's requirements and standards and perhaps the capital value of the item. The record demanded may from extremely comprehensive to minimal. The plant that fails to develop good records cannot possibly have a reasonable maintenance and overhaul program.

Date _____ Location _____

Engine _____ Inventory _____

Component _____

Previous hours _____ Present hours _____

	Tolerance		Dimension at overhaul					
	New	Reject	A_1	B_1	A_2	B_2	A_3	B_3
1								
2								
3								
4								
5								
6								
7								
8								

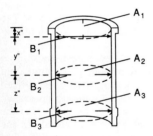

FIGURE 46.2 Cylinder liner wear record.

CHAPTER 47
MEASURES AND CONVERSIONS

47.1 METRIC, IMPERIAL, AND OTHER MEASUREMENT UNITS

With the developing trend toward standardization of measurement units it is common to find power plants with more than one measuring system in use on the various instruments. The following conversion tables are intended to assist the operator in matching original factory test sheet data with the readings observed on the installed equipment or to usefully read from instruments that have been replaced by others of a different scale.

It is recommended that all log sheets, instrumentation test sheets, and replacement instruments be standardized to the SI, or ANSI[1] system wherever and whenever possible. If that is to happen, it is important that the operator become familiar with the new units and then compare the new units with the old.

TABLE 47.1 Conversion Constants, English to SI

Measure	English units	Equivalent SI units
Length	0.039 inch (in)	25.4 millimeters (mm)
	3.280 feet (ft)	0.3048 meter (m)
Mass	28.350 grams (g)	0.035 ounce (oz)
	0.454 kilogram (kg)	2.204 pounds (lb)
	0.907 tonne (T)	1.102 tons (not abbreviated)
Area	0.155 square inch (in^2)	6.45 square centimeters (cm^2)
	10.763 square feet (ft^2)	0.093 square meter (m^2)
Pressure	1 pound per square inch (psi)	0.068 atmosphere (atm)
		2.036 inches of mercury (in Hg)
		27.680 inches of water (in H$_2$O)
	1 atmosphere (atm)	14.700 pounds per square inch (lb/in^2)
		29.920 inches of mercury (in Hg)
		1.860 inches of water (in H$_2$O)
	1 inch mercury (in Hg)	0.491 pounds per square inch (lb/in^2)
		13.580 inches of water (in H$_2$O)
		0.033 atmosphere (atm)
	1 kilopascal (kPa)	0.145 pounds per square inch (lb/in^2)
		0.295 inches of mercury (in Hg)
		0.009 atmosphere (atm)

[1]SI = Societé Internationale; ANSI = American National Standards Institute. The two systems are essentially the same.

TABLE 47.2 Conversion Table: Imperial Gallons to Liters, and Vice Versa*

Imperial gallons	Number of units	Liters
0.219	1	4.54
0.439	2	9.09
0.659	3	13.63
0.879	4	18.18
1.099	5	22.73
1.319	6	27.27
1.539	7	31.82
1.759	8	36.36
1.979	9	40.91
2.199	10	45.46
2.419	11	50.00
2.639	12	54.55
2.859	13	59.09
3.079	14	63.65
3.299	15	68.19
3.519	16	72.73
3.739	17	77.28
3.959	18	81.83
4.179	19	86.37
4.399	20	90.92

*To use the table, first find the number of units to be converted. If from liters to imperial gallons, read to the left; if from imperial gallons to liters, read to the right.

TABLE 47.3 Conversion Table: U.S. Gallons to Liters and Vice Versa*

U.S. gallon	Number of units	Liters
0.26	1	3.79
0.53	2	7.57
0.79	3	11.36
1.06	4	15.14
1.32	5	18.93
1.59	6	22.71
1.85	7	26.50
2.11	8	30.28
2.39	9	34.07
2.64	10	37.85
2.90	11	41.64
3.18	12	45.42
3.44	13	49.21
3.70	14	53.00
3.96	15	56.78
4.22	16	60.56
4.49	17	64.35
4.76	18	68.14
5.02	19	71.92
5.28	20	75.70

*To use the table, first find the number of units to be converted. If from U.S. gallons to liters, read to the right. If from liters to U.S. gallons, read to the left.

MEASURES AND CONVERSIONS

TABLE 47.4 Useful Measures and Conversions

Base units	Conversion Factor	Equivalents
Atmosphere (atm)	1.0132	Bars
	76.0000	Centimeters of mercury (cm Hg) at 0°C
	1033.2600	Centimeters of water (cm H_2O) at 4°C
	33.8995	Feet of water (ft H_2O) at 32°C
	1033.2300	Grams per square centimeter (g/cm^2)
	29.9213	Inches of mercury (in Hg) at 32°F
	1.0332	Kilograms per square centimeter (kg/cm^2)
	760.0000	Millimeters of mercury (mm Hg) at 0°C
	14.6960	Pounds per square inch (lb/in^2)
Barrels, imperial, (imp bbl)	1.3725	U.S. barrels (U.S. bbl)
	36.0000	Imperial gallons (imp gal)
Barrels, U.S. (U.S. bbl)	4.2093	Cubic feet (ft^3)
	26.2292	Imperial gallons (imp gal)
	119.2370	Liters (L)
Bars	0.9869	Atmospheres (atm)
	75.0062	Centimeters of mercury (cm Hg) at 0°C
	33.4883	Feet of water (ft H_2O) at 60°F
	1019.7160	Grams per square centimeter (g/cm^2)
	29.5300	Inches of mercury (in Hg) at 32°F
	1.0197	Kilograms per square centimeter (kg/cm^2)
	14.5038	Pounds per square inch (psi)
Centimeters (cm)	0.0328	Feet (ft)
	0.3937	Inches (in)
	10000.0000	Microns (μ) or micrometers (μm)
	393.7007	Mils
Centimeters of mercury (cm Hg) at 0°C	0.0132	Atmospheres (atm)
	0.0133	Bars
	0.4461	Feet of mercury (in Hg) at 4°C
	0.4465	Feet of mercury (in Hg) at 60°F
	0.3937	Feet of mercury (in Hg) at 0°C
	0.1934	Pounds per square inch (psi or lb/in^2)
Centimeters of water (cm H_2O)	0.0010	Atmospheres (atm)
	0.0142	Pounds per square inch (psi or lb/in^2)
Centimeters per second (cm/s)	1.9685	Feet per minute (ft/min)
	0.0328	Feet per second (ft/s)
Cubic centimeters (cm^3)	0.0610	Cubic inches (in^3)
	0.0352	Imperial ounces (imp oz)
	0.0338	U.S. ounces (U.S. oz)
Cubic feet (ft^3)	28316.8000	Cubic centimeters (cm^3)
	0.0283	Cubic meters (m^3)
	7.4805	U.S. gallons (U.S. gal)
	28.3161	Liters (L)
Cubic feet of water (ft^3 H_2O) at 60°F	63.3663	Pounds of water (lb H_2O)

TABLE 47.4 Useful Measures and Conversions *(Continued)*

Base units	Conversion Factor	Equivalents
Cubic inches (in³)	16.3871	Cubic centimeters (cm³)
	0.0164	Liters (L)
	0.5767	Imperial ounces (imp oz)
Cubic inches of water (in³ H₂O) at 60°F	0.0361	Pounds of water (lb H₂O)
Feet	30.4800	Centimeters (cm)
	0.3048	Meters (m)
Feet of mercury (ft Hg) at 32°F	30.4800	Centimeters of mercury (cm Hg) at 0°C
	163.3020	Inches of water (in H₂O) at 60°F
	5.8939	Pounds per square inch (psi or lb/in²)
Feet of water (ft H₂O) at 4°C	0.0295	Atmospheres (atm)
	2.2419	Centimeters of mercury (cm Hg) at 0°C
	30.4791	Grams per square centimeter (g/cm²)
	0.8826	Inches of mercury (in Hg) at 32°F
	0.4335	Pounds per square inch (psi or lb/in²)
Foot-pound per second (ft · lb/s)	0.0018	Horsepower (hp)
	1.3558	Joules per second (J/s)
	0.0014	Kilowatts (kW)
	1.3558	Watts (W)
Gallons, imperial (imp gal)	4.5450	Liters (L)
	0.0278	Imperial barrels (imp bbl)
	4546.0870	Cubic centimeters (cm³)
	0.1605	Cubic feet (ft³)
	277.4193	Cubic inches (in³)
	1.2009	U.S. gallons (U.S. gal)
Gallons, U.S. (U.S. gal)	3785.4118	Cubic centimeters (cm³)
	0.1334	Cubic feet (ft³)
	231.0000	Cubic inches (in³)
	0.0038	Cubic meters (m³)
	0.8327	Imperial gallons (imp gal)
	8.0000	U.S. pints (U.S. pt)
	4.0000	U.S. quarts (U.S. qt)
Gallons, U.S., of water (U.S. gal H₂O) at 60°F	8.3372	Pounds of water (lb H₂O)
Grams per centimeter (g/cm)	0.0672	Pounds per foot (lb/ft)
	0.0056	Pounds per inch (lb/in)
Horsepower (hp)	2542.4800	Btu hour (Btu · h)
	33000.0000	Foot-pounds per minute (ft · lb/min)
	0.7457	Kilowatts (kW)
	745.7000	Watts (W)
Horsepower-hours (hph)	2546.1400	Btu
	1.98×10^6	Foot-pounds (ft · lb)
	0.7457	Kilowatt-hours (kWh)
Inches (in)	2.5400	Centimeters (cm)
	0.0833	Feet (ft)
	0.0254	Meters (m)

TABLE 47.4 Useful Measures and Conversions *(Continued)*

Base units	Conversion Factor	Equivalents
Inches of mercury (in Hg) at 32°F	0.0334	Atmospheres (atm)
	0.0339	Bars
	34.5316	Grams per square centimeter (g/cm^2)
	345.3160	Kilograms per square meter (kg/m^2)
	25.4000	Millimeters of mercury (mm Hg) at 60°F
Inches of mercury (in Hg) at 60°F	0.0333	Atmospheres (atm)
	34.4343	Grams per square centimeter (g/cm^2)
Inches of Water (in H$_2$O) at 4°F	0.0025	Atmospheres (atm)
	0.0361	Pounds per square inch (lb/in^2)
Kilograms (kg)	35.2740	Ounces (oz)
	2.2046	Pounds (lb)
	0.0010	Long tons
	0.0011	Short tons
Kilograms per square centimeter (kg/cm^2)	0.9678	Atmospheres (atm)
	0.9807	Bars
	73.5559	Centimeters of mercury (cm Hg) at 0°F
	32.8093	Feet of water (ft H$_2$O) at 39.2°F
	28.9590	Inches of mercury (in Hg) at 32°F
	14.2233	Pounds per square inch (psi or lb/in^2)
Kilowatts (kW)	3414.4300	Btu per hour (Btu/h)
	44253.7000	Foot-pounds per minute (ft · lb/min)
	1.3410	Horsepower (hp)
Kilowatt-hours (kWh)	1000.0000	Watt-hours (Wh)
Liters (L)	0.0353	Cubic feet (ft^3)
	61.0255	Cubic inches (in^3)
	0.8799	Imperial quarts (imp qt)
	1.0567	U.S. quarts (U.S. qt)
Milliliters (mL)	1.0000	Cubic centimeters (cm^3)
	0.0610	Cubic inches (in^3)
	0.0010	Liters (L)
	0.0352	Imperial ounces (imp oz)
	0.0338	U.S. ounces U.S. oz)
Millimeters (mm)	0.1000	Centimeters (cm)
	0.0394	Inches (in)
	0.0010	Meters (m)
	1000.0000	Microns (μ) or micrometers (μm)
Mils (mil)	0.0010	Inches (in)
	25.4000	Microns (μ) or micrometers (μm)
	0.0254	Millimeters (mm)
Newtons (N)	0.2248	Pounds (lb)
Newton-meters (N · m)	0.1020	Kilogram-meters (kg · m)
	0.7376	Pound-feet (lb · ft)
Ounces, imperial (imp oz)	28.4131	Cubic centimeters (cm^3)
	1.7339	Cubic inches (in^3)
	28.4123	Milliliters (mL)
	0.9608	U.S. ounces (U.S. oz)

TABLE 47.4 Useful Measures and Conversions *(Continued)*

Base units	Conversion Factor	Equivalents
Ounces, U.S. (U.S. oz)	29.5737	Cubic centimeters (cm^3)
	1.8047	Cubic inches (in^3)
Pints, imperial (imp pt)	568.2601	Cubic centimeters (cm^3)
	0.1250	Imperial gallons (imp gal)
	0.5682	Liters (L)
	1.2009	U.S. pints (U.S. pt)
Pints, U.S. (U.S. pt)	473.1764	Cubic centimeters (cm^3)
	0.0167	Cubic feet (ft^3)
	28.8750	Cubic inches (in^3)
	0.4732	Liters (L)
	0.8327	Imperial pints (imp pt)
Pounds avoirdupois (lb)	453.5923	Grams (g)
	0.4536	Kilograms (kg)
	16.0000	Avoirdupois ounces (oz)
	0.0004	Long tons
	0.0005	Metric tons (t)
	0.0005	Short tons
Pounds per square inch (psi or lb/in^2)	70.3089	Centimeters of water (cm H$_2$O) at 4°C
	70.3069	Grams per square centimeter (g/cm^2)
	2.0360	Inches of mercury (in Hg) at 32°C
	27.6807	Inches of water (in H$_2$O) at 39.2°F
	0.0703	Kilograms per square centimeter (kg/cm^2)
	51.7149	Millimeters of mercury (mm Hg) at 0°C
Quarts, imperial (imp qt)	1136.5200	Cubic centimeters (cm^3)
	69.3548	Cubic inches (in^3)
	0.2500	Imperial gallons
	0.3002	U.S. gallons
	1.1365	Liters (L)
Quarts, U.S. (U.S. qt)	946.3529	Cubic centimeters (cm^3)
	57.7500	Cubic inches (in^3)
	0.2500	U.S. gallons (U.S. gal)
	0.9463	Liters (L)
	32.0000	U.S. ounces (U.S. oz)
	0.8327	Imperial quarts (imp qt)
Square centimeters (cm^2)	0.0011	Square feet (ft^2)
	0.1550	Square inches (in^2)
	0.0001	Square meters (m^2)
	100.0000	Square millimeters (mm^2)
Square feet (ft^2)	929.0304	Square centimeters (cm^2)
	144.0000	Square inches (in^2)
	0.0929	Square meters (m^2)
Square inches (in^2)	6.4516	Square centimeters (cm^2)
	0.0069	Square feet (ft^2)
	0.0006	Square meters (m^2)
	645.1600	Square millimeters (mm^2)
Square meters (m^2)	10.7639	Square feet (ft^2)
	1550.0031	Square inches (in^2)

TABLE 47.4 Useful Measures and Conversions *(Continued)*

Base units	Conversion Factor	Equivalents
Square millimeters (mm^2)	0.0100	Square centimeters (cm^2)
	0.0016	Square inches (in^2)
Square yards (yd^2)	8361.2736	Square centimeters (cm^2)
	9.0000	Square feet (ft^2)
	1296.0000	Square inches (in^2)
Tons, long	1016.0460	Kilograms (kg)
	2240.0000	Pounds (lb)
	1.1060	Metric tons (t)
	1.1200	Short tons
Tons, metric (t)	1000.0000	Kilograms (kg)
	2204.6220	Pounds (lb)
	0.9842	Long tons
	1.1023	Short tons
Tons, short	907.1847	Kilograms (kg)
	32000.0000	Avoirdupois ounces (oz)
	2000.0000	Pounds (lb)
	0.8929	Long tons
	0.9072	Metric tons (t)
Watts	44.2537	Foot-pounds per minute (ft · lb/min)
	1.0000	Joules per second (J/s)
	0.0010	Kilowatts (kW)
Watt-hours (Wh)	3.4144	British thermal units (Btu)
	2655.2200	Foot-pounds (ft · lb)
	0.0013	Horsepower-hours (hph)
	3600.0000	Joules (J)
	0.0010	Kilowatt-hours (kWh)

CHAPTER 48
EQUIPMENT INVENTORY

48.1 INVENTORY PREPARATION

The owner and the plant personnel responsible for parts and material procurement will benefit greatly from the availability and use of a comprehensive plant equipment inventory.

Example sheets of plant equipment inventories are shown in subsequent pages, but an individual plant may find it convenient to adopt a different format to suit its particular convenience and needs. The sheets may be used at some plants as worksheets when the equipment inventory is being committed to a computerized data storage system. The examples are not comprehensive, and the individual superintendent will find it expedient to make his own additions. The assembly of such information into a single reference file or folder will save a great deal of repetitious searching for information of a sort that notoriously becomes lost, defaced, or distorted with the passage of time.

The equipment inventory should be checked off at the time of the recommended annual plant inspection, when all alterations are noted and a revised equipment inventory is subsequently issued to both the headquarters and the power plant. The list will show the basic information relating to each piece of equipment and will also include any detailed remarks that will assist in the full identification of any variance in a given machine. The data will be essential when ordering new parts or having repairs performed.

Many OEMs use different terms for their individual model variations. For example, Caterpillar uses the term *arrangement number*; General Motors uses the term *model number*; Turbocharger makers use *build number*; and so on throughout the industry.

The inventory sheets have many applications, principally for ordering correct replacement parts as a basis for maintenance and overhaul planning and as a guide for the annual inspection. The inventory must be available for the preparation of statistical reports, procurement of parts, forecasting, storekeeping, and so on. In many plants the data will be committed to a computer, in which form it will be far more flexible in its utilization.

Numerous examples of plant equipment inventory sheets follow.

Prime Mover.

    Unit number
    Make
    Model
    Serial number
    Arrangement number
    Commissioning date
    RPM
    Lubricating oil brand
    Lubricating oil change period
    Sump capacity

Starter.

    Make
    Model
    Type
    Serial number (L)
    Serial number (R)
    Voltage or pressure
    Rotation

Starting Batteries.

    Make
    Model
    Type (lead or nickel)
    Ampere-hour rating
    Battery voltage
    Date of installation

Governor.

    Make
    Model
    Serial number
    Arrangement
    Speed motor voltage
    Speed motor type
    Solenoid voltage
    RPM range
    Speeder motor rpm
    Solenoid mode

(Energize to run)
(Energize to S/D)

Fuel Injection Pump.

    Fuel pump number
    Injector body number
    Injector nozzle number
    Injector pressure setting

Turbocharger.

    Make
    Model
    Type
    Arrangement
    Lubricating oil make
    Lubricating oil grade
    Lubricating oil change period

Pressure Charger.

    Make
    Model
    Serial number (R)
    Serial number (L)
    Arrangement

Aftercoolers.

    Make
    Model
    Serial number

Speed Switch.

    Make
    Model
    Serial number
    Switch 1 set point
    Switch 2 set point
    Switch 3 set point

Oil Pressure Switch.

    Make
    Model
    Serial number
    Set point alarm
    Set point shutdown

Lubricating Oil Heater.

    Make
    Model
    Serial number
    Voltage
    Phase
    Watts
    Thermostatic switch
    Setting ON, °C/°F
    Thermostatic switch
    Setting OFF, °C/°F

Jacket Coolant Circulating Pump.

    Make
    Model
    Serial number
    Gallons per hour or minute

Jacket Coolant Circulating Pump Motor.

    Make
    Model
    Serial number
    Voltage
    Amperage
    Phase
    RPM
    Frame

Radiator.

    Make
    Model
    Serial number

EQUIPMENT INVENTORY

Fan Motor.

    Make
    Model
    Serial number
    Voltage
    Amperage
    Phase
    RPM
    Frame
    HP
    Bearing

Radiator Fan Control Thermostat.

    Make
    Model
    Serial number
    Set point ON
    Set point OFF

Raw Water Circulating Pump.

    Make
    Model
    Serial number
    Gallons per hour or min

Raw Water Circulating Pump Motor.

    Make
    Model
    Serial number
    Voltage
    Amperage
    Phase
    RPM
    HP
    Frame
    Bearing

Block (Jacket) Heater.

    Make
    Model

Voltage
Phase
Watts
Temperature switch make
Temperature switch model
ON setting °C/°F
OFF setting °C/°F

Jacket High-Temperature Switch.

Make
Model
Serial number
Shut-down setting, °C/°F

Lubricating Oil Temperature Switch.

Make
Model
Serial number
Shut-down setting, °C/°F

Generator.

Unit number

Make
Model
Serial number
Type
Code
Frame
Kilovolt-amperes
Kilowatts
Volts
Amperes
Power factor
Bearing number

Exciter.

Make
Model
Serial number
Field amperes
Field volts

EQUIPMENT INVENTORY

Automatic Voltage Regulator.

    Make
    Model
    Serial number
    Sensing voltage
    Input voltage

Manual Voltage Regulator.

    Make
    Model
    Serial number
    Input voltage

Circuit Breaker (Generator).

    Make
    Model
    Serial number
    Voltage
    Amperage
    Control voltage

Circuit Breaker (Feeders).

    Make
    Model
    Serial number
    Voltage
    Amperage
    Control voltage

Station Service Batteries.

    Make
    Model
    Type (lead or nickel)
    Specific gravity
    Volts per battery
    Date of installation

Station Service Transformer.

    Make
    Model

Serial number
Kilovolt-amperes
Primary volts
Secondary volts
Connection: High voltage
              Low voltage

Battery Charger.

Make
Model
Serial number
Input voltage
Input amperage
Output voltage
Output amperage
Output breaker
Current rating

Kilowatt-hour Meter Generator.

Make
Model
Serial number
Type
Disk constant (kWH)
Register ratio (Rr)
Meter multiplier (A)
CT ratio (B)
PT ratio (C)
Total reading multiplier (A = B × C)
Element

Kilowatt-hour Meter (Feeder) and Station Service.

Make
Model
Serial number
Type
Disk constant (kWh)
Register ratio (Rr)
Meter multiplier (A)
CT ratio (B)

PT ratio (C)
Total reading multiplier (A = B × C)
Element

Lubricating Oil Separator.

Make
Model
Serial number
Throughput rate

Mechanical Overspeed Switch.

Make
Model
Serial number
Shutdown rpm

Lubricating Oil Circulating Pump.

Make
Model
Serial number
Gallons per hour or minute

Lubricating Oil Circulating Pump Motor.

Make
Model
Serial number
Voltage
Amperage
Phase
RPM
Frame

Fuel Oil Separator.

Make
Model
Serial number
Throughput rate

Air Intake Filters (Curtain Type).

Make
Model

Serial number
Oil type

Fuel Meter at Engine.

Make
Model
Serial number
Format (US or imp gal, liter, etc.)

Water Separating Filter.

Make
Model
Serial number
Filter element type

Coolant Level Switch.

Make
Model
Serial number
Location

Vibration Switch.

Make
Model
Serial number
Operating range
Shut-down setting

Fuel Transfer Pump.

Make
Model
Serial number
Gallons per hour or minute

Fuel Transfer Pump Motor.

Make
Model
Serial number
HP
Volts

Amperage
Phase
RPM
Frame
Bearing

Receiving Fuel Meter.

Make
Model
Type

Fuel Day Tank.

Capacity
Heater watts
Level switch make
Level switch model
Drain fitted

Fuel Storage Tanks.

Capacity
Installation date
Diameter
Height
Bolted or welded

Filter Elements.

OEM part number
Primary fuel
Seconday fuel
Primary lubricating oil (main)
Secondary lubricating oil (bypass)
Coolant
Air

Fuel Meter at Day Tank.

Make
Model
Serial number
Format (U.S. or imp gal, liter, etc.)

Compressor.

    Make
    Model
    Serial number
    Maximum operating pressure
    Cutout setting
    RPM

Compressor Engine.

    Make
    Model
    Serial number
    RPM
    HP
    Fuel pump number
    Injector number
    Fuel filter element
    Lubricating filter element
    Air filter element

Compressor Motor.

    Make
    Model
    Serial number
    Voltage
    Amperage
    Phase
    RPM
    Horsepower
    Frame
    Bearing

Air Receivers.

    Make
    Model
    Serial number
    Certified pressure
    Working pressure
    Volume (or dimensions)
    Relief valve setting

Plant Ventilating Fan Motor.

Make
Model
Serial number
HP
Volts
Amperage
Phase
RPM
Frame
Bearing

Crane.

Make
Model
Serial number
Capacity
Voltage
Hoist motor:			Make
				Model
				Serial number
Cross-traverse motor:	Make
				Model
				Serial number
Travel motor:		Make
				Model
				Serial number

Waste-Heat Boiler.

Make
Model
Serial number
Working pressure
Maximum pounds per hour

Blackstart/Emergency Generator.

Engine make
Engine model
Serial number
RPM

Fuel pump number
Injector body number
Injector nozzle number
Fuel filter element
Lubricating oil filter element
Air filter element

Generator.

Make
Model
Serial number
Kilovolts-amperes
Kilowatts
Volts
Phase
Frame

Index

Abbreviations for drawings, **30**.12–**30**.14
Accesses, inspections of, **45**.3
Acetylene tanks, safety, **41**.13–**41**.14
Additives, **22**.9, **23**.5–**23**.6, **24**.23–**24**.24
Adequacy of power, **34**.9
Adhesives for gaskets, **29**.3
Advisory signals, **3**.10–**3**.11
Aeration of lubricating oil, **24**.19
Aftercooler water pressures, **4**.6
Aftercooler water temperatures, inlet and outlet data on log sheets, **4**.7
Aftercoolers, **17**.1–**17**.4, **19**.13
Air, **23**.14, **23**.26, **23**.27–**23**.29
Air boxes, **20**.1–**20**.3
Air compressors, **38**.1–**38**.4
Air-cooled engines, **23**.33
Air filter differential pressure gauges, **25**.1
Air filters, **19**.10–**19**.15
Air gaps on motors/generators, **35**.5–**35**.6
Air-starting systems, **39**.1–**39**.2, **39**.3
Air-starting valves, **39**.2
Air vessels:
 isolation of, **38**.4
 upkeep of, **38**.3–**38**.4
Alarms, **3**.10–**3**.11
 audible negation of, **3**.11
Alignment, **26**.1–**26**.5
 of shafts in engine-generator sets, **35**.5
Alternator winding thermometers, **25**.6
Alternators, vibration of, **27**.9
Ambient air temperatures, **17**.2
 records of, **25**.1–**25**.2
Ammeters, **25**.1
AMOT (*see* Directional temperature control valves)
Amperage and load, **4**.3–**4**.4
Amplitude, **8**.5
Amplitude (vibration), **27**.3
Analysis:
 of lubricating oil, **24**.21

Analysis (*Cont.*):
 of vibration, **27**.9–**27**.10
ANSI units, **47**.1–**47**.7 (tables)
Antifoam agents, **24**.23
Antifriction bearings, **9**.9–**9**.10, **9**.12
 electric arcing effects on, **9**.10, **9**.11 (dwg.), **9**.12, **35**.6–**35**.7 (dwgs.)
 lubrication of, **35**.7
Antiwear additives, **24**.23
Ash, definition of, **22**.4
Aspirating air, **20**.1–**20**.3
Assets (fixed and portable), inspections of, **45**.6
Authority delegation in maintenance work, **32**.2–**32**.3
Automatic and manual regulation switches, **34**.3–**34**.4
Automatic voltage regulation, relation to quality of service, **34**.9
Automatic voltage regulators, **34**.3–**34**.7
AVR (*see* Automatic voltage regulators)

Barring over of engines, **28**.3
Batteries, **28**.2, **37**.1–**37**.6
Bearing housings, out-of-round, **9**.5
Bearings, **9**.1–**9**.13
 life of, **35**.8
Berms of fuel storage tanks, **22**.23
Block diagrams, uses of, **31**.4
Blowby, **7**.1
Blueprints, **31**.3–**31**.4
Bolts, hold-down, **26**.1, **26**.5 (dwg.)
Boost air pressure gauges, **25**.2
Boost air pressures, **4**.6–**4**.9
Boost air temperature gauges, **25**.2
Boost air temperatures, **4**.7
Boosting, **39**.7
Breaker trip counters, **25**.10
Breakers, protective tripping of, **5**.16

Breathers, crankcases, **7**.3
Brushes, **6**.8
Buildings, inspections of, **45**.5–**45**.6
Bulk fuel records, **46**.3–**46**.4
Buna-N, **29**.4
Buried cables, hazards of, **41**.11–**41**.12

Cable tray fires, **41**.11
Calibration, of instruments, **25**.8–**25**.9
Cams, **14**.2
Camshafts, **14**.1–**14**.2
Cannibalization of equipment, **43**.2
Capacity, selection of, **3**.12–**3**.13
Capacity dispatch, **44**.3–**44**.4
Caps, for radiators, **23**.23, **23**.24 (dwg.)
Carbon deposits, **16**.7, **21**.6–**21**.7
Carbon dioxide, free, **23**.34
Carbon loading of exhaust bellows, **21**.8–**21**.9 (dwgs.)
Carbon residue, definition of, **22**.4
Casings, turbochargers, **16**.7
Cataloging of books/manuals, **31**.5–**31**.6
Catalytic converters, **21**.8–**21**.9
Cavitation of water pumps, **23**.11–**23**.13
Centrifugal filters, **19**.6
   for lubricating oil, **24**.17
Cetane number, definition of, **22**.4–**22**.5
Chains, timing, **14**.7
Changeouts of components, **32**.21
Charging of batteries, **37**.5
Chromate inhibitors, **23**.9
Chrome-plated liners, **10**.3, **10**.4 (dwg.)
Chrome-plated piston rings, **11**.4
Circumstantial information, **46**.3
Cleaning:
   of aftercoolers, **17**.2–**17**.4
   of air boxes, **20**.2–**20**.3
   of air filter elements, **19**.12
   of concrete floors, **6**.2
   of cooling systems, **23**.31–**23**.32
   of crankcases and sumps, **7**.6–**7**.7
   of electric lights, **6**.8
   of electrical apparatus, **6**.6–**6**.7
   of engines, **28**.4
   of mechanical apparatus, **6**.5–**6**.6
   of motors and generators, **35**.4–**35**.5
   of pans, **6**.8–**6**.9
   of radiators, **23**.25
Cleaning agents, selection of, **6**.4
Cleaning fluids, temperature of, **6**.7
Cleaning materials, **6**.4–**6**.5
Cleaning media, disposal of, **6**.7–**6**.8

Cleanliness in power plants, **2**.4
Clearance permits, **41**.5
Clearance procedures, **41**.4–**41**.5
Clearances of cylinder head valves, **13**.1
Climates in power plants, **2**.4–**2**.5
Clothing, **1**.3–**1**.4
   safety types of, **44**.6–**44**.7
Cloud point, definition of, **22**.5
CO_2 hand-held fire extinguishers, **41**.9
Coir mats, **6**.8
Cold-stuck rings, **11**.5
Colors, code for, **6**.9 (table)
Commitment to maintenance, **32**.5–**32**.6
Compensating governor adjustments, **18**.5
Component interchangeability, **33**.3
Compressed air systems, **38**.1–**38**.4
Compression rings, **11**.3
Concrete floors, cracking of, **6**.14
Conformable rings, **11**.4
Connections for pipes, **23**.20–**23**.21
Consignment stock parts, **32**.27–**32**.28
Constant-speed governors, **18**.2
Consumable materials, **33**.4
Consumable parts, **32**.26
Contaminants:
   of coolant fluids, **23**.33–**23**.34
   of lubricating oil, **24**.24
Continuity of power, **34**.8
Coolant fluids, **23**.24–**23**.35, **23**.36
Coolant leakage, **19**.8–**19**.9
Coolant temperature gauges, **25**.2
Coolants, **23**.4–**23**.10
   gassing of, **12**.3–**12**.4, **23**.4
   leaks of, **23**.29–**23**.31
   specific conductance of, **23**.35
Cooling:
   air handling systems for, **23**.26, **23**.27–**23**.28 (dwgs), **23**.29
   of pistons, **11**.1
Cooling fluids, **23**.5–**23**.6
Cooling systems, **23**.1–**23**.36
   overfilling of, **12**.3
Cooling water intakes, **45**.4–**45**.5
Corrosion of impellers, **23**.11
Corrosion inhibitors, **24**.24
Cost of power, **34**.9
Counters, **25**.10–**25**.11
Counting meters, **25**.9–**25**.10
Covers for cylinder heads, **12**.4–**12**.5
Cracking:
   of concrete floors, **6**.14
   of cylinder heads, **12**.1–**12**.3

Cracking (*Cont.*):
 of gear teeth, **14.**6 (dwg.)
Crankcase breathers, **7.**3
Crankcase exhausters, **7.**3
Crankcase paint, **6.**11–**6.**12
Crankcase sludges, **24.**11, **24.**18
Crankcases, **7.**1–**7.**8
Crankshaft diameters, out-of-round diameters, **9.**4–**9.**5
Crankshaft torsional displacement, **8.**5
Crankshafts, **8.**1–**8.**5
Critical speed, **8.**5
Crosscurrent generator compensation, **34.**6
Crude oil, use of as fuel, **22.**15–**22.**16
Crush on varied bearing shells, **9.**3–**9.**4
Cutting, safety rules for, **41.**13
Cylinder head valves, **13.**1–**13.**9
Cylinder heads, **12.**1–**12.**7
 carbon deposits in, **21.**6
Cylinder liners, **10.**1–**10.**5
Cylinders, **12.**6–**12.**7
 flooding of, **10.**3, **10.**4 (dwg.)

Dampers (*see* Torsional vibration in dampers)
Dampness, effects of on electrical machines, **35.**1
Data plates, **25.**5–**25.**6
 on equipment, **2.**6
DC ammeters, **25.**6
Decay of inhibitor coolants, **23.**8–**23.**9
Defect and incident records, **4.**2–**4.**3
Defect rectification, **43.**1–**43.**2
Defective gauges, disposal of, **25.**8
Defects of electric starters, **39.**5, **39.**6 (dwg.), **39.**7
Detergents of metallic origin, **24.**24
Diesel fuel quality, **22.**3–**22.**4
Diesel generator sets, levels of vibration in, **27.**6
Dipsticks, **24.**10, **24.**11 (dwg.), **24.**12, **25.**2–**25.**3
Directional temperature control valves, **23.**18 (dwg.), **23.**19
Dispersants, **24.**24
Displacement (vibration), **27.**3, **27.**4
Disposal:
 of cleaning media, **6.**7–**6.**8
 of defective gauges, **25.**8
 of used lubricating oil, **22.**10
Distillation end point, **22.**5
Distilled water for batteries, **37.**3

Doors, **6.**14
 for crankcases, **7.**5, **7.**6 (dwg.)
Draining of engines, **28.**4
Drip trays, **6.**2
Drive gears, appearances of, **14.**4–**14.**7
Droop settings, **18.**3
Dry chemical fire extinguishers, **41.**9–**41.**10
Duties during shifts, **2.**2–**2.**3

Economics of engine utilization, **44.**4
Economy of effort in maintenance work, **32.**3–**32.**4
Edge filters, **19.**10
Education of operators, **1.**2–**1.**3
Electric arcing, effect on antifriction bearings, **9.**10–**9.**11 (dwg.), **9.**12, **35.**6–**35.**7 (dwgs.)
Electric lights, cleaning of, **6.**8
Electric machines, nomenclature for, **35.**2
Electric starters, **39.**2, **39.**4–**39.**5
 defects of, **39.**5, **39.**6 (dwg.), **39.**7
Electric tools, hazards of portable, **41.**12
Electrical apparatus, cleaning of, **6.**6–**6.**7
Electrical discharges, damage to crankshafts by, **8.**4–**8.**5
Electrical equipment, protection of, **41.**12
Electrical firefighting safety, **41.**11
Electrical safety, rules for, **41.**3–**41.**4
Electrolytes, **37.**3
Emergency handling, **3.**11
Emergency lighting, **41.**6
End-of-day kilowatt-hour readings, **4.**9
Engine coolant overheating, **5.**10
Engine excessive crankcase pressure, **5.**11
Engine faults, types of, **5.**1–**5.**17
Engine fuel dilution of lubricating oil, causes of, **5.**10–**5.**11
Engine gases, causes of, **5.**11
Engine high exhaust temperatures, **5.**15
Engine high lubricating oil pressure, causes of, **5.**12–**5.**13
Engine hour meters, **25.**9–**25.**10
Engine hours, **4.**8–**4.**9
Engine hunts, causes for, **5.**6
Engine knocking, causes of, **5.**7, **5.**8
Engine knowledge and experience, exchange of, **44.**2–**44.**3
Engine log sheets, **4.**1–**4.**9
Engine low lubricating oil pressure, causes of, **5.**12
Engine lubricating oil level rises, **5.**12
Engine lubricating oil too hot, **5.**13

Engine making black smoke, **5.13–5.14**
Engine making blue smoke, **5.14–5.15**
Engine making unusual noises, **5.16**
Engine making white smoke, **5.14**
Engine misfiring, causes of, **5.9–5.10**
Engine not attaining synchronous speed, causes of, **5.8**
Engine not carrying any load after breaker closure, causes of, **5.8**
Engine not carrying full load, **5.9**
Engine not stopping, causes of, **5.16**
Engine overspeeding, causes of, **5.8**
Engine paint, **6.10–6.11**
Engine room temperatures, **25.3**
Engine sludge in sump, causes of, **5.12**
Engine start counters, **25.11**
Engine startup failures, **5.3–5.4**
Engine stoppages, causes for, **5.5–5.6**
Engine timing, **14.2–14.3**
Engine turning, **40.1–40.2**
  safety aspects of, **40.1**
Engine using excessive amount of lubricating oil, causes of, **5.13**
Engine using too much fuel, **5.15–5.16**
Engine vibration, causes of, **5.6–5.7**
Engines:
  air-cooled, **23.33**
  alignment of, **26.1–26.3**
  barring over of, **28.3**
  causes of dirt in, **9.2**
  cleaning of, **28.4**
  damaged by fuel sulfur, **22.11**
  draining of, **28.4**
  fuel supplies for, **28.4**
  inspection of, **28.2**
  instruments for, **25.1–25.5**
  laid-up, **28.3–28.4**
  with large-capacity oil pans, **24.9**
  leaks in, **29.2**
  lubrication of, **28.2**
  monitoring conditions of, **27.1**
  operating temperatures of, **3.4–3.5, 23.1–23.2**
  operation of, **3.1–3.18**
  overcooling of, **23.2**
  overfilling of, **23.4**
  overheating of, **12.3, 23.3**
  painting of, **28.4**
  precautions for restart of, **28.5**
  preservation/storage of, **28.1–28.4**
  protection of, **3.5–3.11**

Engines (*Cont.*):
  restarting of after protective shutdowns, **3.9–3.10**
  returning to service, **28.4–28.5**
  shutdown of, **3.2**
  standby units of, **28.1**
  starting of, **3.2–3.4**
  stopping of, **3.1–3.2**
  storage of, **28.1**
  timing of, **15.9**
  vibration monitoring of, **27.6–27.8, 28.3**
  vibration troubleshooting of, **27.3–27.6**
Engines with exposed flywheels, painting of, **6.11**
English measurement units, **47.1–47.7**
Environmental conditions, **2.4–2.6**
EPDM (*see* Ethylene-propylene terpolymer)
Equipment, cannibalization of, **43.2**
Equipment, inventories of, **48.1–48.14**
Ethylene glycol, **23.6–23.7**
Ethylene-propylene terpolymer (EPDM), **29.4**
Exchange of engine knowledge, **44.2–44.3**
Exchange parts, **32.26**
Exhaust bellows, **21.8–21.9** (dwgs.)
Exhaust emissions, **21.4**
Exhaust expansion pieces, **21.8**
Exhaust gas pressures, **21.7**
Exhaust gas temperatures, **4.8**
Exhaust manifold gaskets, **21.7–21.8**
Exhaust passages, carbon deposits in, **21.6–21.7**
Exhaust smoke:
  black, **21.4–21.5**
  white, **21.5**
Exhaust sparking, **21.4**
Exhaust systems, **21.1–21.10**
  catalytic converters for, **21.8–21.9**
Exhaust temperature gauges, **25.3**
Exhaust temperatures, **21.1–21.4**
Exhaust valves, **13.6–13.8**
Exhausters, crankcases, **7.3**
Exit doors, **6.14**
Expansion tanks (*see* Header tanks)
Eyewash stations, **41.17**

Failures:
  of bearings, **9.2**
  causes of in instruments, **25.8**
  of cylinder head valves, **13.5, 13.6**
  of water pumps, **23.10–23.11, 23.12**

INDEX

Fan belts, **23.25–23.26**
Fan shrouds for radiators, **23.25**
Fans for radiators, **23.24**
Fasteners, **43.2–43.3**
Faults (*see* Engine faults)
Feeder panels, instruments, **25.6–25.7**
Fillet ride, **9.5**
Filters, **19.1–19.15, 23.20**
    fuel waxing of, **22.12**
    for lubricating oil, **24.12, 24.13–24.14**
        (dwgs.), **24.15–24.17**
    (*See also* Air filters; Centrifugal filters; Fuel filters; Oil filters)
Fire extinguishers, **41.7–41.10**
Fire protection, **41.6–41.7**
Fires, **41.10–41.11**
First-aid kits, **41.16–41.17**
Fittings for fuel pipe systems, **22.19**
FKM (*see* Fluoroelastomer)
Flaking of gear teeth, **14.5** (dwg.)
Flash point, definition of, **22.5**
Flex disk balance weights for generators, **26.9, 26.10** (dwg.)
Flexible fuel lines, **22.19**
Flexible hoses, **23.21–23.22**
Flexible pipe connections, **23.20–23.21**
Flooding of cylinders, **10.3, 10.4** (dwg.)
Floor painting, **6.13–6.14**
Floors, **6.2, 6.14**
Floors and foundations, checking of, **45.2**
Fluoroelastomer (FKM), **29.4**
Flushing systems for lubricating oil, **24.27–24.28**
Foreign particles, effects of on bearing shells, **9.2**
Frequency (vibration), **27.3, 27.4** (dwg.)
Frequency meters, **25.6**
Frequency of power, **34.9**
Fresh water, **23.6**
Friction clutch adjustment for governors, **18.8–18.9**
Frost-protection coolants, **23.6–23.8**
Fuel:
    heating of, **22.15**
    unfiltered, **19.10**
    water and sediment in, **22.7**
    water content of, **22.12–22.13, 22.14**
Fuel consumption:
    comparison of, **3.15–3.18** (tables)
    effect of utilization factor on, **22.2–22.3**
Fuel day tanks, **22.16, 22.17–22.18**

Fuel dilution, **19.9**
    of lubricating oil, **24.19**
Fuel economy, **3.13–3.15** (graphs)
Fuel filters, selection of, **19.9–19.10**
Fuel gauges, **25.3**
Fuel injectors, **15.1–15.12**
Fuel lines, flexible, **22.19**
Fuel metering, **22.2**
Fuel meters, **4.9, 25.10–25.11**
Fuel oil, **22.1–22.23**
    terminology of, **22.4**
Fuel pipe systems, fittings for, **22.19**
Fuel pressure gauges, **25.3**
Fuel pressures, data on log sheets, **4.6**
Fuel pumps (*see* Pumps, fuel injectors)
Fuel service tanks, fuel temperatures in, **22.10–22.11, 22.12** (table)
Fuel spills, **22.19–22.20**
    cleanup of, **6.2–6.3**
    procedures for handling, **22.20–22.21**
    use of peat moss in, **22.21**
Fuel storage areas, inspections of, **45.4**
Fuel storage tanks, **22.16, 22.17–22.22**
Fuel supplies for engines, **28.4**
Fuel supply pipe systems, **45.5**
Fuel systems, security of, **22.21**
Fuel temperatures, data on log sheets, **4.6**
Fuel transfer pumps (*see* Pumps, fuel injectors)
Fuel usage, **22.1–22.2**
Fuel use, records of, **46.4**
Fuel waxing in filters, **22.12**
Fuels, **22.7–22.16**
    temperature of heavy types, **20.2**
Fuses, **36.16**

Gases in cooling systems, **23.14**
Gaskets, **29.2–29.3**
    for crankcase doors, **7.6** (dwg.)
    storage of, **29.5, 33.3–33.4**
Gassing of coolants, **12.3–12.4, 23.4**
Gate valves, **30.2** (dwg.)
Gauges, disposal of defective units, **25.8**
Generator sets:
    allocating/dispatching of, **3.12–3.13**
    balance of, **26.9–26.10**
    startup of, **3.3–3.4**
Generators, operation of, **34.1–34.9**
    alignment of, **26.1–26.3**
    overhaul and upkeep of, **35.1–35.9**
    fires in, **41.11**

Generators, operation of (*Cont.*):
  flex disk balance weights for, **26.9, 26.10** (dwg.)
  grounding straps for, **35.**9
  instruments for, **25.**6
  operating hours records, **46.**3
  workbooks and manuals for, **31.3–31.**5
Glazed cylinder liners, **10.2–10.**3
Globe valves, **30.**3 (dwg.)
Glossary, lubrication terms, **24.28–24.**30
Governors, **18.1–18.**9
  safety rules concerning, **3.**6
Grease, application guidelines for, **24.**28
Greasing of antifriction bearings, **9.**10
Ground fault indication, **34.7–34.**8
Grounding straps, **35.**9
Guards on machines, **41.5–41.**6

Hand tools, safety rules for, **41.**14
Hardness of coolant fluids, **23.**35
Hazards:
  of buried cables, **41.11–41.**12
  of liquid fire extinguishers, **41.**11
  of portable electric tools, **41.**12
Hazy atmospheres in power plants, **2.**4
Header tanks, **23.3–23.**5
  ejection of coolants from, **12.3–12.**4
Heat-sensitive crayons, **21.**3
Heavy fuels, **22.13–22.**15
High jacket coolant temperature shutdowns, **23.2–23.**3
Hold-down bolts, **26.1, 26.**5 (dwg.)
Hot-stuck rings, **11.**5
Hour meters, **25.3–25.4, 25.**6
Housekeeping, **6.1–6.**14
Hunt, definition of, **18.**2
Hydrometers, **37.4–37.**6

Identity numbers and marks, **43.**2
Idle speeds, **3.**5
Impellers, corrosion of, **23.**11
Indicator cocks for cylinders, **12.**7
Indicator lights, **25.**6
Information, circumstantial, **46.**3
Information plates for lubricating oil systems, **24.**16
Information provision in maintenance work, **32.4–32.**5
Inhibitors, **23.8–23.**9
Injection, timing of, **14.**4
Injector nozzle changeouts, **32.**21
Inlet valves, rehabilitation of, **13.6–13.**7

Inspection certificates, statutory, **44.**6
Inspections:
  of bearings, **9.**12
  of crankcases, **7.7–7.**8
  of facilities, **45.1–45.**6
  of machinery, **3.7–3.9, 28.**2
  of flexible hoses, **23.21–23.**22
  of fuel storage tanks, **22.**22
  of mufflers, **21.9–21.10, 45.**3
  of oil filters, **19.6–19.**7
  post-overhaul of motors and generators, **35.**8
  of raw water intakes, **23.32–23.**33
  of rubber-bonded torsional dampers, **8.6, 8.8–8.**9
Installation of antifriction bearings, **9.**12
Instrumentation, **25.1–25.**11
Instruments, **25.1, 25.8–25.**9
Insulation:
  testing of, **35.3–35.**4
  of turbochargers, **16.**8
Internal combustion engines, torsional vibration dampers in, **8.**6
Inventories, **48.1–48.**14
Isochronous governors, **18.**2
Isolating valves for air compressors, **38.**4
Isolators, vibration, **26.3–26.**7

Jacket water expansion tank level gauges, **25.**4
Jacket water heaters, **23.19–23.**20
Jacket water inlet/outlet temperatures, **3.7–3.8, 4.**8
Jiggle, definition of, **18.**2
Journeyman notes, **43.1–43.**4
Jump starting, **39.**7
Jumper cables, **39.7–39.**8

Kilowatt-hour meters, **25.**6
Kilowatt hours, **4.5–4.**6
Kilowatts per liter, **4.**9

Ladders, safety rules for, **41.15–41.**16
Laid-up engines, **28.3–28.**4
Lead-acid batteries, care of, **37.3–37.**4
Leaking air inlets valves, **13.5–13.**6
Leaks, **23.29–23.31, 23.37, 29.**2
Lifting equipment, safety, **41.14–41.**15
Lighting, emergency, **41.**6
Liquid fire extinguishers, **41.**11
Load and amperage, data on log sheets, **4.3–4.**4

INDEX

Load division among generators, **3.4, 34.5–34.6**
Load limit controls, **18.3**
Load-limiting governors, **18.1**
Load trips off and engine continuing to run, causes of, **5.16**
Log sheets, daily review of, **44.5–44.6**
Logs (*see* Engine log sheets)
Low temperatures, effects of on batteries, **37.2** (graph)
Lubricants:
  contaminants in, **19.5**
  for seals, **29.3**
  for thread, **43.4**
Lubricating oil, **24.1–24.28**
Lubricating oil coolers, **24.25–24.26**
Lubricating oil filter differential pressure gauges, **25.4**
Lubricating oil filters (*see* Oil filters)
Lubricating oil level sight glasses, **25.4**
Lubricating oil meters, **25.11**
Lubricating oil pressure gauges, **25.4–25.5**
Lubricating oil pressure shutdowns, **3.6–3.7**
Lubricating oil pressures, data on log sheets, **4.7**
Lubricating oil temperature gauges, **25.5**
Lubricating oil temperatures, inlet and outlet data on log sheets, **4.7**
Lubrication:
  of bearings, **9.6, 35.7**
  of engines, **28.2**
  of turbochargers, **16.2, 16.3** (dwg.)
Lubrication systems, ancillary, **24.10**
Lubrication terms, glossary of, **24.28–24.30**

Machine guards, **41.5–41.6**
Machine vibration signatures, **27.10**
Magnetic plugs, **24.18**
Maintenance:
  of air compressors, **38.1–38.2**
  of batteries, **37.5**
  of motors and generators, **35.3, 35.4**
  of vehicles, **44.8**
  unplanned, **32.21–32.22**
  (*See also* Planned maintenance; Preventive maintenance)
Maintenance in plants, **2.7–2.11**
Maintenance schedules:
  for 1.0-mW unit, **32.12–32.13**
  for 300-kW unit, **32.11–32.12**

Maintenance systems, records of, **46.4–46.5, 46.6** (sample form)
Manifold gaskets, exhaust, **21.7–21.8**
Manifolds for aspirating air, **20.1**
Manuals for operators, **1.3**
Marine growth, **23.13–23.14**
Materials handling, safety, **41.15**
Measurement of vibration, **27.2–27.3**
Measurement units, **47.1–47.7** (tables)
Mechanical apparatus, cleaning, **6.5–6.6**
Metering, of fuel, **22.2**
Meters, **25.9–25.11**
Methanol, for diesels, **22.9–22.10**
Methyl alcohol (*see* Methanol)
Metric units, **47.1–47.7** (tables)
Moisture, effects of on electrical machines, **35.1**
  condensation of, **19.8–19.9**
Mops, **6.8**
Motor vehicles, **44.7–44.8**
Motors, care and running of, **35.3–35.9**
Mufflers:
  exhaust stacks of, **45.3**
  inspections of, **21.9–21.10, 45.3**

NBR (*see* Nitrate butadiene rubber)
Needle valve adjustments, **18.5**
Negative pressure in crankcases, **7.5**
Neoprene, **29.4**
Nickel-cadmium batteries, **37.3**
Nimonic exhaust valves, **13.4**
Nitrate butadiene rubber (NBR), **29.4**
Noise of turbochargers, **16.6** (dwg.), **16.7**
Nonslip paint, **6.13**
Nozzles, fuel injectors, **15.4–15.6**

O-rings, **29.3–29.5**
Odors in power plants, **2.3–2.4**
OEM, definition of, **1.4**
Oil barrels, handling of, **24.26–24.27**
Oil contamination, excessive, **19.7**
Oil control rings, **11.3**
Oil coolers, **24.25–24.26**
Oil filters, **19.2** (dwgs.), **19.3–19.9**
Operating hours, records of, **44.6, 46.4**
Operating temperatures of engines, **23.1–23.2**
Operation of engines, **3.1–3.18**
Operation of machines, efficiency, **1.1**
Operator-in-charge, duties of, **2.1–2.2**
Operators, **2.1–2.12**
  duties of, **1.1, 2.2–2.3**

Operators (*Cont.*):
   relation to power outages, **34**.8
Outages:
   due to overspeed shutdowns, **3**.8–**3**.9
   review of, **5**.16–**5**.17
Outdoor and indoor air temperatures, **4**.8
Outfalls, inspections of, **45**.4
Overcooled lubricating oil, **24**.25
Overcooling of engines, **23**.2
Overfilling, **12**.3, **23**.4
Overhaul, completeness of, **32**.23–**32**.24
Overhaul of machinery, **32**.1
   classifications of, **32**.15–**32**.16
   purpose of, **32**.14–**32**.15
   schedule variations, **32**.20–**32**.21
   schedules for, **32**.16–**32**.20
Overhaul parts, **32**.27
Overhauls, manpower requirements for, **32**.28–**32**.29
Overheating of engines, **12**.3–**23**.3
Overload, **3**.12
Overpressure in crankcases, **7**.3–**7**.5
Overspeed, and valve problems, **13**.8, **13**.9
Overspeed shutdowns, **3**.8–**3**.9
Oxidation inhibitors, **24**.24
Oxygen tanks, safety, **41**.13–**41**.14

Pails, **6**.8
Paint, **6**.9–**6**.14
Painting, **6**.9–**6**.14, **28**.4
Pans, cleaning of, **6**.8–**6**.9
Parasitic radiator loads, **23**.22–**23**.23
Parking areas, inspections of, **45**.5
Part numbers, **33**.1–**33**.2
Parts books, uses of, **31**.2–**31**.3
Parts classification, **32**.25–**32**.27
Parts control (stocktaking), **33**.4
Peat moss, use in fuel spills, **22**.21
Pedestal bearing temperatures, **4**.8
Permafrost area inspection, **45**.1–**45**.2
Personal health, **41**.16
Personal protection during cleaning, **6**.5
Personnel dangers of working alone, **41**.16
pH of cooling systems, **23**.34–**23**.35
Phase (vibration), **27**.3
Phase amperage of generators, **34**.6, **34**.7
Pipe connections, flexible, **23**.20–**23**.21
Pipe diagram symbols, **30**.5–**30**.11
Pipes, **30**.1–**30**.14
Pipework, **30**.3, **30**.5–**30**.11 (dwgs.)
Piston crowns, scuffing, scratching, or scoring of, **11**.2

Piston rings, **11**.3–**11**.4
Piston skirts, scuffing of, **11**.2
Pistons, **11.1–11**.5
Pitting of gear teeth, **14**.4–**14**.6
Planned maintenance, **32**.2–**32**.6
   at set frequencies, **32**.22–**32**.23
Portable electric tools, hazards of, **41**.12
Positional definition of engines, **2**.6–**2**.7
Pour point, definition of, **22**.6
Pour point depressants, **24**.24
Powders for fire extinguishers, **41**.10
Power, quality of, **34**.6, **34**.8–**34**.9
Power factor meters, **25**.6
Power factors, data on log sheets, **4**.5
Power outages, **34**.8
   safety instructions for, **41**.12–**41**.15
Power plant design, doors/windows, **2**.12
Power plants:
   inspections of, **45**.1–**45**.6
   management of, **44**.1–**44**.8
   physical environment in, **2**.3–**2**.6
Prepared parts lists, **32**.25
Pressure in crankcases, **7**.1–**7**.3
Pressure gauges:
   for cylinders, **12**.7
   failures of, **25**.7
Pressure relief valves:
   crankcases, **7**.4 (dwg.)
   for lubricating oil, **24**.19–**24**.20
Preventive maintenance, **32**.6–**32**.14
Propylene glycol, **23**.7–**23**.8
Protection:
   of electrical equipment **41**.12
   of engines, **3**.5–**3**.11
Protective devices, **3**.5–**3**.6
   high jacket coolant temperatures, **23**.2–**23**.3
Protective tripping of breakers, **5**.16
Pumps:
   fuel injectors, **15**.9, **15**.10 (dwg.), **15.11–15**.12
   for lubricating oil, **24**.24–**24**.25

Quality control, **32**.25
   of produced power, **34**.6, **34**.8–**34**.9

Radiator return flex connection collapse, **23**.22
Radiators, **23**.22–**23**.25
Raw water intakes, **23**.32–**23**.33
Raw water pressure gauges, **25**.5
Raw water temperature gauges, **25**.5

INDEX

Rebuilding of components, **32**.21
Rebuilt parts, **32**.26–**32**.27
Record keeping, **46**.1–**46**.6
  for maintenance work, **32**.6–**32**.8
Regulators (*see* Thermostats)
Repetitive maintenance work, comparison of, **32**.24
Replacement parts, **32**.27
Residual fuels, **22**.13–**22**.15
Response time, definition of, **18**.2
Restart of engines:
  precautions for, **28**.5
  after protective shutdowns, **3**.9–**3**.10
Reuse of antifriction bearings, **9**.12
Reverse engine turning, **40**.1, **40**.2 (dwg.)
Ring gumming, **7**.2
Roofs, inspections of, **45**.2
Rotors of generators, **35**.1
Rubber-bonded torsional dampers, **8**.6–**8**.10
Rubber-bonded torsional vibration dampers, condition identification of, **8**.9–**8**.10
Rubber O-rings, storage of, **29**.5
Rust-inhibiting additives, **24**.24

Safeguards of machines, **1**.1
Safety, **41**.1–**41**.17, **44**.8
  during cleaning, **6**.5
Safety clothing, **44**.6–**44**.7
Safety meetings, **44**.7
Safety rules, **41**.3–**41**.16
Safety shutdown settings, **3**.6
Safety training, **44**.7
Salt, **23**.35
Sampling of lubricating oil, **24**.21, **24**.22
Scavenge trunks (*see* Air boxes)
Schematic diagram symbols, uses of, **31**.5
Schematic diagrams, uses of, **31**.4
Scraper rings, **11**.3, **11**.4–**11**.5
Scuffing:
  of gear teeth, **14**.6 (dwg.)
  of pistons, **11**.2
Seals, **29**.1–**29**.2
Security:
  of fuel systems, **22**.21
  of vehicles, **44**.8
Sediment in fuel, **22**.7
Selection of O-rings, **33**.3
Separators for water, **22**.13, **22**.14 (dwg.)
Service bulletins, uses of, **31**.3
Shaft seals, **29**.1–**29**.2

Shift conduct, **2**.2–**2**.3
Shift operators, identification of, **4**.8
Shifts, changes in, **44**.4–**44**.5
Shop manuals, uses of, **31**.2
Shutdown devices, **3**.5–**3**.10
Shutdowns, **23**.2–**23**.3, **24**.20–**24**.21
  of engines, **3**.2
SI units, **47**.1–**47**.7 (tables)
Silicates, in coolants, **23**.9
Silicone rubber, **29**.4
Single-bearing generators, alignment of, **26**.2 (dwg.), **26**.3–**26**.5 (dwgs.)
Single-line diagrams, **31**.4–**31**.5
Slobber, **21**.5–**21**.6
Sludge-breaking additives, **22**.10
Sludges (*see* Crankcase sludges)
Solids, in coolant fluids, **23**.34
Sound levels in power plants, **2**.6
Spalling of gear teeth, **14**.5 (dwg.)
Spare parts, **32**.27, **33**.1, **33**.2
Spark erosion of bearings, **9**.7–**9**.8
Specific gravity, **22**.5–**22**.6
Speed adjusting controls, **18**.3
Speed droop, **18**.2–**18**.3
Speed-limiting governors, **18**.2
Spills (*see* Fuel spills)
Spin-on elements for oil filters, **19**.5–**19**.6
Split pins, **43**.3–**43**.4
Stability, definition of, **18**.2
Stacks, inspection of, **21**.9–**21**.10
Standard device numbers lists, **36**.1, **36**.2–**36**.9 (tables)
Standards, pipes and valves, **30**.5–**30**.11 (dwgs.)
Standards of operation, need for, **1**.1
Standby units of engines, **28**.1
Standing orders, **44**.5
Start-up, post-overhaul of motors and generators, **35**.8–**35**.9
Starters, electric (*see* Electric starters)
Starting aids, **39**.8–**39**.9
Starting air pressures, **4**.6
Starting engines, **3**.2–**3**.4
Starting problems, standby engines, **39**.8
Starting systems, **39**.1–**39**.9
Startup check of generator sets, **3**.3
Startup sequence of generator, **3**.3–**3**.4
Stators of generators, **35**.1
Statutory inspection certificates, **44**.6
Stem seizure of cylinder head valves, **13**.7
Sticking rings, **11**.5
Stopping engines, **3**.1–**3**.2

Storage of parts, **9**.12, **28**.1, **29**.5, **33**.3–**33**.4
Storage tanks (*see* Fuel storage tanks)
Strainers, **23**.20, **24**.16–**24**.17
Sulfates, **23**.35
Sulfur, **22**.6, **22**.11–**22**.12
Sumps, cleaning of, **7**.6–**7**.7
Superintendents of plants, **44**.1–**44**.2
Surface fatigue of bearings, **9**.5–**9**.6
Surface preparation, paint, **6**.12–**6**.13
Surge, definition of, **18**.2
Surge tanks (*see* Header tanks)
Switchgear, **36**.1–**36**.16
Switchgear device function numbers, **36**.1
Symbols, piping and valves, **30**.5–**30**.11
Symptoms and causes of faults, **5**.3–**5**.16
Synchronizer controls, **18**.3
Synthetic lubricating oil, **24**.2

Tachometers, **25**.5
Tappet clearance, **13**.1
Temperature control valves, **23**.19
Temperature gauging, **2**.12
Temperatures, **23**.1–**23**.3, **23**.10
  of aspirating air, **20**.1
  of cleaning fluids, **6**.7
  exhaust (*see* Exhaust temperatures)
  in fuel service tanks, **22**.10–**22**.12
  of lubricating oil, **24**.4
  for operation of engines, **3**.4–**3**.5
  of pistons, **11**.1
  in power plants, **2**.4–**2**.5
  rise of in machines, **35**.1
Terminology, **42**.1–**42**.9
Test equipment, fuel injectors, **15**.2–**15**.4
Testing:
  of coolant fluids, **23**.35, **23**.36 (chart)
  of generator and motor winding insulation, **35**.3–**35**.4
Textured paint, **6**.13
Thermal shock, **12**.2
Thermocouple leads, **25**.9
Thermostats, **23**.14–**23**.15, **23**.16–**23**.17 (dwgs.), **24**.25
Thread lubricants, **43**.4
Timing, **14**.3–**14**.4, **15**.9
Timing chains, adjustment of, **14**.7
Timing gears, **14**.3–**14**.6 (dwgs.), **14**.7
Tools for shift use, **2**.11–**2**.12
Torsional vibration dampers, **8**.5–**8**.11
  (*See also* Rubber-bonded torsional vibration dampers)
Training of operators, **1**.2–**1**.3

Troubleshooting, **5**.1–**5**.2, **23**.19
Turbocharger exhaust entries, **4**.7
Turbochargers, **9**.9, **16**.1–**16**.8, **24**.28

Undercutting of gear teeth, **14**.6 (dwg.)
Unfiltered fuel, damage done by, **19**.10
Unplanned maintenance, **32**.21–**32**.25
Urethane rubber, **29**.4
Used spare parts, **33**.2
Utilization factor, **22**.2–**22**.3
Utilization of engines, **44**.4

V-belts, selection/installation of, **26**.8
Valve diagram symbols, **30**.5–**30**.11
Valve stems, lubrication of, **13**.8
Valves, **30**.1–**30**.14
  air-starting, **39**.2
  for directional temperature control, **23**.18 (dwg.), **23**.19
  (*See also* Cylinder head valves)
Variant parts, **33**.2–**33**.3
Vehicles, **44**.8
Velocity (vibration), **27**.3, **27**.4 (dwg.)
Ventilation, **8**.11, **19**.14–**19**.15
  air handling systems for, **23**.26, **23**.27–**23**.28 (dwgs.), **23**.29
Ventilation systems, inspections of, **45**.3
Vibration, **27**.1–**27**.11, **28**.3
Vibration isolators, **26**.3–**26**.5, **26**.6–**26**.7 (dwgs.)
Vibration levels, floor and foundation, **26**.10 (table), **26**.11 (dwg.)
Viscosity, definition of, **22**.6–**22**.7
Viscosity index modifiers, **24**.24
Viscous fluid life in torsional vibration dampers, **8**.10–**8**.11
Voltage, **4**.4–**4**.5, **34**.8–**34**.9
Voltage regulators (*see* Automatic voltage regulators)

Waste-heat recovery, **21**.10
Water, **22**.12–**22**.14
  in fuel, **22**.7
  minerals and impurities in, **23**.6
  (*See also* Raw water)
Water heaters, jacket, **23**.19–**23**.20
Water pumps, **23**.10–**23**.13
Water treatment chemicals, **23**.34–**23**.35
Wattmeters, **25**.6
Wear:
  in bearings, **9**.1–**9**.2
  evidences of in lubricating oil, **24**.23

Wear (*Cont.*):
   in thermostats, **23**.16–**23**.17
Welding, **6**.12, **41**.13
Winding insulation, **35**.3–**35**.4
Windings, **35**.1–**35**.2
Wiping-down routines, **6**.1

Wiping rags, **6**.3
Wiring diagrams, uses of, **31**.5
Work motivation in maintenance work, **32**.3
Work practices, proper and safe, **43**.1
Workbooks, uses of, **31**.1–**31**.2

ABOUT THE AUTHOR

Clive T. Jones is a diesel power plant consultant and author with many years of experience in the field of diesel-powered generation, beginning with the Petter Oil Engine Company.

In 1972, he became Area Superintendent of Northern Canada Power Commission and later Mechanical Maintenance Superintendent at the Edmonton head office where he was responsible for the maintenance, overhaul, and repair of some 200 generator sets. Currently, Clive Jones resides in Edmonton, Alberta, where he is President of his own consulting company.